Eckhard Worch

Adsorption Technology in Water Treatment

Also of interest

Drinking Water Treatment.
An Introduction
E. Worch, 2019
ISBN 978-3-11-055154-9, e-ISBN 978-3-11-055155-6,
e-ISBN (EPUB) 978-3-11-055166-2

Hydrochemistry.
Basic Concepts and Exercises
E. Worch, 2015
ISBN 978-3-11-031553-0, e-ISBN 978-3-11-031556-1,
e-ISBN (EPUB) 978-3-11-038202-0

Nanomaterials for Water Remediation
A. K. Mishra, C. M. Hussain, S. B. Mishra (Eds.), 2020
ISBN 978-3-11-064336-7, e-ISBN 978-3-11-065060-0,
e-ISBN (EPUB) 978-3-11-063455-6

Green Chemistry.
Water and its Treatment
M. A. Benvenuto, H. Plaumann (Eds.), 2021
ISBN 978-3-11-059730-1, e-ISBN 978-3-11-059782-0,
e-ISBN (EPUB) 978-3-11-059806-3

Ultrafiltration Membrane Cleaning Processes.
Optimization in Seawater Desalination Plants
G. Gilabert-Oriol, 2021
ISBN 978-3-11-071507-1, e-ISBN 978-3-11-071514-9,
e-ISBN (EPUB) 978-3-11-071516-3

Eckhard Worch

Adsorption Technology in Water Treatment

Fundamentals, Processes, and Modeling

2nd, Revised Edition

DE GRUYTER

Author
Prof. Dr. Eckhard Worch
Dresden University of Technology
Institute of Water Chemistry
01062 Dresden
Germany

ISBN 978-3-11-071542-2
e-ISBN (PDF) 978-3-11-071550-7
e-ISBN (EPUB) 978-3-11-071560-6

Library of Congress Control Number: 2020950607

Bibliographic information published by the Deutsche Nationalbibliothek
The Deutsche Nationalbibliothek lists this publication in the Deutsche Nationalbibliografie; detailed bibliographic data are available on the Internet at http://dnb.dnb.de.

Cover image: krystiannawrocki/E+/Getty Images
Typesetting: Integra Software Services Pvt. Ltd.
Printing and binding: CPI books GmbH, Leck

www.degruyter.com

Preface to the second edition

The first edition of *Adsorption Technology in Water Treatment* was published in 2012. Gratifyingly, the book has been well received by researchers, practitioners, and students. Therefore, the author, encouraged by the publisher, has decided to provide an improved and extended edition of the book.

The current edition includes extended presentations of the basics of adsorption modeling with a special focus on isotherm prediction, multicomponent adsorption modeling, and fixed-bed adsorber design by scale-up methods. Furthermore, some practical aspects, such as activated carbon application in drinking water and waste-water treatment and regeneration of oxidic adsorbents, are discussed in more detail and under consideration of recent developments.

A number of new and revised figures as well as additional items in the appendix complement the updated text passages of the book.

I hope the current edition will be a helpful tool and a source of know-how that enables users to better understand and successfully apply adsorption processes for water treatment.

Eckhard Worch
November 2020

https://doi.org/10.1515/9783110715507-202

Preface to the first edition

The principle of adsorption and the ability of certain solid materials to remove dissolved substances from water have long been known. For about 100 years, adsorption technology has been used to a broader extent for water treatment, and during this time, it has not lost its relevance. On the contrary, new application fields, besides the conventional application in drinking water treatment, have been added in recent decades, such as groundwater remediation or enhanced wastewater treatment.

The presented monograph treats the theoretical fundamentals of adsorption technology for water treatment. In particular, it presents the most important basics needed for planning and evaluation of experimental adsorption studies as well as for process modeling and adsorber design. The intention is to provide general basics, which can be adapted to the respective requirements, rather than specific application examples for selected adsorbents or adsorbates. As a practice-oriented book, it focuses more on the macroscopic processes in the reactors than on the microscopic processes at the molecular level.

The book begins with an introduction to basic concepts and an overview of adsorption processes in water treatment, followed by a chapter on adsorbents and their characterization. The main chapters of the book deal with the three constituents of the practice-related adsorption theory: adsorption equilibria, adsorption kinetics, and adsorption dynamics in fixed-bed columns. Single-solute systems as well as multicomponent systems of known and unknown composition are considered. A special emphasis is given to the competitive adsorption of micropollutants and organic background compounds due to the high relevance for micropollutant removal from different types of water. The treatment of engineered processes ends with a chapter on the restoration of the adsorbent capacity by regeneration and reactivation. The contents of the book are completed by an outlook on geosorption processes, which play an important role in seminatural treatment processes such as bank filtration or groundwater recharge.

It was in the mid-1970s, at the beginning of my PhD studies, when I was first faced with the theme of adsorption. Although I have broadened my research field during my scientific career, adsorption has always remained the focus of my interests. I would be pleased if this book, which is based on my long-term experience in the field of adsorption, would help readers to find an easy access to the fundamentals of this important water treatment process.

I would like to thank all those who contributed to this book by some means or other, in particular my PhD students as well as numerous partners in different adsorption projects.

Eckhard Worch
January 2012

https://doi.org/10.1515/9783110715507-203

Contents

1 Introduction

1.1 Basic concepts and definitions

1.1.1 Adsorption as a surface process

Adsorption is a phase transfer process that is widely used in practice to remove substances from fluid phases (gases or liquids). It can also be observed as a natural process in different environmental compartments. The most general definition describes adsorption as an enrichment of chemical species from a fluid phase on the surface of a liquid or a solid. In water treatment, adsorption has been proved as an efficient removal process for a multiplicity of solutes. Here, molecules or ions are removed from the aqueous solution by adsorption onto solid surfaces.

Solid surfaces are characterized by active, energy-rich sites that are able to interact with solutes in the adjacent aqueous phase due to their specific electronic and spatial properties. Typically, the active sites have different energies, or – in other words – the surface is energetically heterogeneous.

In adsorption theory, the basic terms shown in Figure 1.1 are used. The solid material that provides the surface for adsorption is referred to as the adsorbent; the species that will be adsorbed are named adsorbate. By changing the properties of the liquid phase (e.g. concentration, temperature, pH) adsorbed species can be released from the surface and transferred back into the liquid phase. This reverse process is referred to as desorption.

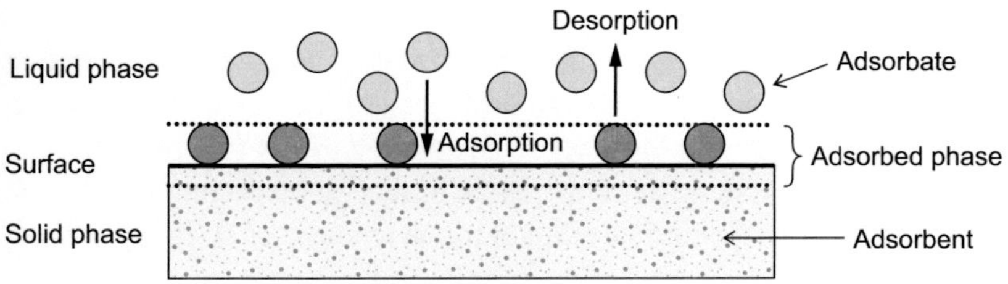

Figure 1.1: Basic terms of adsorption.

Since adsorption is a surface process, the surface area is a key quality parameter of adsorbents. Engineered adsorbents are typically highly porous materials with surface areas in the range between 10^2 and 10^3 m^2/g. Their porosity allows realizing such large surfaces as internal surfaces constituted by the pore walls. In contrast, the external surface is typically below 1 m^2/g and therefore of minor relevance. As an example, the external surface of powdered activated carbon with a particle density of 0.6 g/cm^3 and a particle radius of 0.02 mm is calculated to be only 0.25 m^2/g,

https://doi.org/10.1515/9783110715507-001

whereas typical values of the internal surface of activated carbons are in the range between 600 and 1,200 m^2/g.

1.1.2 Some general thermodynamic considerations

In thermodynamics, the state of a system is described by fundamental equations for the thermodynamic potentials. The Gibbs free energy, G, is one of these thermodynamic potentials. In surface processes, the Gibbs free energy is not only a function of temperature (T), pressure (p), and composition of the system (number of moles, n_i) but also a function of the surface area, A. Its change is given by the fundamental equation

$$\mathrm{d}G = -S\,\mathrm{d}T + V\,\mathrm{d}p + \sum_i \mu_i\,\mathrm{d}n_i + \sigma\,\mathrm{d}A \tag{1.1}$$

where S is the entropy, V is the volume, μ is the chemical potential, and σ is the surface free energy, also referred to as surface tension

$$\sigma = \left(\frac{\partial G}{\partial A}\right)_{T,p,n_i} \tag{1.2}$$

If adsorption takes place, the surface free energy is reduced from the initial value σ_{ws} (surface tension at the water-solid interface) to the value σ_{as} (surface tension at the interface between adsorbate solution and solid). The difference between σ_{ws} and σ_{as} depends on the adsorbed amount and is referred to as spreading pressure, π,

$$\sigma_{ws} - \sigma_{as} = \pi > 0 \tag{1.3}$$

The Gibbs fundamental equation (eq. (1.1)) and the relationship between spreading pressure and adsorbent loading provide the basis for the most frequently applied competitive adsorption model, the ideal adsorbed solution theory (Chapter 4).

Conclusions on the heat of adsorption can be drawn by inspecting the change of the free energy of adsorption and its relation to the changes of enthalpy and entropy of adsorption. The general precondition for a spontaneously proceeding reaction is that the change of free energy of reaction has a negative value. Considering the relationship between free energy, enthalpy, and entropy of adsorption, the respective condition for a spontaneous adsorption process reads

$$\Delta G_{ads} = \Delta H_{ads} - T\,\Delta S_{ads} < 0 \tag{1.4}$$

The change of the adsorption entropy describes the change in the degree of disorder in the considered system. Typically, the immobilization of the adsorbate leads to a decrease of disorder in the adsorbate/adsorbent system, which means that the change of the entropy is negative ($\Delta S_{ads} < 0$). Exceptions could be caused by dissociation during adsorption or by displacement processes where more species are desorbed than adsorbed. Given that ΔS_{ads} is negative, it follows from eq. (1.4) that adsorption must be an exothermic process ($\Delta H_{ads} < 0$).

Depending on the value of the adsorption enthalpy, adsorption can be categorized as physical adsorption (physisorption) or chemical adsorption (chemisorption). The physical adsorption is caused by van der Waals forces (dipole-dipole interactions, dispersion forces, induction forces), which are relatively weak interactions. The adsorption enthalpy in the case of physisorption is mostly lower than 50 kJ/mol. Chemisorption is based on chemical reactions between the adsorbate and the surface sites, and the interaction energies are therefore in the order of magnitude of reaction enthalpies (>50 kJ/mol). It has to be noted that the differentiation between physisorption and chemisorption is widely arbitrary and the boundaries are fluid.

1.1.3 Adsorption versus absorption

As explained before, the term "adsorption" describes the enrichment of adsorbates on the surface of an adsorbent. In contrast, absorption is defined as transfer of a substance from one bulk phase to another bulk phase. Here, the substance is enriched within the receiving phase and not only on its surface. The dissolution of gases in liquids is a typical example of absorption.

In natural systems, some materials with complex structure can bind substances from the aqueous phase on their surface but also in the interior of the material. The uptake of organic solutes by the organic fractions of soils, sediments, or aquifer materials is a typical example of such complex binding mechanisms. In such cases, it is not easy to distinguish between adsorption and absorption. Therefore, the more general term "sorption" is preferred to describe the phase transfer between the liquid and the solid in natural systems. The term "sorption" comprises adsorption and absorption. Moreover, the general term "sorption" is also used for ion exchange processes on mineral surfaces.

1.1.4 Mathematical description of adsorption processes: The structure of the adsorption theory

In accordance with the character of adsorption as a surface process, it would be reasonable to express the adsorbate uptake by the adsorbent surface as surface concentration, Γ (in mol/m^2), which is the quotient of the adsorbed amount, n_a, and the adsorbent surface area, A,

$$\Gamma = \frac{n_a}{A} \tag{1.5}$$

However, since the surface area, A, cannot be determined as exactly as the adsorbent mass, in practice the mass-related adsorbed amount, q, is typically used instead of the surface concentration, Γ,

$$q = \frac{n_a}{m_A} \tag{1.6}$$

where m_A is the adsorbent mass. The amount adsorbed per mass adsorbent is also referred to as adsorbent loading or simply loading.

In view of the practical application of adsorption, it is important to study the dependences of the adsorbed amount on the characteristic process parameters and to describe these dependences on a theoretical basis. The practice-oriented adsorption theory consists of three main elements: the adsorption equilibrium, the adsorption kinetics, and the adsorption dynamics. The adsorption equilibrium describes the dependence of the adsorbed amount on the adsorbate concentration and the temperature

$$q = \mathrm{f}(c, T) \tag{1.7}$$

For the sake of simplicity, the equilibrium relationship is typically considered at constant temperature and expressed in the form of the adsorption isotherm

$$q = \mathrm{f}(c) \quad T = \text{constant} \tag{1.8}$$

The adsorption kinetics describes the time dependence of the adsorption process, which means the increase of the adsorbent loading with time or, alternatively, the decrease of liquid-phase concentration with time

$$q = \mathrm{f}(t), \quad c = \mathrm{f}(t) \tag{1.9}$$

The adsorption rate is typically determined by slow mass transfer processes from the liquid to the solid phase.

Adsorption within the frequently used fixed-bed adsorbers is not only a time-dependent but also a spatial-dependent process. The dependence on time (t) and space (z) is referred to as adsorption dynamics or column dynamics

$$q = \mathrm{f}(t, z), \quad c = \mathrm{f}(t, z) \tag{1.10}$$

Figure 1.2 shows the main constituents of the practice-oriented adsorption theory and their interdependences. The adsorption equilibrium is the basis of all adsorption models. Knowledge about the adsorption equilibrium is a precondition for the application of both kinetic and dynamic adsorption models. To predict adsorption dynamics, information about adsorption equilibrium as well as about adsorption kinetics is required.

These general principles of adsorption theory are not only valid for single-solute adsorption but also for multisolute adsorption, which is characterized by competition of the adsorbates for the available adsorption sites and, in particular in

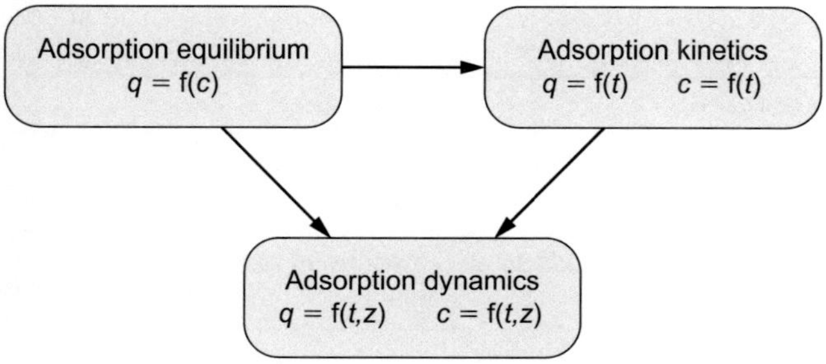

Figure 1.2: Elements of the adsorption theory.

fixed-bed adsorbers, by displacement processes. The prediction of multisolute adsorption behavior from single-solute data is an additional challenge in practice-oriented adsorption modeling.

Adsorption equilibria in single-solute and multisolute systems will be considered in detail in Chapters 3 and 4. Chapter 5 focuses on adsorption kinetics, whereas Chapters 6 and 7 deal with adsorption dynamics in fixed-bed adsorbers.

1.2 Engineered adsorption processes in water treatment

1.2.1 Overview

Adsorption processes are widely used in water treatment. Table 1.1 gives an overview of typical application fields and treatment objectives. Depending on the adsorbent type applied, organic substances as well as inorganic ions can be removed from the aqueous phase. A detailed characterization of the different adsorbents can be found in Chapter 2.

Table 1.1: Adsorption processes in water treatment.

Application field	Objective	Adsorbent
Drinking water treatment	Removal of dissolved organic matter	Activated carbon
	Removal of organic micropollutants	Activated carbon
	Removal of arsenic	Aluminum oxide, iron hydroxide
Municipal wastewater treatment	Removal of phosphate	Aluminum oxide, iron hydroxide
	Removal of micropollutants	Activated carbon

Table 1.1 (continued)

Application field	Objective	Adsorbent
Industrial wastewater treatment	Removal or recycling of specific chemicals	Activated carbon, polymeric adsorbents
Swimming-pool water treatment	Removal of organic substances	Activated carbon
Groundwater remediation	Removal of organic substances	Activated carbon
Treatment of landfill leachate	Removal of organic substances	Activated carbon
Aquarium water treatment	Removal of organic substances	Activated carbon

Activated carbon is the most important engineered adsorbent applied in water treatment. It is widely used to remove organic substances from different types of water such as drinking water, wastewater, groundwater, landfill leachate, swimming-pool water, and aquarium water. Other adsorbents are less often applied. Their application is restricted to special adsorbates or types of water.

1.2.2 Drinking water treatment

For nearly 100 years, adsorption processes with activated carbon as adsorbent have been used in drinking water treatment to remove organic solutes. At the beginning, taste and odor compounds were the main target solutes, whereas later the application of activated carbon was proved to be efficient for removal of a wide range of other organic micropollutants, such as phenols, chlorinated hydrocarbons, pesticides, pharmaceuticals, personal care products, corrosion inhibitors, and so on. Activated carbon adsorption is mainly applied in cases where surface water or riverbank filtrate is used as raw water source. However, its application can also be necessary in groundwater treatment if the groundwater is polluted with harmful compounds.

Since natural organic matter (NOM, measured as dissolved organic carbon, DOC) is present in all raw waters and often not completely removed by upstream processes, it is always adsorbed together with the organic micropollutants. Since activated carbon is not very selective in view of the adsorption of organic substances, the competitive NOM adsorption and the resulting capacity loss for micropollutants cannot be avoided. The competition effect is often relatively strong not least due to the different concentration levels of DOC and micropollutants. The typical DOC concentrations in raw waters are in the lower mg/L range, whereas the concentrations of organic micropollutants are in the ng/L or µg/L range. On the other hand, the NOM removal also has a positive aspect. NOM is known as a precursor for the formation of disinfection by-products (DBPs) during the final disinfection with chlorine or chlorine dioxide. Therefore, removal of NOM during the adsorption process helps to reduce the formation of DBPs.

Activated carbon can be applied as powdered activated carbon (PAC) in slurry reactors or as granular activated carbon (GAC) in fixed-bed adsorbers. The particle sizes of powdered activated carbons are in the medium micrometer range, whereas the GAC particles have diameters in the lower millimeter range.

PAC is often temporarily applied to control the seasonal occurrence of taste and odor compounds or pesticides in surface waters. In such cases, the PAC suspension is added to the existing pipes or contactors. Figure 1.3 shows the possible PAC introduction points in the typical treatment train used for surface water treatment. In most cases, PAC is added to the rapid mix tank that is used for coagulation. The contact time, determined by the residence time in the downstream flocculation basin, ranges between 10 and 45 minutes. The addition before the rapid mix tank allows somewhat longer contact times but can lead to an increased competition because at this point compounds that can be removed by coagulation/flocculation are still present in the water. However, studies suggest that the effect of coagulation/flocculation on the competitive adsorption of micropollutants and natural organic matter is not very strong because coagulation/flocculation preferentially removes larger molecules of the natural organic matter that are not able to strongly compete with the micropollutants in the micropores (Zoschke et al. 2011). If the PAC is added directly before the granular filtration, it is retained longer within the filter, which increases the adsorption efficiency. However, attention has to be paid to the possible breakthrough of fine PAC particles.

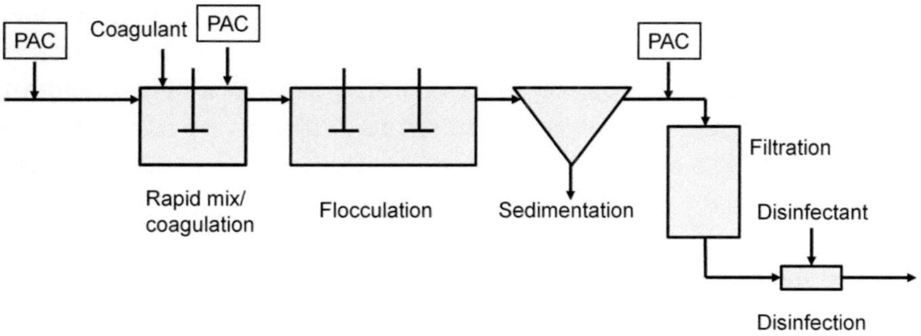

Figure 1.3: Flow sheet of a surface water treatment train with possible PAC introduction points.

GAC adsorbers as integral parts of the treatment train are typically arranged behind granular filtration and just prior to final disinfection (Figure 1.4). In a special process variant, adsorption is combined with biodegradation of the organic compounds. To enhance the biological activity in the GAC adsorber, ozonation is applied prior to adsorption. Ozonation breaks down the NOM molecules and makes them better assimilable and microbially oxidizable. The degradation of NOM constituents in the GAC adsorber

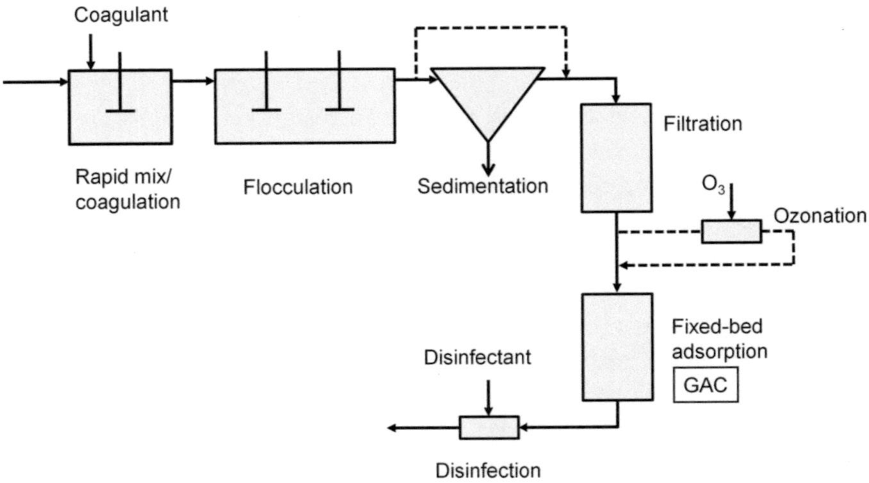

Figure 1.4: Arrangement of GAC adsorption within the drinking water treatment train.

can lead to weaker competition with nondegradable micropollutants. This combination of biodegradation and adsorption is known as biological activated carbon (BAC) process or biologically enhanced activated carbon process (see Chapter 7, Section 7.6.2).

In recent years, the problem of arsenic in drinking water has increasingly attracted public and scientific interest. In accordance with the recommendations of the World Health Organization (WHO), many countries have reduced their limit value for arsenic in drinking water to 10 µg/L. As a consequence, a number of waterworks, in particular in areas with high geogenic arsenic concentrations in groundwater and surface water have to upgrade their technologies by introducing an additional arsenic removal process. Adsorption processes with oxidic adsorbents such as ferric hydroxide or aluminum oxide have been proved to remove arsenate very efficiently. The same adsorbents are also expected to remove anionic uranium and selenium species.

1.2.3 Wastewater treatment

The conventional wastewater treatment process includes mechanical and biological treatment (primary and secondary treatment). In order to further increase the effluent quality and to protect the receiving environment, more and more often a tertiary treatment step is introduced into the treatment train. A main objective of the tertiary treatment is to remove nutrients, which are responsible for eutrophication of lakes and rivers. To remove the problematic phosphate, a number of different processes are in use (e.g. biological and precipitation processes). Adsorption of phosphate onto ferric hydroxide or aluminum oxide is an interesting alternative in particular for smaller decentralized treatment plants. A further aspect is that adsorption allows for recycling the phosphate, which is a valuable raw material, for instance, for fertilizer production.

In recent years, the focus has been directed to persistent micropollutants, which are not degraded during the activated sludge process. To avoid their input in water bodies, additional treatment steps are in discussion and, in some cases, already realized. Besides oxidation with ozone, adsorption onto activated carbon is considered a promising additional treatment process because its suitability to remove organic substances is well known from drinking water treatment. An advantage of activated carbon adsorption in comparison to ozonation is that no by-products are formed. However, some of the emerging compounds are polar and therefore only weakly adsorbable. Furthermore, the micropollutant adsorption is strongly influenced by competition effects, in this case between the micropollutants and the effluent organic matter (EfOM). This is a comparable effect as in drinking water treatment where NOM and micropollutants compete for the adsorption sites.

As in drinking water treatment, activated carbon can be applied as powdered activated carbon (PAC) or granular activated carbon (GAC). If PAC should be added, different introduction points are possible (Figure 1.5): directly into the aerated activated sludge tank (alternatively into the upstream anoxic denitrification stage), prior to a filter, or into a separate contactor.

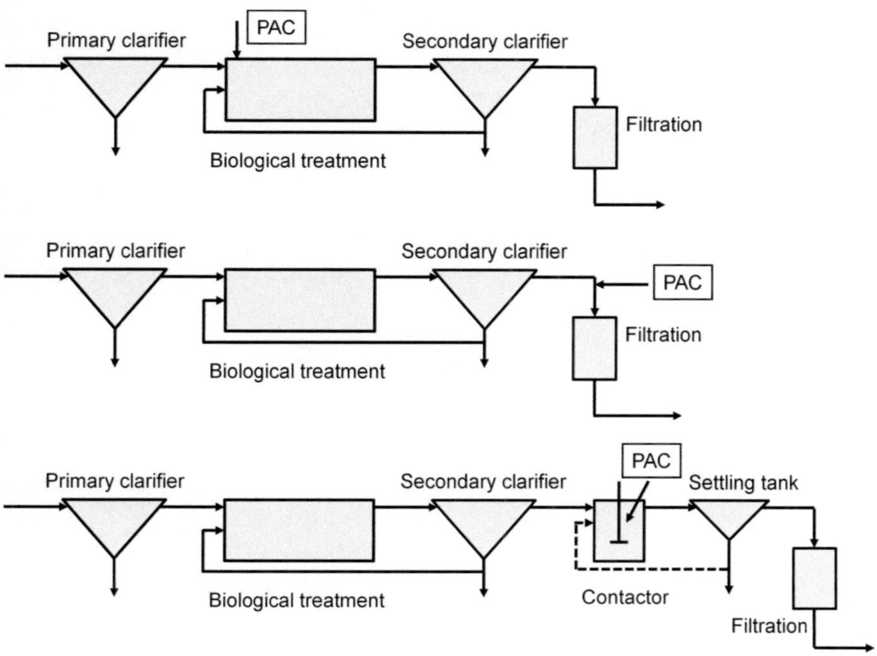

Figure 1.5: Removal of organic micropollutants in wastewater treatment: Different variants of PAC application.

In the first and third variant, the main part of the loaded carbon is removed by sedimentation, in the first variant together with the activated sludge, and in the third variant in a separate settling tank, optionally supported by coagulants. It was found that coagulation has only a minor effect on the adsorption of organic micropollutants because coagulation mainly affects EfOM fractions that are irrelevant for competitive adsorption. Furthermore, incorporation of PAC into the flocs does not appear to impede the mass transfer of the micropollutants to the PAC surface (Altmann et al. 2015). Since the separation by sedimentation is not complete, an additional filtration over sand or anthracite is necessary to avoid a carbon particle discharge into the receiving water body. In the third variant, the efficiency of the adsorption process can be improved by recirculation of the settled carbon particles to the contactor.

In the second variant, filtration over sand or anthracite serves as the only particle separation stage. Here, frequent backwashing is necessary to avoid a particle breakthrough. In this case, too, addition of coagulants can support the separation of fine PAC particles.

GAC is applied in fixed-bed adsorbers that are arranged after the secondary clarifier (Figure 1.6). GAC adsorption can also be combined with ozonation. Besides the direct oxidation of micropollutants, ozonation reduces the aromaticity, molecular size, and hydrophobicity of the effluent organic matter, leading to a decreased adsorbability and in turn to a weaker competition with organic micropollutants (Zietzschmann et al. 2015).

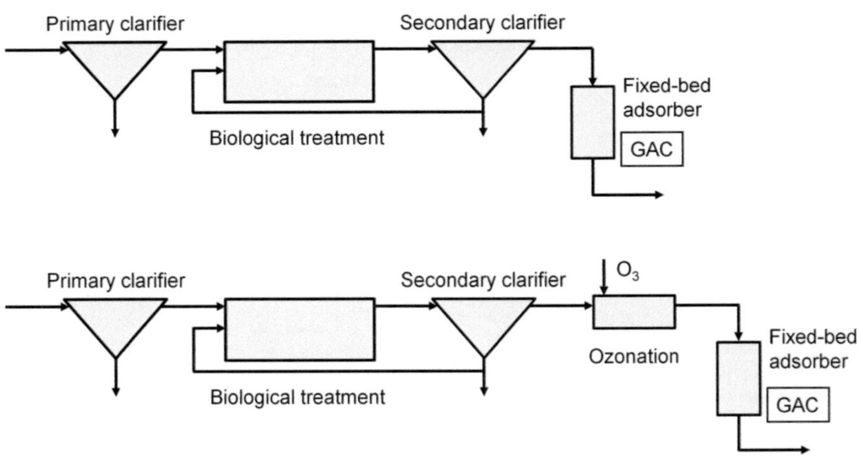

Figure 1.6: Removal of micropollutants in wastewater treatment: GAC adsorption and combined ozonation and GAC adsorption.

In industrial wastewater treatment, adsorption processes are also an interesting option, in particular for removal or recycling of organic substances. If the treatment objective is only removal of organics from the wastewater, activated carbon is an appropriate adsorbent. On the other hand, if the focus is more on the recycling of valuable chemicals, alternative adsorbents (e.g. polymeric adsorbents), which allow an easier desorption (e.g. by solvents), can be used.

1.2.4 Hybrid processes in water treatment

Adsorbents can also be used in other water treatment processes to support these processes by synergistic effects. Mainly activated carbon, in particular powdered activated carbon (PAC), is used in such hybrid processes. As in other activated carbon applications, the target compounds of the removal processes are organic substances.

Addition of PAC to the activated sludge process is a measure that has been well known for a long time. Here, activated carbon increases the removal efficiency by adsorbing substances that are not biodegradable or inhibit biological processes (see also Section 1.2.3). Furthermore, activated carbon provides an attachment surface for the microorganisms. The high biomass concentration at the carbon surface allows for an enhanced degrading of initially adsorbed substances. Due to the biological degradation of the adsorbed species, the activated carbon is permanently regenerated during the process. This effect is referred to as bioregeneration. Since activated carbon acts as a kind of buffer against substances that would disturb the biodegradation due to their toxicity or high concentrations, this process was originally developed for the treatment of highly contaminated industrial wastewaters or landfill leachates. In recent years, it is also considered an option for micropollutant removal from domestic wastewaters (Section 1.2.3).

The combination of activated carbon adsorption with membrane processes is a current development in water treatment. In particular, the application of PAC in ultrafiltration (UF) and nanofiltration (NF) processes is under discussion and in some cases already implemented.

Ultrafiltration membranes are able to remove particles and large molecules from water. By adding PAC to the membrane system, dissolved low-molecular-weight organic substances can be adsorbed and removed together with the PAC and other particles. As an additional effect, a reduction of membrane fouling can be expected because the concentration of organic matter is decreased by adsorption.

Addition of PAC to nanofiltration systems is also proposed, although nanofiltration itself is able to remove dissolved substances, including small molecules. Nevertheless, a number of benefits of an NF/PAC hybrid process can be expected. The high solute concentrations on the concentrate side of the membrane provide favorable conditions for adsorption so that high adsorbent loadings can be achieved. Furthermore, the removal of organics on the concentrate side by adsorption decreases the organic membrane fouling. Additionally, abrasion caused by the activated carbon particles reduces the coating of the membrane surface.

UF/PAC or NF/PAC hybrid processes can be used for different purposes, such as drinking water treatment, wastewater treatment, landfill leachate treatment, or groundwater remediation.

1.3 Natural sorption processes in water treatment

Sorption processes (adsorption, absorption, ion exchange) may occur in many natural systems. In principle, sorption can take place at all interfaces where an aqueous phase is in contact with natural solid material. Table 1.2 gives some typical examples. Sorption processes are able to remove dissolved species from the aqueous phase and lead to accumulation and retardation of these species. These processes are part of the self-purification within the water cycle. Natural sorbents are often referred to as geosorbents. Accordingly, the term "geosorption" is sometimes used for natural sorption processes.

Table 1.2: Examples of natural sorption systems.

Natural solid material acting as sorbent	Liquid phase in contact with the solid
Lake and river sediments	Surface water
Suspended matter in groundwater and surface water	Groundwater or surface water
Soil (vadose zone)	Seepage water, infiltrate
Aquifer material (saturated zone)	Groundwater, bank filtrate, infiltrate

Under certain conditions, sorption processes in special environmental compartments can be utilized for water treatment purposes. In drinking water treatment, bank filtration or infiltration (Figure 1.7) are typical examples of utilizing the attenuation potential of natural sorption processes. Bank filtration is a pretreatment option if polluted surface water has to be used for drinking water production. In this case, the raw

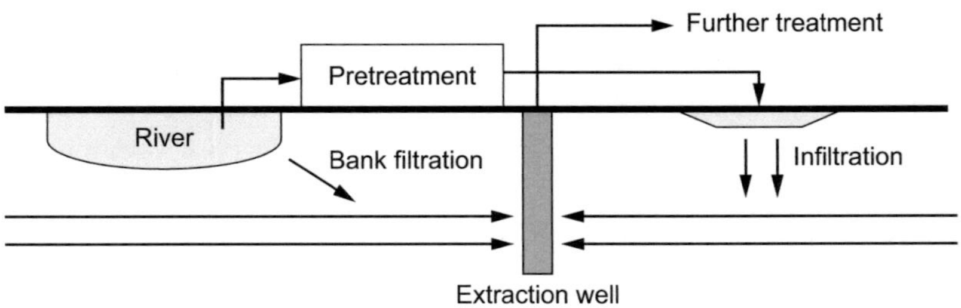

Figure 1.7: Schematic representations of the processes bank filtration and infiltration as part of drinking water treatment.

water is not extracted directly from the river or lake but from extraction wells located a certain distance from the bank. Due to the hydraulic gradient between the river and the extraction wells caused by pumping, the water flows in the direction of the extraction wells. During the subsurface transport, complex attenuation processes take place. Although biodegradation is the most important process, in particular in the first part of the flow path, sorption onto the aquifer material during the subsurface transport also can contribute to the purification of the raw water, in particular by retardation of nondegradable or poorly degradable solutes.

Infiltration of surface water, typically pretreated by engineered processes (e.g. flocculation, sedimentation), is based on analogous principles. During the contact of the infiltrated water with the soil and the aquifer material, biodegradation and sorption processes can take place leading to an improvement of the water quality. The infiltrated water is then extracted by extraction wells and further treated by engineered processes.

Natural processes are not only used in drinking water treatment but also for reuse of wastewater. Particularly in regions with water scarcity, the use of reclaimed wastewater for artificial groundwater recharge becomes increasingly important. In this case, wastewater, treated by advanced processes, is infiltrated into the subsurface where in principle the same attenuation processes as during bank filtration or surface water infiltration take place. Since the degradable water constituents are already removed to a high extent in the wastewater treatment plant, it can be expected that sorption of nondegradable or poorly degradable solutes is of particular relevance as purification process during wastewater infiltration.

In all these cases, the solid material that acts as sorbent is of complex composition. Therefore, different types of interactions are possible. It is well known from a multitude of studies that the organic fraction of the solid material is of particular importance for the binding of neutral organic solutes. Other components such as clay minerals or oxidic surfaces are mainly relevant for ionic species. More details of natural sorption processes are discussed in Chapter 9.

It has to be noted that the application of the above-mentioned processes for water treatment is only possible if the subsurface layers at the considered sites show sufficient permeability.

2 Adsorbents and adsorbent characterization

2.1 Introduction and adsorbent classification

Adsorbents used for water treatment are either of natural origin or the result of an industrial production and/or activation process. Typical natural adsorbents are clay minerals, natural zeolites, oxides, or biopolymers. Engineered adsorbents can be classified into carbonaceous adsorbents, polymeric adsorbents, oxidic adsorbents, and zeolite molecular sieves. Activated carbons produced from carbonaceous material by chemical activation or gas activation are the most widely applied adsorbents in water treatment. Polymeric adsorbents made by copolymerization of nonpolar or weakly polar monomers show adsorption properties comparable to activated carbons, but high material costs and costly regeneration have prevented a broader application to date. Oxides and zeolites are adsorbents with stronger hydrophilic surface properties. The removal of polar, in particular ionic, compounds is therefore their preferred field of application. In recent decades, an increasing interest in using wastes and by-products as alternative low-cost adsorbents (LCAs) can be observed.

In general, engineered adsorbents exhibit the highest adsorption capacities. They are produced under strict quality control and show nearly constant properties. In most cases, the adsorption behavior towards a broad variety of adsorbates is well known, and recommendations for application can be derived from scientific studies and producers' information. On the other hand, engineered adsorbents are often very expensive. In contrast, the adsorption capacities of natural and other LCAs are much lower and the properties are subject to stronger variations. They might be interesting due to their low prices, but in most cases, the studies about LCAs are limited to very specific applications and not enough information is available for a generalization of the experiences and for a final assessment.

To guarantee the safety of drinking water, adsorbents for use in drinking water treatment have to fulfill high quality standards and typically must be certified. Therefore, the number of possible adsorbents is limited and comprises basically commercial activated carbons and oxidic adsorbents. The other adsorbents, including the LCAs, are rather suitable for wastewater treatment.

Since adsorption is a surface process, the surface area of the adsorbent is of great importance for the extent of adsorption and is therefore a key quality parameter. In general, natural adsorbents have much smaller surface areas than highly porous, engineered adsorbents. The largest surface areas can be found for activated carbons and special polymeric adsorbents. A precondition for high surface area is high porosity of the material, which enables a large internal surface constituted by the pore walls. The internal surface of engineered adsorbents is much larger than their external particle surface. As a rule, the larger the pore system and the finer the pores, the higher is the internal surface. On the other hand, a certain fraction of

https://doi.org/10.1515/9783110715507-002

larger pores is necessary to enable fast adsorbate transport to the adsorption sites. Therefore, pore-size distribution is a further important quality aspect. Besides the texture, the surface chemistry may also be of interest, in particular for chemisorption processes.

In this chapter, the most important adsorbents and frequently used methods for their characterization are presented.

2.2 Engineered adsorbents

2.2.1 Activated carbons

The adsorption properties of carbon-rich materials (e.g. wood charcoal, bone charcoal) have been known for millennia, but only since the beginning of the twentieth century has this material been improved by special activation processes. Activated carbons can be produced from different carbon-containing raw materials and by different activation processes. The most common raw materials are wood, wood charcoal, peat, lignite and lignite coke, hard coal and coke, bituminous coal, petrol coke as well as residual materials, such as coconut shells, sawdust, or plastic residuals.

For organic raw materials like wood, sawdust, peat, or coconut shells, a preliminary carbonization process is necessary to transform the cellulose structures into a carbonaceous material. Such cellulose structures contain a number of oxygen- and hydrogen-containing functional groups, which can be removed by dehydrating chemicals. The dehydration is typically carried out at elevated temperatures under pyrolytic conditions and leads to a destruction of the cellulose structures with the result that the carbon skeleton is left. This process, referred to as chemical activation, combines carbonization and activation processes. Typical dehydrating chemicals are zinc chloride and phosphoric acid. After cooling the product, the activation agent has to be extracted. Since the extraction is often not complete, residuals of the activation chemicals remain in the activated carbon and might be leached during the application. This is in particular critical for drinking water treatment. Furthermore, the application of chemicals in the activation process requires an expensive recycling, and the products of chemical activation are typically powders with low densities and low content of micropores. For these reasons, most of the activated carbons used in drinking water treatment are produced by an alternative process termed physical, thermal, or gas activation.

In gas activation, carbonized materials such as coals or cokes are used as raw materials. These carbon-rich materials already have a certain porosity. For activation, the raw material is brought in contact with an activation gas (steam, carbon dioxide, air) at elevated temperatures (800 °C–1,000 °C). During the activation, the activation gas reacts with the solid carbon to form gaseous products. In this manner, closed pores are opened and existing pores are enlarged. The reactions cause a mass loss of the solid material. Since the development of the pore system and the surface area are correlated with the burn-off, an optimum for the extent of the activation has to be found. This optimum depends on the material and is often in the range of 40% to 50% burn-off. Higher burn-off degrees lead to a decrease of net surface area because no more new pores are opened, but existing pore walls are burned away.

The main reaction equations for the chemical processes during gas activation together with the related reaction enthalpies are given below. The reactions in gas activation are the same as in the carbochemical process of coal gasification for synthesis gas production, but in contrast to it, the gasification is not complete. A positive sign of the reaction enthalpy indicates an endothermic process, whereas a negative sign indicates an exothermic process

$$C + H_2O \rightleftharpoons CO + H_2 \qquad \Delta_R H = +131\,\text{kJ/mol}$$

$$C + CO_2 \rightleftharpoons 2\,CO \qquad \Delta_R H = +172\,\text{kJ/mol}$$

$$2\,C + O_2 \rightleftharpoons 2\,CO \qquad \Delta_R H = -111\,\text{kJ/mol}$$

$$CO + 0.5\,O_2 \rightleftharpoons CO_2 \qquad \Delta_R H = -285\,\text{kJ/mol}$$

$$CO + 3\,H_2 \rightleftharpoons CH_4 + H_2O \qquad \Delta_R H = -210\,\text{kJ/mol}$$

$$CO + H_2O \rightleftharpoons CO_2 + H_2 \qquad \Delta_R H = -41\,\text{kJ/mol}$$

$$H_2 + 0.5\,O_2 \rightleftharpoons H_2O \qquad \Delta_R H = -242\,\text{kJ/mol}$$

The products of gas activation mainly occur in granulated form. Different particle sizes can be obtained by grinding and sieving. Gas activation processes can also be used for a further activation of chemically activated carbons.

Activated carbons are applied in two different forms, as granular activated carbon (GAC) with particle sizes in the range of 0.5 to 4 mm and powdered activated carbon (PAC) with particle sizes <40 µm. The different particle sizes are related to different application techniques: slurry reactors for PAC application and fixed-bed adsorbers for GAC. More technological details are discussed in the following chapters.

Activated carbons show a broad variety of internal surface areas ranging from some hundreds m^2/g to more than a thousand m^2/g depending on the raw material and the activation process used. Activated carbon for water treatment should not have pores that are too fine so that larger molecules are also allowed to enter the pore system and to adsorb onto the inner surface. Internal surface areas of activated carbons applied for water treatment are typically in the range of 800–1,000 m^2/g.

The activated carbon structure consists of crystallites with a strongly disturbed graphite structure (Figure 2.1). In graphite, the carbon atoms are located in layers and are connected by covalent bonds (sp^2 hybridization). Graphite possesses a delocalized π-electron system that is able to interact with aromatic structures in the adsorbate molecules. The graphite crystallites in activated carbons are randomly

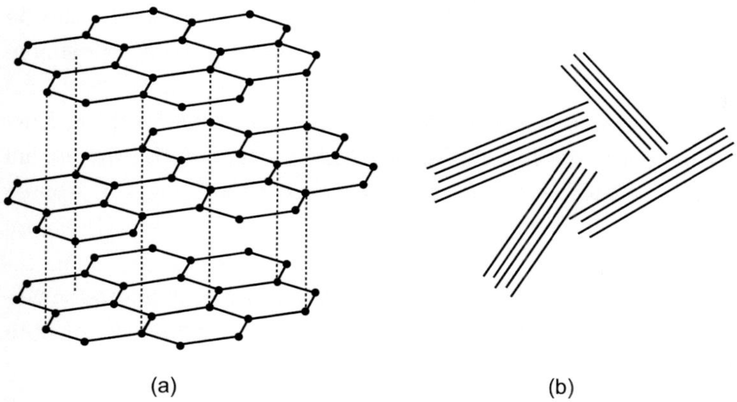

Figure 2.1: Structural elements of activated carbons: (a) graphite structure, (b) randomly oriented graphite microcrystallites.

oriented and interconnected by carbon cross-links. The micropores are formed by the voids between the crystallites and are therefore typically of irregular shape. Frequently, slit-like pores are found.

Activated carbons are able to adsorb a multiplicity of organic substances mainly by weak intermolecular interactions (van der Waals forces), in particular dispersion forces. These attraction forces can be superimposed by π-π interactions in the case of aromatic adsorbates or by electrostatic interactions between surface oxide groups (Section 2.5.6) and ionic adsorbates. Their high adsorption capacities make activated carbons to the preferred adsorbents in all water treatment processes where organic impurities should be removed. Besides trace pollutants (micropollutants), background dissolved organic matter (DOM) can also be efficiently removed by activated carbon.

Below, some general trends in activated carbon adsorption are listed.

- The adsorption increases with increasing internal surface (increasing micropore volume) of the adsorbent.
- The adsorption increases with increasing molecule size of the adsorbates as long as no size exclusion hinders the adsorbate molecules from entering the pore system.
- The adsorption decreases with increasing temperature because (physical) adsorption is an exothermic process (see Chapter 1).
- The adsorbability of organic substances onto activated carbon increases with decreasing polarity (solubility, hydrophilicity) of the adsorbate.
- Aromatic compounds are better adsorbed than aliphatic compounds of comparable size.
- Organic ions (e.g. phenolates or protonated amines) are not adsorbed as strongly as the corresponding neutral compounds (pH dependence of the adsorption of weak acids and bases).

- In multicomponent systems, competitive adsorption takes place, resulting in decreased adsorption of a considered compound in comparison with its single-solute adsorption.
- Inorganic ions (e.g. metal ions) can be adsorbed by interactions with the functional groups of the adsorbent surface (Section 2.5.6) but to a much lower extent than organic substances, which are adsorbed by dispersion forces and hydrophobic interactions.

To assess the performance of activated carbons, often practice-oriented test methods are used. In these tests, the adsorption strength of model adsorbates is determined under defined conditions. More details are given in Section 2.6.

Loaded GAC is typically regenerated by thermal processes (Chapter 8). In most cases, the activated carbon is reactivated analogously to the gas activation process. Reactivation causes a mass loss due to burn-off. PAC cannot be reactivated and is therefore used as a one-way adsorbent and has to be burned or deposited after application.

A relatively new development in the field of activated carbon manufacturing is the production of synthetic activated carbons on the basis of polymer materials. For example, porous styrene-divinylbenzene copolymers (ion exchange resins or polymeric adsorbents, see Section 2.2.2) can be used as educts. The porous polymer balls are pyrolyzed to transform the organic material into carbon, and the carbonized balls are then further activated, typically with gas activation. The advantages of polymer-based spherical activated carbons are their reproducible quality, their regular shape, and the possibility to adjust the pore system to specific requirements. In contrast to the polymeric adsorbents, they can be reactivated in a comparable way as the conventional activated carbons. The disadvantage is that they are made from comparatively expensive starting materials.

2.2.2 Polymeric adsorbents

Polymeric adsorbents, also referred to as adsorbent resins, are porous solids with considerable surface areas and distinctive adsorption capacities for organic molecules. They are produced by copolymerization of styrene, or sometimes also acrylic acid esters, with divinylbenzene as a cross-linking agent. Their structure is comparable to that of ion exchange resins, but in contrast to ion exchangers, the adsorbent resins have no or only few functional groups and are nonpolar or only weakly polar. To obtain a high porosity, the polymerization is carried out in the presence of an inert medium that is miscible with the monomer and does not strongly influence the chain growth. After polymerization, the inert medium is removed from the polymerizate by extraction or evaporation. Polymeric adsorbent materials tailored for particular needs can be produced by variation of the type and the concentration of the inert compound, the monomer concentration, the fraction of divinylbenzene, the concentration of polar monomers, and the reaction conditions. Figure 2.2 shows the typical structure of a styrene-divinylbenzene copolymer. The conventional polymeric adsorbents have surface areas up to 800 m^2/g. Polymeric adsorbents typically show a narrow pore-size distribution, and the surface is relatively homogeneous. With increasing degree of cross-linking, the pore size becomes smaller and the surface area increases.

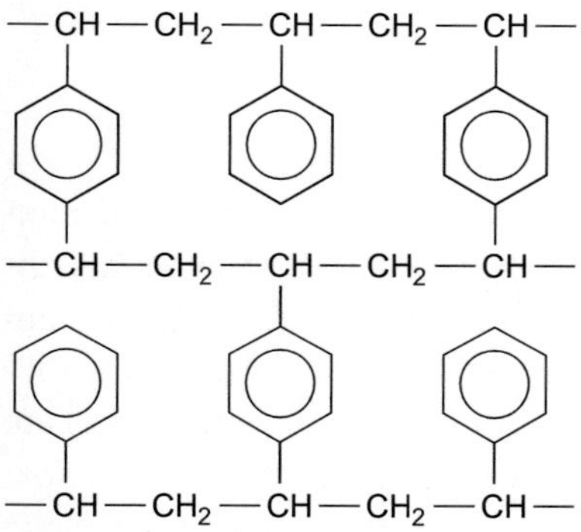

Figure 2.2: Structure of a styrene-divinylbenzene copolymer.

By specific post-cross-linking reactions, such as chloromethylation with subsequent dehydrochlorination (Figure 2.3), the pore size can be further reduced and large surface areas, up to 1,200 m^2/g and more, can be received.

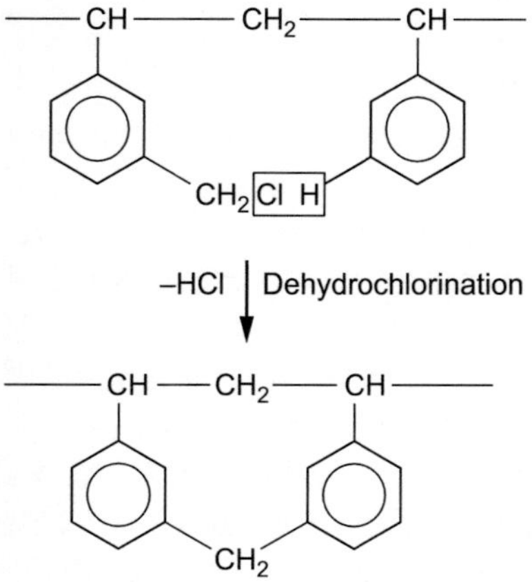

Figure 2.3: Principle of post-cross-linking of polymeric networks by chloromethylation and subsequent dehydrochlorination.

Highly cross-linked polymeric adsorbents show adsorption capacities that are comparable to that of activated carbons. Desorption of the adsorbed organic compounds is possible by extraction with solvents, in particular alcohols, such as methanol or isopropanol. The much higher costs for the polymeric materials in comparison to activated carbons and the need for extractive regeneration by solvents make the polymeric adsorbents unsuitable for treatment of large amounts of water with complex composition – for instance, for drinking water treatment or treatment of municipal wastewater effluents. Instead of that, polymeric adsorbents can be beneficially applied for recycling of valuable chemicals from process wastewaters. To separate the solvent from the desorbed compounds, an additional process step – for instance, distillation – is necessary (see also Chapter 8).

2.2.3 Oxidic adsorbents

The term "oxidic adsorbents" comprises solid hydroxides, hydrated oxides, and oxides. Among the engineered oxidic adsorbents, aluminum and iron materials are the most important. The general production process is based on the precipitation of hydroxides followed by a partial dehydration at elevated temperatures. The hydroxide products are thermodynamically metastable. Further strong heating would result in a transformation to stable oxides with only small surface areas.

The dehydration process of a trivalent metal (Me) hydroxide can be described in a simplified manner as

$$Me(OH)_3 \rightarrow MeO(OH) + H_2O$$

$$2\,MeO(OH) \rightarrow Me_2O_3 + H_2O$$

In these reactions, species with different water contents can occur as intermediates. The nomenclature used in practice for the different species is not always precise. Independent of the real water content, the hydrated materials are often simply referred to as oxides or hydroxides. In the following text, the established names for the materials will be used.

The oxidic adsorbents exhibit a relatively large number of surface OH groups, which substantially determine their adsorption properties. The polar character of the surface together with the possible protonation or deprotonation of the OH groups (Section 2.5.6) makes the oxidic adsorbents ideally suited for the removal of ionic compounds, such as phosphate, arsenate, fluoride, or heavy metal species.

Activated aluminum oxide (γ-aluminum oxide, γ-Al_2O_3) can be used for the removal of arsenate and fluoride from drinking water or for the removal of phosphate from wastewater. The surface areas are in the range of 150–350 m^2/g. Activated aluminum oxide is produced in different particle sizes, ranging from about 0.1 to 10 mm.

Recently, iron(III) hydroxide (ferric hydroxide) in granulated form finds increasing interest, in particular as an efficient adsorbent for arsenate (Driehaus et al. 1998), but also for phosphate (Sperlich 2010) and other oxyanions. Different products are available with crystal structures according to α-FeOOH (goethite) and β-FeOOH (akaganeite). The surface areas are comparable to that found for aluminum oxide and range from 150 to 350 m^2/g. Typical particle sizes are between 0.3 and 3 mm.

Ion adsorption onto oxidic adsorbents strongly depends on the pH value of the water to be treated. This can be explained by the influence of the pH on the surface charge (Section 2.5.6). This pH effect provides the opportunity to desorb the ions from the adsorbent by changing the pH. In the case of activated aluminum oxide and granular ferric hydroxide, the surface charge is positive up to pH values of about 8. Therefore, anions are preferentially adsorbed in the neutral pH range. The regeneration of the adsorbent (desorption of the anions) can be done by increasing the pH (e.g. by NaOH addition) to values where the surface charge becomes negative. However, it was found that the presence of calcium in the water to be treated complicates the desorption due to different effects, such as surface complexation and precipitation. Therefore, an additional acidic conditioning prior to the desorption step is recommended. More details on this specific aspect are presented in Chapter 8, Section 8.2.2.

2.2.4 Synthetic zeolites

Zeolites occur in nature in high diversity. For practical applications, however, often synthetic zeolites are used. Synthetic zeolites can be manufactured from alkaline aqueous solutions of silicium and aluminum compounds under hydrothermal conditions.

Zeolites are alumosilicates with the general formula $(Me^{II},Me^{I}_2)O \cdot Al_2O_3 \cdot n\ SiO_2 \cdot p\ H_2O$. In the alumosilicate structure, tetrahedral AlO_4 and SiO_4 groups are connected via joint oxygen atoms. Zeolites are tectosilicates (framework silicates) with a porous structure characterized by windows and caves of defined sizes. Zeolites can be considered as derivatives of silicates where Si is partially substituted by Al. As a consequence of the different number of valence electrons of Si (4) and Al (3), the zeolite framework carries negative charges, which are compensated by metal cations. Depending on the molar SiO_2/Al_2O_3 ratio (modulus n), different classes can be distinguished – for instance, the well-known types A ($n = 1.5 \ldots 2.5$), X ($n = 2.2 \ldots 3.0$), and Y ($n = 3.0 \ldots 6.0$). These classical zeolites are hydrophilic. They are in particular suitable for ion exchange processes (e.g. softening) but not for the adsorption of neutral organic substances. The hydrophobicity of zeolites increases with increasing modulus. High-silica zeolites with $n > 10$ are more hydrophobic and are therefore potential adsorbents for organic compounds. Although some promising experimental results for several organic adsorbates were published in the past, zeolites have not found broad application as adsorbents in water treatment until now.

2.3 Natural and low-cost adsorbents

Among the natural and low-cost adsorbents (LCAs), clay minerals have a special position. The application of natural clay minerals as adsorbents has been studied for a relatively long time. The adsorption properties of clay minerals or mineral mixtures such as bentonite (main component: montmorillonite) or Fuller's earth (attapulgite and montmorillonite varieties) are related to the net negative charge of the mineral structure. This property allows clays to adsorb positively charged species – for instance, heavy metal cations such as Cu^{2+}, Zn^{2+}, or Cd^{2+}. Relatively high adsorption capacities were also reported for organic dyes during treatment of textile wastewater. To improve the sorption capacity, clay minerals can be modified by organic cations to make them more organophilic.

In recent decades, a growing interest in LCAs has been observed, and, in addition to clay, other potential adsorbents have gained increasing interest. This can be seen, for instance, from the strongly increasing number of published studies in this field. This ongoing development is driven by the fast industrial growth in some regions of the world (e.g. in Asia) accompanied by increasing environmental pollution and the search for low-cost solutions to these problems. In these regions, often natural materials as well as wastes from agricultural and industrial processes are available, which come into consideration as potential adsorbents. Review articles about LCAs were published by Gupta et al. (2009) and Crini et al. (2019). Based on these reviews, the scheme shown in Figure 2.4 was derived, which illustrates the broad variety of possible LCAs. The adsorbents are mainly used untreated, but in some cases, physical and

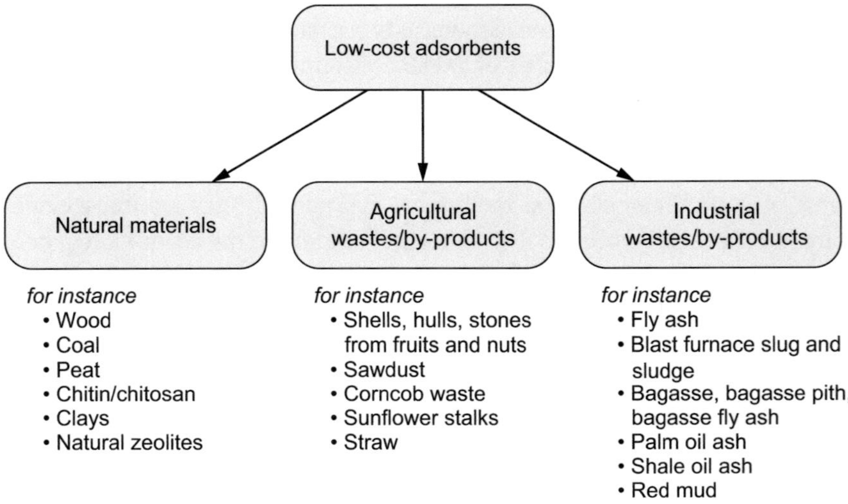

Figure 2.4: Selected low-cost adsorbents.

chemical pretreatment processes, such as heating or treatment with hydrolyzing chemicals, were also proposed. The studies on the adsorption properties of the alternative LCAs were mainly directed to the removal of problematic pollutants from industrial wastewaters, in particular heavy metals from electroplating wastewaters and dyes from textile wastewaters. In some studies, phenols were also considered.

In recent years, some studies were carried out that deals with the application of biochar as an alternative to coal-based activated carbon for organic micropollutant removal from wastewater (Thompson et al. 2016, Kearns et al. 2020). Biochar is pyrolyzed (carbonized) biomass (e.g. wood or biosolids from wastewater treatment) not subjected to further physical or chemical activation.

Despite the increasing number of studies on the application of LCAs, there is still a lack of systematic investigations, including in-depth studies on the adsorption mechanisms on a strict theoretical basis, and also a lack of comparative studies under defined conditions. Therefore, it is not easy to evaluate the practical importance of the different alternative adsorbents for wastewater treatment.

2.4 Geosorbents in environmental compartments

As already mentioned in Chapter 1, Section 1.3, sorption processes in certain environmental compartments can be utilized for water treatment purposes. Bank filtration and artificial groundwater recharge by infiltration of pretreated surface water or wastewater are typical examples of using the sorption capacity of natural sorbents to remove poorly biodegradable or nonbiodegradable substances from water. In these cases, soil and/or the aquifer materials act as sorbents. They are also referred to as geosorbents, and therefore the process of accumulating solutes by these solids can be termed “geosorption”. The geosorbents are typically heterogeneous solids consisting of mineral and organic components. The mineral components are mainly oxidic substances and clay minerals. Due to their surface charge (Sections 2.2.3, 2.3, and 2.5.6), they preferentially adsorb ionic species, including ionized organic compounds. In contrast, the organic fractions of the geosorbents (sorbent organic matter (SOM)) are able to bind neutral organic solutes, in particular hydrophobic compounds. The high affinity of hydrophobic solutes to the hydrophobic organic material can be explained by the effect of hydrophobic interactions, which is an entropy-driven process that induces the hydrophobic solute to leave the aqueous solution and to aggregate with other hydrophobic material. In accordance with the sorption mechanism, it can be expected that the sorption of neutral organic solutes increases with increasing hydrophobicity.

In contrast to the properties of engineered and low-cost adsorbents, which can be influenced by technical measures, the properties of the geosorbents in a considered environmental compartment are fixed and have to be accepted as they are. However, it is possible to take samples of the material and to study the composition and the characteristic sorption properties. The latter is typically done in column experiments with an experimental setup comparable to that used for fixed-bed adsorption studies with engineered adsorbents (for more details, see Chapter 9).

Due to the relevance of the sorbent organic matter for the uptake of organic solutes, the content of organic material in the solid is a very important quality parameter. To make the sorption properties of different solid materials comparable, the characteristic sorption coefficients are frequently normalized to the organic carbon content, given as fraction f_{oc}. Aquifer materials often have organic carbon fractions much lower than 1%, but these small fractions are already enough to sorb considerable amounts of organic solutes, in particular if these solutes are hydrophobic.

2.5 Adsorbent characterization

2.5.1 Densities

Since adsorbents are porous solids, different densities can be defined depending on the volume used as reference. A distinction can be made between material density, particle density (apparent density), and bulk (bed) density.

Material density

The material density, ρ_M, is the true density of the solid material (skeletal density). It is defined as the quotient of the adsorbent mass, m_A, and the volume of the solid material without pores, V_{mat},

$$\rho_M = \frac{m_A}{V_{mat}} \tag{2.1}$$

The material volume can be measured by means of a pycnometer. A pycnometer is a measuring cell that allows for determining the volume of a gas or liquid that is displaced after introducing the adsorbent. To find the material volume, compounds with small atom or molecule sizes that are able to fill nearly the total pore volume have to be used as the measuring gas or liquid. In this case, the displaced volume can be set equal to the material volume. A well-known method is based on the application of helium, which has an effective atom diameter of 0.2 nm. The displaced helium volume is indirectly determined in a special cell by measuring temperature and pressure. The material density determined by this method is also referred to as helium density.

An easier method is based on the application of liquid methanol in a conventional glass pycnometer (methanol density). Since the methanol molecule is larger than the helium atom, the finest pores are possibly not filled, and therefore the estimated material volume might be slightly too high.

Particle density

The particle density, ρ_P, is defined as the ratio of the adsorbent mass, m_A, and the adsorbent volume including pores, V_A,

$$\rho_P = \frac{m_A}{V_A} = \frac{m_A}{V_{mat} + V_{pore}} \tag{2.2}$$

where V_{pore} is the pore volume. The particle density is also referred to as apparent density.

Like the material density, the particle density can be determined in a pycnometer. However, in contrast to the determination of the material density, mercury is used as the pycnometer liquid because it cannot enter the pores and the displaced volume can be set equal to the sum of material volume and pore volume ($V_A = V_{mat} + V_{pore}$). For an exact determination of the particle density, it is necessary that the adsorbent particles are fully immersed in the liquid mercury. However, this is hard to realize in practice due to the high density difference between the adsorbent particles and mercury. To minimize the experimental error, a high number of parallel determinations have to be carried out. A careful determination of the particle density is necessary because ρ_P as well as further parameters derived from ρ_P are important data for adsorber design. The particle density determined in this way is also referred to as mercury density.

Sontheimer et al. (1988) have described a simple method for determining ρ_M as well as ρ_P for activated carbons by using water as the pycnometer liquid. A representative sample of the dry adsorbent is weighed (m_A), and then the pore system of the adsorbent is filled with water. This can be done by boiling an aqueous adsorbent suspension or by placing the suspension in a vacuum. After the wetting (filling of inner pores with water), the water is carefully removed from the outer surface of the wet adsorbent particles by centrifugation or rolling the particles on a paper towel. The wet adsorbent is then put into an empty pycnometer of known volume (V_{pyc}) and mass (m_{pyc}). The pycnometer with the wet adsorbent is weighed (m_1), completely filled with water, and weighed again (m_2). The mass of the wet carbon is given by

$$m_{wet} = m_1 - m_{pyc} \tag{2.3}$$

and the volume of the wet carbon is

$$V_{wet} = V_{pyc} - \frac{m_2 - m_1}{\rho_W} \tag{2.4}$$

where ρ_W is the density of water. Given that the volume of the wet carbon is equal to the sum of material volume and pore volume, the particle density can be found from

$$\rho_P = \frac{m_A}{V_A} = \frac{m_A}{V_{mat} + V_{pore}} = \frac{m_A}{V_{wet}} \tag{2.5}$$

To find the material volume necessary for the estimation of the material density, the volume of the water within the pores must be subtracted from the volume of the wet adsorbent

$$\rho_M = \frac{m_A}{V_{mat}} = \frac{m_A}{V_{wet} - \dfrac{m_{wet} - m_A}{\rho_W}} \tag{2.6}$$

Bulk density (bed density)

The bulk density, ρ_B, is an important parameter for characterizing the mass/volume ratio in adsorbers. It is defined as the ratio of the adsorbent mass and the total reactor volume filled with liquid and solid, V_R. V_R includes the adsorbent volume, V_A, and the volume of liquid that fills the space between the adsorbent particles, V_L,

$$\rho_B = \frac{m_A}{V_R} = \frac{m_A}{V_L + V_A} = \frac{m_A}{V_L + V_{pore} + V_{mat}} \tag{2.7}$$

In batch reactors, typically low amounts of adsorbent are dispersed in large volumes of liquid. Thus, the bulk density has the character of a mass concentration of the solid particles rather than that of a conventional density.

In fixed-bed adsorbers, the adsorbent particles are arranged in an adsorbent bed. Consequently, the proportion of the void volume is much lower than in the case of the batch reactor. For fixed-bed adsorbers, the term "bed density" is often used instead of bulk density. It has to be noted that in fixed-bed adsorbers the void volume and therefore the bed density may change during filter operation, in particular after filter backwashing and adsorbent resettling.

In the laboratory, the determination of the bed density can be carried out by filling a defined mass of adsorbent particles in a graduated cylinder and reading the occupied volume. To approximate the situation in practice, it is often recommended that the bed density be determined after a certain compaction by shaking or vibrating. On the other hand, too strong a compaction can lead to an experimentally determined bed density that is higher than the bed density under practical conditions. An alternative method is to determine the bed density from a full-scale adsorber.

2.5.2 Porosities

Generally, the porosity specifies the fraction of void space on the total volume. Depending on the total volume considered, a distinction can be made between the particle porosity, ε_P, and the bulk (bed) porosity, ε_B. Both porosities can be derived from the densities.

Particle porosity

The particle porosity (also referred to as internal porosity) gives the void volume fraction of the adsorbent particle. It is therefore defined as the ratio of the pore volume, V_{pore}, and the volume of the adsorbent particle, V_A,

$$\varepsilon_P = \frac{V_{pore}}{V_A} = \frac{V_{pore}}{V_{mat} + V_{pore}} \tag{2.8}$$

The particle porosity is related to the particle density and the material density by

$$\varepsilon_P = \frac{V_{pore}}{V_A} = \frac{V_A - V_{mat}}{V_A} = 1 - \frac{V_{mat}}{V_A} = 1 - \frac{\rho_P}{\rho_M} \tag{2.9}$$

Bulk porosity

The bulk porosity (external void fraction), ε_B, is defined as the ratio of the liquid-filled void volume between the adsorbent particles, V_L, and the reactor volume, V_R,

$$\varepsilon_B = \frac{V_L}{V_R} = \frac{V_L}{V_A + V_L} \tag{2.10}$$

The bulk porosity is related to the particle density and the bulk density by

$$\varepsilon_B = \frac{V_L}{V_R} = \frac{V_R - V_A}{V_R} = 1 - \frac{V_A}{V_R} = 1 - \frac{\rho_B}{\rho_P} \tag{2.11}$$

In the case of fixed-bed adsorption, the term "bed porosity" is often used instead of bulk porosity.

The bulk porosity can be used to express different volume/volume or solid/volume ratios, which are characteristic for the conditions given in an adsorber and therefore often occur in adsorber design equations. Table 2.1 summarizes the most important ratios.

Table 2.1: Ratios characterizing the conditions in adsorbers.

Ratio	Expression
$\frac{\text{volume of liquid}}{\text{total reactor volume}}$	$\frac{V_L}{V_R} = \varepsilon_B$
$\frac{\text{volume of adsorbent}}{\text{total reactor volume}}$	$\frac{V_A}{V_R} = 1 - \varepsilon_B$
$\frac{\text{volume of adsorbent}}{\text{volume of liquid}}$	$\frac{V_A}{V_L} = \frac{1 - \varepsilon_B}{\varepsilon_B}$
$\frac{\text{mass of adsorbent}}{\text{volume of liquid}}$	$\frac{m_A}{V_L} = \rho_P \frac{1 - \varepsilon_B}{\varepsilon_B}$

2.5.3 External surface area

According to the general mass transfer equation

$$\textit{mass transfer rate} = \textit{mass transfer coeffcient} \times \textit{area available for mass transfer} \times \textit{driving force} \tag{2.12}$$

the external surface area has a strong influence on the rate of the mass transfer during adsorption. In the case of porous adsorbents, a distinction has to be made between external and internal mass transfer (Chapter 5).

The external mass transfer is the mass transfer through the hydrodynamic boundary layer around the adsorbent particle. Given that the boundary layer is very thin, the area available for mass transfer in the mass transfer equation can be approximated by the external adsorbent surface area.

The internal mass transfer occurs through intraparticle diffusion processes. If the internal mass transfer is approximately described by a mass transfer equation according to eq. (2.12) (linear driving force approach, Chapter 5, Section 5.4.6), the area available for mass transfer is also given by the external adsorbent surface area.

The external surface area can be determined by the counting-weighing method. In this method, the number of the adsorbent particles in a representative sample (Z_S) is counted after weighing the sample ($m_{A,S}$). The average mass of a single adsorbent particle, $m_{A,P}$, is then given by

$$m_{A,P} = \frac{m_{A,S}}{Z_S} \tag{2.13}$$

If the particles can be assumed to be spherical, the average radius is

$$r_P = \sqrt[3]{\frac{3\, m_{A,P}}{4\pi \rho_P}} \tag{2.14}$$

where ρ_P is the particle density. For irregular particles, the radius r_P represents the equivalent radius of the sphere having the same volume. The external surface area of a spherical adsorbent particle, $A_{s,P}$, is given by

$$A_{s,P} = 4\pi r_P^2 \tag{2.15}$$

and the total surface area available for mass transfer, A_s, is

$$A_s = Z_T\, A_{s,P} \tag{2.16}$$

where Z_T is the total number of the adsorbent particles applied, which can be calculated from the total mass applied, m_A, and the mass of a single adsorbent particle, $m_{A,P}$,

$$Z_T = \frac{m_A}{m_{A,P}} \tag{2.17}$$

For spherical particles, the external surface area can also be calculated by using the bulk density, ρ_B, and the bulk porosity, ε_B. With

$$m_A = V_R \rho_B \tag{2.18}$$

and

$$m_{A,P} = \frac{4}{3} \pi r_P^3 \rho_P \tag{2.19}$$

the following equation can be derived from eqs. (2.15) to (2.19):

$$A_s = Z_T A_{s,P} = \frac{3 V_R \rho_B}{4 \pi r_P^3 \rho_P} \cdot 4 \pi r_P^2 = \frac{3 V_R \rho_B}{r_P \rho_P} \tag{2.20}$$

Introducing the relationship between the bulk porosity, ε_B, and the densities ρ_B and ρ_P (eq. (2.11)) finally leads to

$$A_s = \frac{3 V_R (1 - \varepsilon_B)}{r_P} \tag{2.21}$$

In the mass transfer equations, the surface area available for mass transfer is often applied in form of a volume-related surface area – for instance, related to the adsorber volume, V_R,

$$a_{VR} = \frac{A_s}{V_R} = \frac{3 (1 - \varepsilon_B)}{r_P} \tag{2.22}$$

or to the adsorbent volume, V_A. For the latter case, the respective equation can be found from eq. (2.21) and replacing $(1 - \varepsilon_B)$ by the volume ratio V_A/V_R (see Table 2.1)

$$a_{VA} = \frac{A_s}{V_A} = \frac{3 V_R (1 - \varepsilon_B)}{V_A r_P} = \frac{3}{r_P} \tag{2.23}$$

Alternatively, the external surface area can also be expressed as mass-related surface area, a_m. Dividing eq. (2.20) by m_A and replacing ρ_B by m_A/V_R gives

$$a_m = \frac{A_s}{m_A} = \frac{3 V_R \rho_B}{m_A r_P \rho_P} = \frac{3}{r_P \rho_P} \tag{2.24}$$

From eqs. (2.22–2.24), the following equivalence relationship can be derived:

$$a_{VR} = a_{VA} (1 - \varepsilon_B) = a_m \rho_P (1 - \varepsilon_B) \tag{2.25}$$

2.5.4 Internal surface area

Porous adsorbents typically have internal surface areas that exceed the external surface areas many times over. In particular, engineered adsorbents possess extremely large internal surface areas. Therefore, nearly the whole adsorption capacity is provided by the internal surface area. Hence, the internal surface area is a very important quality parameter of an adsorbent. On the other hand, it has to be noted that the internal surface area alone is not sufficient to characterize or predict the adsorption capacity of an adsorbent, because the strength of adsorption is additionally influenced by a number of other adsorbent- and adsorbate-related properties.

The standard method for the determination of the internal surface area is based on low-temperature gas adsorption (typically nitrogen adsorption at 77 K) and subsequent application of the Brunauer-Emmett-Teller (BET) isotherm. This method is referred to as the BET method, and the internal surface area determined by this method is often referred to as the BET surface area, A_{BET}.

The BET model (Brunauer et al. 1938) is based on the assumption of a multilayer adsorption onto a nonporous adsorbent with energetically homogeneous surface without lateral interactions between the adsorbed molecules. Under these assumptions, the following isotherm equation for an infinite number of adsorbate layers has been derived:

$$q = q_{mono} \frac{C_B\, p}{(p_0 + (C_B - 1)p)\,(1 - p/p_0)} \tag{2.26}$$

which can also be written in the form

$$q = q_{mono} \frac{C_B\, p/p_0}{(1 - p/p_0)\,(1 - p/p_0 + C_B\, p/p_0)} \tag{2.27}$$

where q is the adsorbed amount, q_{mono} is the adsorbed amount in the first layer (monomolecular surface coverage), p is the partial pressure of the adsorbate, p_0 is the saturation vapor pressure, and C_B is a constant. Instead of the adsorbed amounts, q and q_{mono}, the adsorbed volumes, V_{ads} and $V_{ads,mono}$, can also be used in eqs. (2.26) or (2.27) because the amount adsorbed per mass (mol/g) can be expressed as the quotient of the volume adsorbed per mass (cm^3/g) and the molar volume (cm^3/mol); the ratio q/q_{mono} is therefore equal to the ratio $V_{ads}/V_{ads,mono}$. Although derived for nonporous solids, eqs. (2.26) and (2.27) are commonly used to determine the internal surface of porous adsorbents.

The typical isotherm form corresponding to eq. (2.26) is shown in Figure 2.5. The inflection point indicates the transition from monomolecular to multilayer coverage. From the adsorbed amount (or adsorbed volume) in the monolayer and the molecule size of the adsorbate, the surface area can be calculated. Since this isotherm shape is frequently found for the adsorption of nitrogen or noble gases at low temperatures, these adsorbates are commonly used in practice for surface area determination.

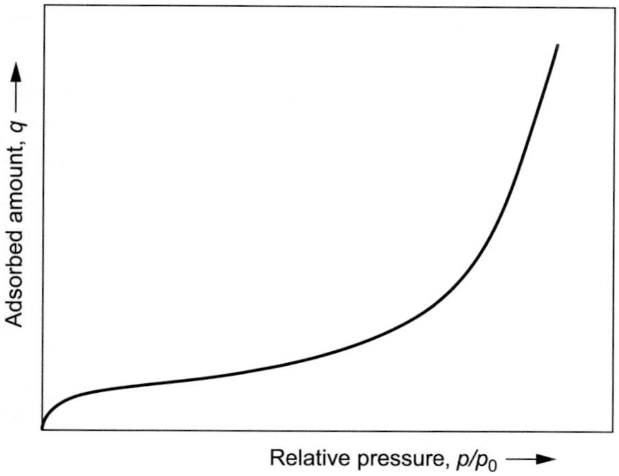

Figure 2.5: Characteristic shape of a BET isotherm.

However, it has to be noted that deviations from the ideal BET isotherm form may occur at low and high relative pressures, in particular in the case of highly porous adsorbents. At low relative pressures, deviations from the assumption of an energetically homogeneous adsorbent surface become noticeable, whereas at high relative pressures the multilayer formation is limited by the pore size and capillary condensation may occur in narrow pores. Furthermore, interactions between the molecules cannot be excluded at high relative pressures (high adsorbent loadings). For these reasons, the evaluation of the isotherm data is typically carried out in the range $0.05 < p/p_0 < 0.3$.

The adsorbed amount in the monomolecular layer, q_{mono}, can be determined from the linearized BET isotherm equation

$$\frac{p}{q\,(p_0 - p)} = \frac{1}{q_{mono}\,C_B} + \frac{C_B - 1}{q_{mono}\,C_B} \cdot \frac{p}{p_0} \tag{2.28}$$

With q_{mono} and the area occupied by the adsorbate molecule, A_M, the surface area can be estimated from

$$A_{BET} = q_{mono}\,N_A\,A_M \tag{2.29}$$

where N_A is Avogadro's number ($6.022 \cdot 10^{23}$ mol^{-1}). For nitrogen adsorbed at 77 K, A_M is $16.2 \cdot 10^{-20}$ m^2.

The parameter C_B is an indicator of the strength of adsorption. In the case of highly porous adsorbents, C_B can reach values >100. Under this condition, the estimation of the monolayer capacity can be simplified. If C_B is much greater than 1 (guide value $C_B > 50$), the BET isotherm equation reduces to

$$q = \frac{q_{mono}}{(1 - p/p_0)} \tag{2.30}$$

and the monolayer capacity, q_{mono}, can be received from only a single isotherm point (single-point BET method; Haul and Dümbgen 1960, 1963).

The size of the internal surface area is related to the dimension of the pore system. Strongly microporous adsorbents as used for water treatment possess large internal surface areas. Typical ranges of BET surface areas for common adsorbents are given in Table 2.2.

Table 2.2: Typical ranges of the specific internal surface area for different adsorbents.

Adsorbents	A_{BET} in m^2/g
Activated carbons	600 ... 1,200
Polymeric adsorbents	300 ... 1,400
Aluminum oxides	150 ... 350
Granular ferric hydroxides	150 ... 350
Zeolites	400 ... 900

To determine the BET surface area, a commercial BET surface area analyzer is required. As a cost-saving alternative to the BET surface area determination, the iodine number can be used to characterize the surface area. The iodine number can be easily determined without expensive equipment. The determination is based on an adsorption experiment with iodine as adsorbate and with defined initial and residual concentrations (0.1 M and 0.02 M, respectively). To meet the defined residual concentration, the adsorbent dose has to be varied. With the adsorbent dose found from the variation and the initial and residual concentrations, the amount adsorbed can be calculated by means of the material balance equation for isotherm tests (Chapter 3). The adsorbed amount expressed in mg/g is referred to as the iodine number. Since the numerical value of the iodine number is approximately equal to the numerical value of the BET surface area, the iodine number can be used as a compensatory parameter to characterize the internal surface area – for instance, for comparison of different adsorbent types.

2.5.5 Pore-size distribution

Most of the engineered adsorbents with large internal surface areas possess a multitude of pores with different shapes and sizes. According to the definition of the International Union of Pure and Applied Chemistry (IUPAC), three types of pores can be distinguished: macropores, mesopores, and micropores (Table 2.3). The macropores

Table 2.3: Classification of pores according to the IUPAC definition.

Pore class	Range of pore radius
Macropores	>25 nm
Mesopores	1 nm ... 25 nm
Micropores	<1 nm

and the mesopores are primarily relevant for the mass transfer into the interior of the adsorbent particles, whereas the micropore volume mainly determines the size of the internal surface and therefore the adsorbent capacity. As a rule, the internal surface area increases with increasing micropore volume. In principle, the higher the micropore volume, the larger the amount of adsorbate that can be adsorbed. However, it has to be considered that in the case of very fine pores and large adsorbate molecules, there may be a limitation of the extent of adsorption by size exclusion. Such size exclusion can be found, for instance, in the case of the adsorption of high-molecular-weight natural organic matter onto microporous adsorbents.

Due to the relevance of the pore system for both adsorption kinetics and adsorption equilibrium, it is interesting to get information about the frequency of occurrence of different pore sizes in the considered adsorbent. However, the analysis of the pore-size distribution is not a simple matter. Among others, the following problems are related to the pore-size analysis:

- All pore-size analyses are based on models that are subject to simplifications and restricted validation.
- Since the shape of the pores is typically irregular, simplifying assumptions in view of the pore geometry have to be made.
- There is no single method that can be used for all ranges of pore sizes.
- The measurement procedures and the data analysis are laborious.
- The results of the different methods are often not comparable.

An extensive discussion of the multitude of different methods proposed for pore-size analyses is not possible within the framework of this book. Therefore, only some general methods will be discussed. For detailed information, the monograph of Lowell et al. (2010) is recommended.

Mercury intrusion

During mercury intrusion, also referred to as mercury porosimetry, mercury is pressed under increasing pressure into the pore system. Because mercury does not wet the solid material, and any other effects that could allow for spontaneously penetrating the pore system are also absent, the volume of pore space filled with mercury is directly related to the pressure applied. This relationship is known as Washburn's equation and is given by

$$r_{pore} = -\frac{2\sigma \cos \Theta}{p} \tag{2.31}$$

where r_{pore} is the lowest radius to which the pores are filled under the applied pressure p, σ is the surface tension of mercury (0.48 N/m at 20 °C), and Θ is the contact angle. The contact angle depends on the adsorbent type and varies between 110° and 142°. For $\Theta = 140°$, the following relationship holds:

$$r_{pore} = \frac{0.7354\,\text{N/m}}{p} \tag{2.32}$$

Mercury porosimeters are working with pressures up to 400 MPa (1 Pa = 1 N/m^2). According to eq. (2.32), this maximum pressure corresponds to the minimum pore radius $r_{pore} = 1.8$ nm. Thus, the working range of mercury intrusion comprises only macropores and mesopores. Furthermore, eq. (2.31) was developed under the assumption that the pores are of cylindrical shape, which is not realistic for most of the adsorbents used in water treatment, in particular for activated carbons. Even though the results of mercury porosimetry are therefore not absolutely precise, this method has proved worthwhile for adsorbent characterization, in particular for comparison of adsorbents.

During the mercury intrusion, at first the volume pressed into the pore space of the given adsorbent sample, V_{pore}, is recorded as a function of the applied pressure, p. Then, the corresponding radii can be calculated from eqs. (2.31) or (2.32) to get the cumulative pore-size distribution (Figure 2.6a)

$$\sum \Delta V_{pore} = \mathrm{f}(\log r_{pore}) \tag{2.33}$$

or the differential pore-size distribution (Figure 2.6b)

$$\frac{\mathrm{d}V_{pore}}{\mathrm{d}(\log r_{pore})} = \mathrm{f}(\log r_{pore}) \tag{2.34}$$

Note that in pore-size distribution diagrams, typically the logarithm of the pore radius is used.

Taking into consideration eq. (2.31), the surface of the macropores and mesopores can be calculated by

$$A(\text{macropores} + \text{mesopores}) = \int \frac{2}{r_{pore}} \left(\frac{\mathrm{d}V_{pore}}{\mathrm{d}r_{pore}}\right) \mathrm{d}\,r_{pore} = -\frac{1}{\sigma \cos \Theta} \int p\, \mathrm{d}V_{pore} \tag{2.35}$$

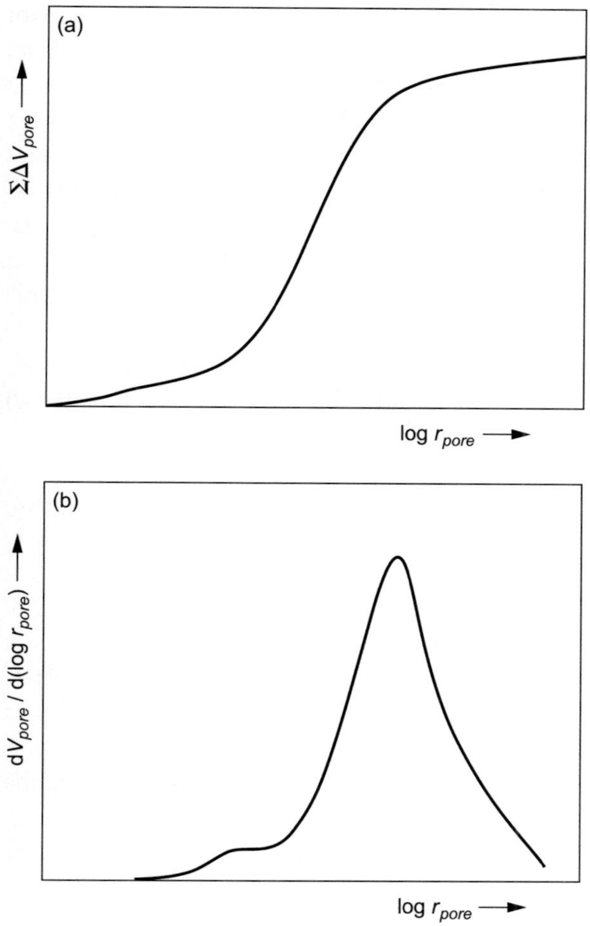

Figure 2.6: Schematic representation of cumulative (a) and differential (b) pore-size distribution.

Gas or vapor adsorption

Determination of pore-size distributions by gas or vapor adsorption measurements is based on the fact that the adsorption in porous adsorbents at very low and high relative pressures is not a gradual accumulation of adsorbate layers as assumed, for example, in the BET model. Instead of that, the gas adsorption in mesopores at medium and high relative pressures is dominated by capillary condensation, whereas the adsorption of gases and vapors in micropores can be explained by the theory of volume filling of micropores (TVFM; see also Chapter 3, Section 3.3.3).

The capillary condensation can be described by the Kelvin equation

$$r_{pore,K} = -\frac{2\,\sigma\,V_m \cos\Theta}{R\,T\,\ln(p/p_0)} \tag{2.36}$$

where σ is the surface tension of the condensed phase, V_m is the molar volume of the condensed phase, R is the gas constant, T is the absolute (Kelvin) temperature, Θ is the contact angle between the condensed phase and the solid surface, p/p_0 is the relative adsorbate pressure, and $r_{pore,K}$ is the radius up to which the pores are filled with liquid. The Kelvin equation is often used in a modified form that considers the fact that a layer of thickness t is already adsorbed before capillary condensation takes place. The corrected form of the Kelvin equation is therefore

$$r_{pore} = r_{pore,K} + t = -\frac{2\sigma V_m \cos\Theta}{RT\ln(p/p_0)} + t \tag{2.37}$$

The statistical thickness of the adsorbed layer depends on the relative pressure. Several relationships were proposed to determine the value of t as a function of p/p_0 – for instance, the Halsey equation (Halsey 1948)

$$t(\text{nm}) = 0.354\left[\frac{-5}{\ln(p/p_0)}\right]^{1/3} \tag{2.38}$$

or the Harkins-Jura equation (Harkins and Jura 1944a, 1944b)

$$t(\text{nm}) = 0.1\left[\frac{13.99}{0.034 - \log(p/p_0)}\right]^{1/2} \tag{2.39}$$

Both equations are valid for nitrogen adsorption at 77 K. Figure 2.7 shows a comparison of the adsorbed layer thickness as a function of relative pressure as calculated from eqs. (2.38) and (2.39).

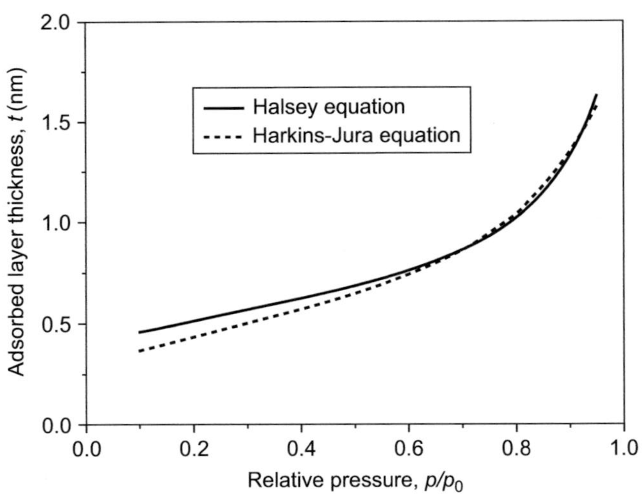

Figure 2.7: Adsorbed layer thickness as a function of relative pressure according to the Halsey and Harkins-Jura equations.

Since eq. (2.37) relates the relative pressure to the pore radius and the corresponding volume of the filled pores is available from the isotherm, the pore-size distribution can be derived in the same manner as described for mercury porosimetry. In most cases, nitrogen is used as the adsorbate, and the isotherm measurement is done under the same condition as for surface area determination (77 K). Alternatively, argon can be used at the temperature of liquid nitrogen (77 K) or liquid argon (87 K). The contact angle in eqs. (2.36) and (2.37), respectively, is typically assumed to be 0° ($\cos \Theta = 1$).

For most adsorbents, different isotherm forms are found if the equilibration is carried out by stepwise pressure increase (adsorption) or by stepwise pressure decrease (desorption). This effect is referred to as adsorption hysteresis (Figure 2.8). The occurrence of a hysteresis can be explained by different forms of the meniscus of the liquid during filling and emptying of the pores. It is an indicator for capillary condensation in mesopores. The shape of the hysteresis is related to the texture of the pore system (Lowell et al. 2010). Although not obligatory, it is often recommended that the desorption branch of the isotherm be used for evaluating the pore-size distribution.

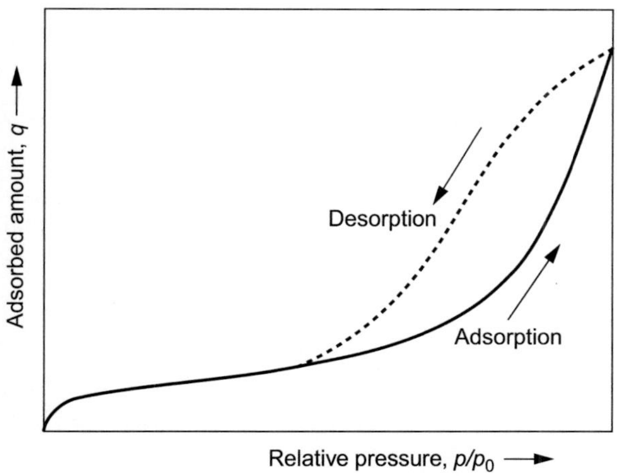

Figure 2.8: Adsorption hysteresis.

The transition to the micropore range ($r_P < 1$ nm) is considered the lower limit for the application of the Kelvin equation because the definition of a liquid meniscus loses its sense in the range of molecular dimensions.

Adsorption of gases or vapors within the micropores can be better described by the TVFM, especially by the well-known Dubinin-Radushkevich isotherm equation (Chapter 3, Section 3.3), here given in the linearized form

$$\ln V_{ads} = \ln V_0 - \frac{1}{E_C^2}\left(R\,T\ln\frac{p_0}{p}\right)^2 \tag{2.40}$$

where V_{ads} is the volume adsorbed (which equals the filled pore volume, V_{pore}) per unit adsorbent mass, V_0 is the micropore volume, and E_C is a characteristic adsorption energy (isotherm parameter). The term within the brackets is also referred to as adsorption potential, ε,

$$\varepsilon = R\,T \ln \frac{p_0}{p} \tag{2.41}$$

Plotting the isotherm data as ln V_{ads} over ε^2 allows estimating the micropore volume from the intercept of the resulting line. Gases (e.g. N_2, Ar) as well as organic vapors (benzene) are frequently used as adsorbates. V_{ads} and the adsorbed amount, q, are related by

$$q = \frac{V_{ads}}{V_m} \tag{2.42}$$

where V_m is the molar volume of the adsorbate.

Figure 2.9 shows exemplarily the Dubinin-Radushkevich plot of a benzene isotherm measured on activated carbon.

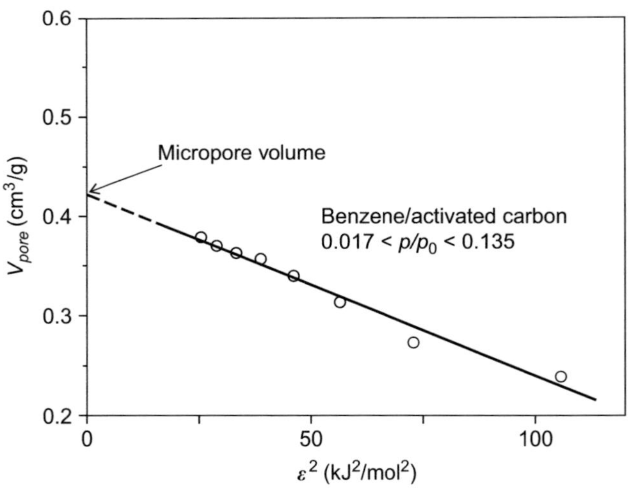

Figure 2.9: Dubinin-Radushkevich plot of a benzene isotherm measured on activated carbon.

To obtain the pore-size distribution for micropores, Jüntgen and Seewald (1975) extrapolated the Kelvin equation by means of the Dubinin-Radushkevich isotherm. Combining these equations under the assumption of a contact angle $\Theta = 0°$ leads to

$$V_{pore} = V_0 \exp\left[-\left(\frac{2\,\sigma\,V_m}{E_C\, r_{pore}} \right)^2 \right] \tag{2.43}$$

Since eq. (2.43) relates the isotherm data to the pore radius, the pore-size distribution can be calculated in the same manner as shown for the above-mentioned methods.

Besides the conventional methods already discussed, new approaches have been developed in recent years that are based on microscopic treatment of sorption phenomena on a molecular level by statistical mechanics. Methods like the density functional theory (DFT) or Monte Carlo (MC) simulation methods provide microscopic models of adsorption and a realistic description of the thermodynamic properties of the pore fluid. Complex mathematical modeling of different interactions and geometrical considerations concerning the pore geometry leads to density profiles for the confined fluid as a function of temperature and pressure. From these density profiles, the amount adsorbed can be derived to get theoretical isotherms for different materials, pore geometries, and analysis conditions. Under the assumption that the total isotherm consists of a number of individual single-pore isotherms multiplied by their relative distribution, the theoretical total isotherm can be fitted to the experimental isotherm data to find the pore-size distribution of the tested adsorbent. These methods are applicable for pore-size analysis of both the micropore and mesopore size ranges. Detailed information about these complex models and their application for pore-size characterization can be found elsewhere (e.g. Lowell et al. 2010).

2.5.6 Surface chemistry

Depending on the type of adsorbent, surface chemistry may affect the adsorbate/adsorbent interaction. This is especially true for the adsorption of ions onto oxidic adsorbents, but in special cases, it can be also relevant for the adsorption onto activated carbons.

Oxidic adsorbents

Manufactured or naturally occurring oxides or hydroxides such as aluminum oxide, ferric hydroxide, or silicon dioxide are characterized by specific structures where positively charged metal or metalloid ions and negatively charged oxygen or hydroxide ions are arranged in such a manner that the different charges compensate each other. At the surface, this regular structure is disturbed and the charges have to be compensated by other ions. In aqueous solutions, the negative charges of the surface oxygen ions are neutralized by protons, whereas the positive charges of the surface metal or metalloid ions are neutralized by hydroxide ions. As a result, the surface of oxidic adsorbents is covered with surface OH groups. These groups are subject to protonation or deprotonation depending on the pH value of the solution

$$\equiv\text{S–OH} + \text{H}^+ \rightleftharpoons \equiv\text{S–OH}_2{}^+$$

$$\equiv\text{S–OH} \rightleftharpoons \equiv\text{S–O}^- + \text{H}^+$$

In these equations, the symbol $\equiv$S stands for the surface of the solid material.

It follows from the equations that the surface is positively charged at low pH values and negatively charged at high pH values. Between these regions, a pH value exists at which the sum of negative charges equals the sum of positive charges and the net charge of the surface is zero. This point is referred to as the point of zero charge (pzc). The pH_{pzc} is an important adsorbent parameter that aids in understanding the adsorption of charged species and the influence of pH on the adsorption process. Generally, the adsorption of charged species onto charged surfaces can be expected to be strongly influenced by electrostatic attraction or repulsion forces.

The surface charge as a function of pH and the pH_{pzc} can be determined by titration of an adsorbent suspension with strong acids and bases (e.g. HCl and NaOH) at a specified ionic strength. For each point of the titration curve, the surface charge, Q_S, can be calculated from the general mass/charge balance

$$Q_S = q(\mathrm{H}^+) - q(\mathrm{OH}^-) = \frac{V_L}{m_A}(c_a - c_b - c(\mathrm{H}^+) + c(\mathrm{OH}^-)) \tag{2.44}$$

where $q(\mathrm{H}^+)$ is the surface loading with H^+, $q(\mathrm{OH}^-)$ is the surface loading with OH^-, c_a is the molar concentration of the acid used for titration, c_b is the molar concentration of the base used for titration, $c(\mathrm{H}^+)$ is the proton concentration after equilibration (measured as pH), $c(\mathrm{OH}^-)$ is the OH^- concentration after equilibration (calculated from the measured pH), V_L is the volume of the solution, and m_A is the adsorbent mass. The surface charge is given in mmol/g or mol/kg. The surface charge density, σ_S (in C/m^2), can be calculated from Q_S by

$$\sigma_S = \frac{Q_S F}{A_m} \tag{2.45}$$

where F is the Faraday constant (96,485 C/mol) and A_m is the specific surface area (m^2/kg). Plotting Q_S or σ_S over pH illustrates the influence of the pH on the surface charge (Figure 2.10). The intersection with the line at $\sigma_S = 0$ gives the pH_{pzc}. Table 2.4 lists points of zero charge for some oxides and hydroxides.

Dissolved ions can be bound to the surface OH groups by different mechanisms: specific adsorption (i.e. surface complex formation) and nonspecific adsorption. These mechanisms are strongly related to the structure of the electric double layer that surrounds the charged solid particle. The term "surface complex formation" includes two different types of reactions, the formation of inner-sphere complexes and the formation of outer-sphere complexes (Figure 2.11).

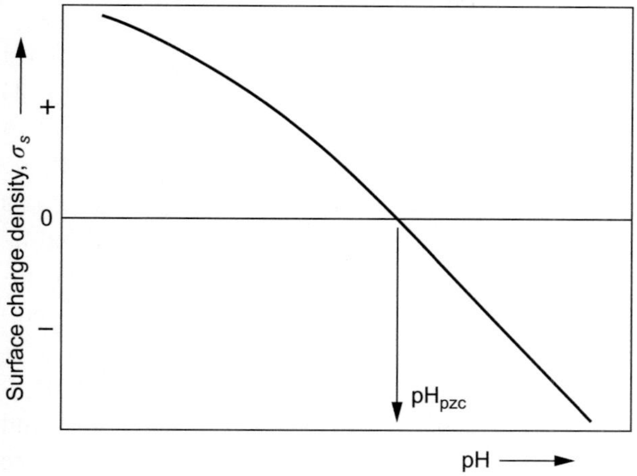

Figure 2.10: Schematic representation of a surface titration curve.

Table 2.4: Points of zero charge for selected oxides, oxohydrates, and hydroxides (Stumm 1992).

Oxidic material	pH_{pzc}
α-Al_2O_3	9.1
α-Fe_2O_3	8.5
$Fe(OH)_3$	8.5
γ-AlOOH	8.2
α-FeOOH	7.8
α-$Al(OH)_3$	5.0
SiO_2	2.0

In the case of inner-sphere complexes, the adsorbate ions without the water molecules of the hydration sphere are directly bound to the surface site by ligand exchange. Cations replace the protons of the surface OH groups as shown in the following reaction equations for a bivalent cation (M^{2+}):

$$\equiv S{-}OH + M^{2+} \rightleftharpoons \equiv S{-}OM^{+} + H^{+}$$

$$2 \equiv S{-}OH + M^{2+} \rightleftharpoons (\equiv S{-}O)_2M + 2\,H^{+}$$

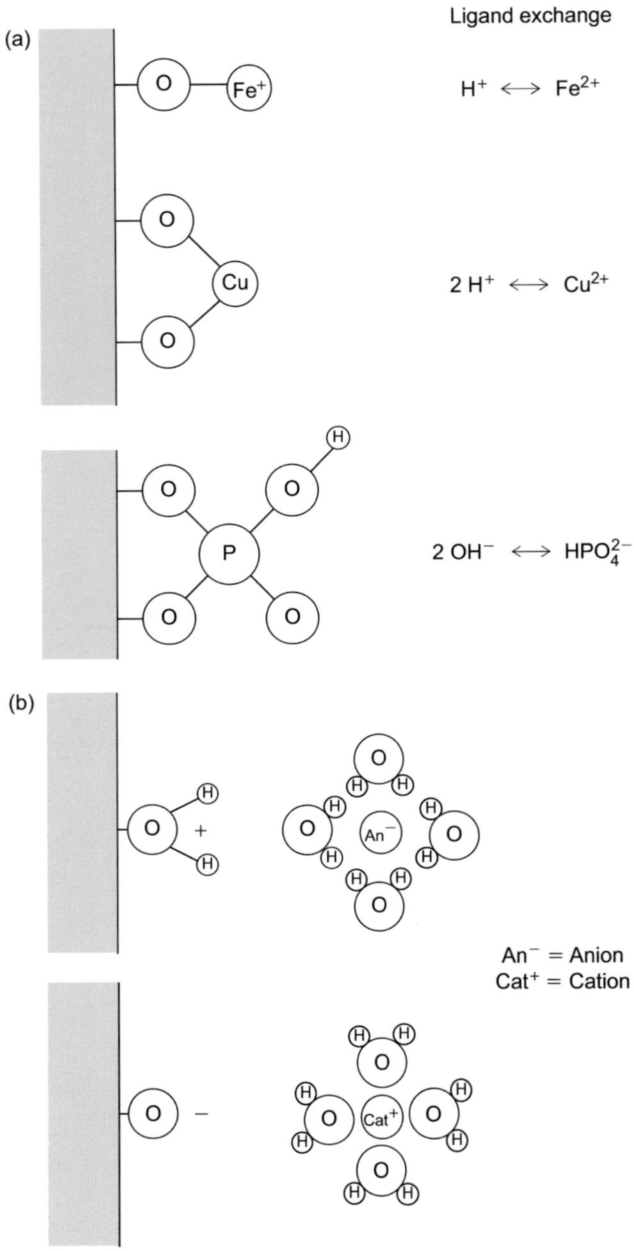

Figure 2.11: Inner-sphere (a) and outer-sphere (b) complexes (adapted from Sigg and Stumm 2011).

In the case of anions (here A^{2-}), the OH groups are replaced – for instance,

$$\equiv S{-}OH + A^{2-} \rightleftharpoons \equiv S{-}A^{-} + OH^{-}$$

$$2 \equiv S{-}OH + A^{2-} \rightleftharpoons (\equiv S)_2A + 2\ OH^{-}$$

According to the equations given, the adsorption of cations increases with increasing pH, whereas the adsorption of anions increases with decreasing pH. If the adsorbate is a weak acid, the pH-dependent acid/base equilibrium also influences the adsorption and has to be additionally considered in equilibrium calculations.

The adsorbed ions are strongly bound and located in a compact layer directly attached to the surface. This first part of the electric double layer is also referred to as the surface layer. As can be seen from the reaction equations, the adsorption of ions can lead to neutral or charged surface complexes depending on the ion charge and the number of surface sites that take part in the reaction.

The model of outer-sphere complex formation presumes that ions can also be bound to the surface sites by chemical bonds without losing their hydration water. That means that water molecules are located between the ions and the adsorption sites. Therefore, the distance to the surface is larger and the binding strength is weaker in comparison to inner-sphere complex formation. The layer where outer-sphere complexation takes place is referred to as the beta (β) layer. The beta layer is also a part of the compact layer within the double-layer model.

Beyond the beta layer, a diffuse layer exists where an excess concentration of counter ions (ions charged oppositely to the charge of the surface layer) compensates the remaining surface charge. Throughout the diffuse layer, the concentration of the counter ions decreases with increasing distance from the surface until in the bulk liquid the equivalent concentrations of cations equal the equivalent concentrations of the anions. The enrichment of counter ions in the diffuse layer is a result of electrostatic interactions and can be considered a nonspecific adsorption.

A number of models were developed to characterize the charge distribution and the accumulation of counter ions. In particular, the constant capacitance model, the diffuse layer model, and the triple-layer model are widely used to describe the adsorption onto oxidic adsorbents (Benjamin 2002, Sigg and Stumm 2011, Worch 2015). The models differ mainly in the assumptions concerning the charge distribution and the location of the adsorbed species (Figure 2.12, Table 2.5).

In principle, the protonation/deprotonation and the complex formation can be described by laws of mass action with respective equilibrium constants. However, it has to be noted that the equilibrium constants depend on the surface charge due to attraction or repulsion caused by the charged surface groups. Generally, the apparent equilibrium constant, K^{app}, can be expressed as the product of a constant that is

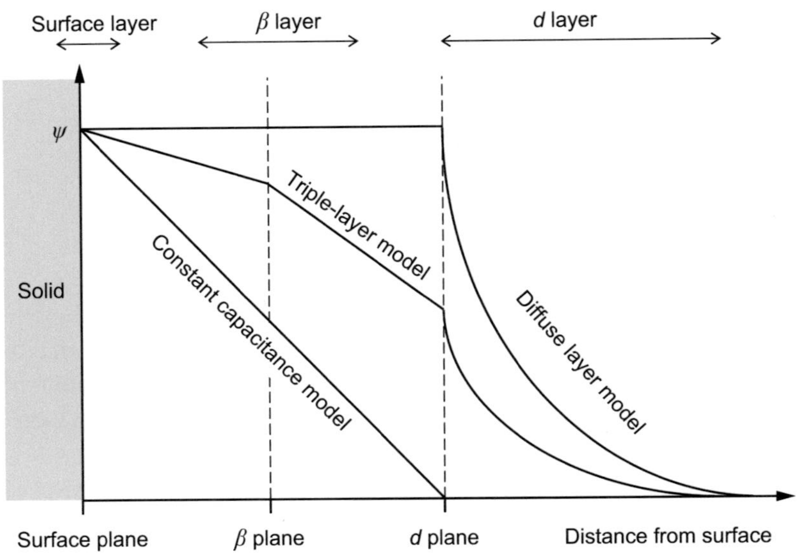

Figure 2.12: Surface potential as a function of distance from the surface as assumed in different surface complexation models (adapted from Benjamin 2002).

Table 2.5: Frequently used models to describe surface complex formation (adapted from Benjamin 2002).

Model	Assumptions
Constant capacitance	– Specifically adsorbed species in the surface plane – No β layer – Charge opposite to that in the surface plane provided by nonspecifically adsorbed ions, all located in the *d* plane
Diffuse layer	– Specifically adsorbed species in the surface plane – No β layer – Charge opposite to that in the surface plane provided by nonspecifically adsorbed ions, starting in the *d* plane and distributed throughout the *d* layer
Triple-layer	– Dehydrated specifically adsorbed species in the surface plane – Hydrated specifically adsorbed species in the β plane – Charge opposite to that in the combined surface and β planes provided by nonspecifically adsorbed ions, starting in the *d* plane and distributed throughout the *d* layer

independent of the electrical charge (intrinsic constant, K^{int}) and a term describing the influence of the surface potential

$$K^{app} = K^{int} \exp\left(\frac{-\Delta z_S \, F \, \psi_S}{R\,T}\right) \tag{2.46}$$

Here, F is the Faraday constant, ψ_s is the surface potential, Δz_s is the change of the charge in the surface layer during the considered reaction, R is the gas constant, and T is the absolute temperature. In eq. (2.46), it is assumed that the reaction takes place only in the surface layer. If it is assumed that adsorption affects the charges in both the surface and beta layers, the apparent equilibrium constant is given by

$$K^{app} = K^{\text{int}} \exp\left(\frac{-\left(\Delta z_s\,\psi_s + \Delta z_\beta\,\psi_\beta\right) F}{R\,T}\right) \tag{2.47}$$

where Δz_β is the change of the charge in the beta layer, and ψ_β is the potential of the beta plane.

To apply a surface complexation model for describing the adsorption of ions onto a charged surface, a multitude of equations has to be combined – in particular, material balances for all species, laws of mass action for all reactions in all considered layers, charge balances in each layer, and charge-potential relationships for all considered layers. To reduce the number of equations, simplifying assumptions can be made – for instance, neglecting the beta layer.

Relatively strong limitations of such equilibrium models result from uncertainties concerning the model assumptions, the need for simplifications, and the problems in parameter determination as well as from the increasing complexity if a large number of ions are present in the water. Therefore, for practical purposes, the conventional adsorption isotherm equations are frequently used to describe the adsorption equilibria instead of applying a surface complex formation model. Nevertheless, a qualitative characterization of surface chemistry – in particular, knowledge about the pH-dependent charges and the location of pH_{pzc} – is helpful for interpreting adsorption processes on oxidic or other charged surfaces.

Besides inorganic ions, organic ions can also be bound to the charged surfaces of oxidic adsorbents. Organic compounds form ions when basic functional groups are protonated or acidic functional groups are deprotonated. Accordingly, the ionic forms of the organic compounds occur only in specific pH ranges (protonated basic groups at low pH values, deprotonated acidic groups at high pH values). The exact pH range, in which the charged species occur, depends on the acidity (or basicity) constant. In general, significant adsorption can be expected in a pH range where the organic ion and the adsorbent surface carry opposite charges.

Under practical conditions, the adsorption of organic ions onto oxidic adsorbents is strongly influenced by the competition from inorganic ions, which typically occur in much higher concentrations than the organic compounds. As an example, Figure 2.13 shows the adsorption isotherms of the drug metoprolol on silica gel (amorphous SiO_2) at different concentrations of the competing calcium ions. The experiments were carried out in a pH range between 5.5 and 7. In this range, metoprolol is completely cationic and the silica gel surface is negatively charged. The remaining sorption capacity for metoprolol at very high concentrations of the competitor

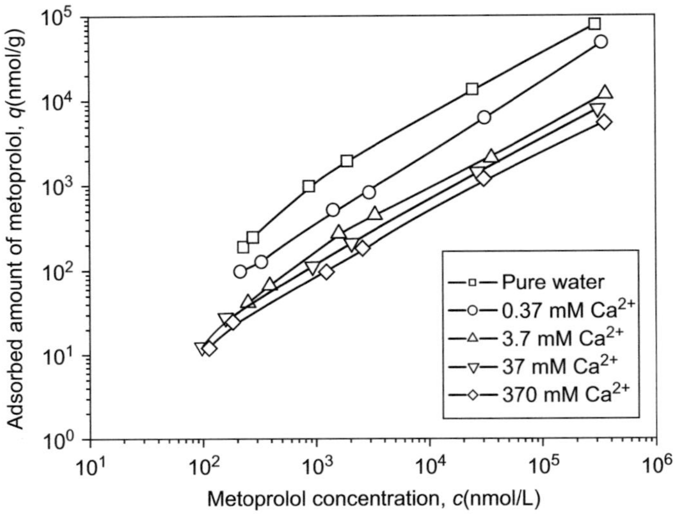

Figure 2.13: Influence of competing calcium ions on the adsorption of metoprolol on silica gel.

Ca^{2+} is an indication of the existence of additional binding mechanisms, such as hydrophobic interactions or hydrogen bonding (Kutzner et al. 2014, 2016).

Activated carbons

It is well known that oxygen-containing functional groups exist at the surface of carbonaceous adsorbents. These groups, also referred to as surface oxides, show either acidic or basic character. They are thought to occur at the cross-links and edges of the graphite crystallites (Section 2.2.1). Their occurrence can be proved by spectroscopic methods and simple titrations with acids and bases (Boehm 2002). The nature of acidic groups is quite well understood. Typical functional groups are carboxyl groups, carboxylic anhydrides, lactone and lactol groups as well as phenolic hydroxyl groups. Less is known about the nature of basic groups, although their existence can be shown by titration. It has been hypothized that pyrone-type groups coupled with polycyclic aromatic structures could be responsible for the basic properties, but this point is still under discussion. Figure 2.14 shows possible acidic and basic surface groups.

The surface charge dependence on pH can be determined with the same method as describe for oxidic adsorbents – that is, acid/base titration and application of eqs. (2.44) and (2.45). The point of zero charge varies with the carbon type and is frequently found to be in the range $5.5 < pH_{pzc} < 7.2$.

To differentiate between the main acidic groups, Boehm (1966, 1974) has proposed a selective neutralization technique with bases of different strengths. In this still frequently used method, the carbon is equilibrated with solutions of sodium bicarbonate ($NaHCO_3$), sodium carbonate (Na_2CO_3), and sodium hydroxide (NaOH),

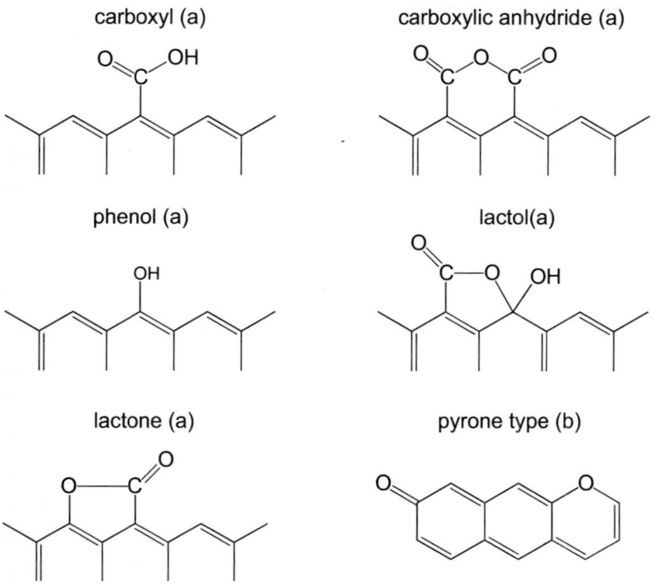

Figure 2.14: Possible acidic (a) and basic (b) groups on activated carbon surfaces.

and the supernatant is back-titrated with hydrochloric acid (HCl). The differences in the reduction of the amount of each base allow calculation of the surface concentrations of strongly carboxylic, weakly carboxylic, and phenolic groups.

At activated carbon surfaces, the number of sites that can be protonated and deprotonated is about one to two orders of magnitude lower than that at oxidic surfaces. Consequently, these groups are less relevant for adsorption. Ionic adsorbates are therefore adsorbed onto activated carbon only to a small extent. On the other hand, in the case of organic ions the physical adsorption, mainly based on dispersion forces, can be superimposed by ionic interactions (attraction or repulsion) between the acidic or basic surface groups and the adsorbate, with the consequence that the adsorption is increased or decreased in comparison to that of the related neutral species (see also Section 4.6 in Chapter 4).

An indirect characterization of the surface functionalities is possible by applying a thermogravimetric analysis combined with an analysis of the decomposition gas by gas chromatography/mass spectrometry (GC/MS). In the thermogravimetric analysis, the activated carbon is heated in an inert gas stream up to 600 °C with a constant heating rate. In this temperature range, the functional groups on the carbon surface are transformed to gaseous products (H_2O, CO_2, CO, H_2) that can be identified and quantified by GC/MS. Under the assumption that each type of functional group decomposes to a defined product, the specific product composition can be used to characterize the surface properties (Boehm 2002). However, due to possible side reactions that can change the concentration ratios, the results are not always unambiguous.

2.6 Practice-oriented test methods for assessing the performance of adsorbents

Pragmatic test methods for assessing the performance of adsorbents are required both for adsorbent selection and for quality control in large-scale applications. In such tests, the adsorption behavior of model substances is studied under defined conditions (initial or residual concentration) and given as mass adsorbent necessary to reduce the test substance by a defined percentage or as concentration decrease at a given adsorbent dosage (see also the material balance equation given in Chapter 3, Section 3.2). The choice of the test solutes depends on the adsorbent type and the purpose of the adsorption process. Oxidic or polymeric adsorbents are typically applied to remove specific single solutes. Here, the target solute can be used for the test. By contrast, the most widely used adsorbent activated carbon, applied in both drinking water and wastewater treatment, adsorbs a multiplicity of organic substances, and therefore test substances have to be selected that are representative for the widest possible range of adsorbates. In the past, phenol and methylene blue were widely used as test substances. Phenol was chosen as a typical pollutant and methylene blue (M = 319.85 g/mol) as a representative of medium-molecular-weight substances. Both substances could be determined well with the analytical possibilities of the time. Today, nitrobenzene is often used as a test substance. The nitrobenzene number (amount of activated carbon necessary to reduce the nitrobenzene concentration under defined conditions by 90%) is a measure of the adsorption capacity for strongly adsorbable compounds. In recent years, new pollutants (emerging pollutants), such as pharmaceuticals, personal care products, pesticides, industrial and household products, industrial additives, or solvents, have come into focus both in drinking water and wastewater treatment. Since in most cases, nitrobenzene does not reflect the adsorption behavior of these compounds satisfactorily, it was necessary to extend the spectrum of test substances to representatives of these compound groups. Benzotriazole, carbamazepine, diclofenac, methylbenzotriazole, and metoprolol are examples of such new test substances.

Single-solute adsorption tests are particularly relevant for the comparison of adsorbents but cannot give information about the adsorption behavior in the presence of background organic matter (natural organic matter in drinking water treatment or effluent organic matter in wastewater treatment). On the other hand, the experimental determination of micropollutant adsorption in the presence of background organic matter requires a relatively high analytical effort. Therefore, simple test methods were developed that use empirical relationships between the relative removal of surrogate parameters, which are easier to measure, and the relative removal of micropollutants. For example, UV absorption at 254 nm (Zietzschmann et al. 2014, Anumol et al. 2015) and total fluorescence (Anumol et al. 2015) were proposed as surrogates. Dittmar et al. (2018) could show that methyl orange can be used not only as a surrogate parameter for competitive MP removal but also as a surrogate competitor replacing wastewater effluent organic matter in laboratory tests.

3 Adsorption equilibrium I: General aspects and single-solute adsorption

3.1 Introduction

Within the framework of the adsorption theory, the adsorption equilibrium and its mathematical description are of outstanding significance. The knowledge of adsorption equilibrium data provides the basis for assessing the adsorption processes and, in particular, for adsorber design. Information about the equilibrium in a considered adsorbate/adsorbent system is necessary, for instance, to characterize the adsorbability of water pollutants, to select an appropriate adsorbent, and to design batch, flow-through, or fixed-bed adsorbers. The equilibrium position in a considered system depends on the strength of the adsorbate/adsorbent interactions and is significantly affected by the properties of the adsorbate and the adsorbent but also by properties of the aqueous solution, such as temperature, pH value, and occurrence of competing adsorbates.

Although single-solute adsorption is rather the exceptional case than the typical situation in water treatment practice, it is reasonable to begin with a deeper look at single-solute adsorption. Some general aspects of adsorption processes can be explained more clearly for the simple case where only one adsorbate has to be considered. Furthermore, to compare adsorbabilities of solutes or capacities of adsorbents, it is sufficient to characterize the adsorption of single solutes. Last but not least, the models for mathematical description or prediction of multisolute adsorption equilibria are typically based on single-solute adsorption isotherms. Therefore, the basics of single-solute adsorption will be discussed in this chapter. Multisolute adsorption equilibrium is the subject matter of Chapter 4.

Each adsorption equilibrium state is uniquely defined by the variables adsorbate concentration, adsorbed amount (also referred to as adsorbent loading), and temperature. For a single-solute system, the equilibrium relationship can be described in its general form as

$$q_{eq} = \mathrm{f}(c_{eq}, T) \tag{3.1}$$

where c_{eq} is the adsorbate concentration in the state of equilibrium, q_{eq} is the adsorbed amount (adsorbent loading) in the state of equilibrium, and T is the temperature.

It is common practice to keep the temperature constant and to express the equilibrium relationship in the form of an adsorption isotherm (Figure 3.1)

$$q_{eq} = \mathrm{f}(c_{eq}), \quad T = \text{constant} \tag{3.2}$$

https://doi.org/10.1515/9783110715507-003

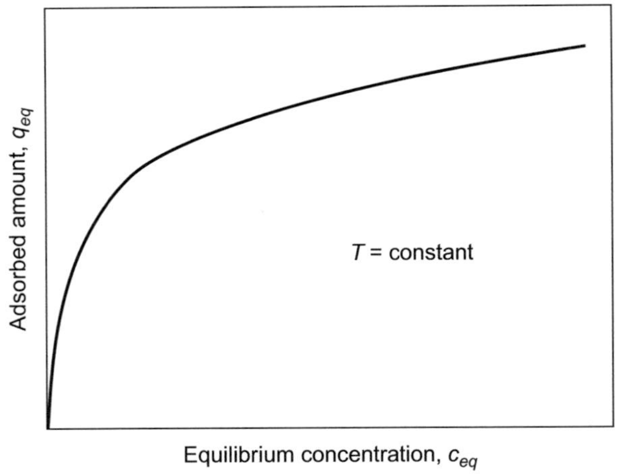

Figure 3.1: Adsorption isotherm.

Typically, the dependence of the adsorbed amount on the equilibrium concentration is determined experimentally at constant temperature, and the measured data are subsequently described by an appropriate isotherm equation. In particular, for the application of equilibrium data in more complex adsorption models (e.g. kinetic models, breakthrough curve models), it is indispensable to describe the data by a mathematical equation. The multitude of isotherm equations proposed in the literature can be classified by the number of the included parameters that have to be determined from the experimental data.

After discussing some general aspects of the experimental determination of equilibrium data (Section 3.2), the most important isotherms and their applications and limits will be presented in Section 3.3 following the above-mentioned classification principle. Further sections deal with model approaches for predicting adsorption data (Section 3.4) and with the effect of temperature on adsorption (Section 3.5). The chapter ends with some practical aspects of the application of equilibrium data (Sections 3.6 and 3.7).

3.2 Experimental determination of equilibrium data

3.2.1 Basics

Provided that an appropriate analytical method for the adsorbate is available, the experimental determination of single-solute adsorption data usually causes no problems. By contrast, the determination of equilibrium data for multisolute (competitive) adsorption is much more problematic due to the multidimensional character of the adsorption isotherms and is therefore in most cases not possible, in particular if the number of components is very high. For natural organic matter (NOM), which is typically present in all natural waters and has the character of a multicomponent mixture with unknown composition, only total isotherms (total adsorbent loading as function of total concentration) based on sum parameters (e.g. dissolved organic carbon (DOC)) can be determined. The same situation occurs in wastewater treatment where effluent organic matter (EfOM) is also a multicomponent mixture with unknown composition. The specific problems of multisolute adsorption will be considered separately in Chapter 4.

To determine equilibrium data, the bottle-point method is usually applied. In this method, a set of bottles is used to determine a larger number of isotherm points in parallel. A minimum number of 8 to 10 is recommended to get enough data points for the subsequent isotherm fitting. Each bottle is filled with the adsorbate solution of known volume, V_L, and known initial concentration, c_0. After adding a defined adsorbent mass, m_A, the solution is shaken or stirred until the state of equilibrium is reached (Figure 3.2). The time required to reach the equilibrium is typically between some hours and some weeks. Besides the type of adsorbent and adsorbate, in particular, the adsorbent particle diameter has a strong influence on the required equilibration time. The problem of finding the appropriate equilibration time will be discussed later in more detail.

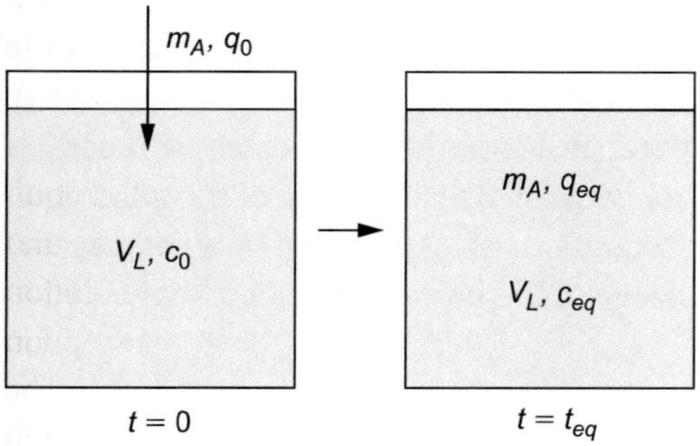

Figure 3.2: Experimental determination of adsorption equilibrium data.

After the equilibrium is established, the residual (equilibrium) concentration, c_{eq}, has to be measured. Then, the adsorbed amount, q_{eq}, can be calculated by using the material balance equation for the batch adsorption process. Under the condition that other elimination processes (e.g. degradation, evaporation, adsorption onto the bottle walls) can be excluded and only adsorption onto the adsorbent particles takes place, the mass Δm^l removed from the liquid phase must be the same as the mass adsorbed onto the adsorbent, Δm^a,

$$\Delta m^l = \Delta m^a \tag{3.3}$$

or explicitly with the starting and end values (subscripts 0 and *eq*)

$$m_0^l - m_{eq}^l = m_{eq}^a - m_0^a \tag{3.4}$$

With the definitions of the mass concentration, c, and the adsorbent loading, q,

$$c = \frac{m^l}{V_L} \tag{3.5}$$

$$q = \frac{m^a}{m_A} \tag{3.6}$$

the material balance (eq. (3.4)) can be written in the form

$$V_L(c_0 - c_{eq}) = m_A(q_{eq} - q_0) \tag{3.7}$$

Because fresh (i.e. not preloaded) adsorbent is usually used in equilibrium measurements ($q_0 = 0$), the balance equation reduces to

$$q_{eq} = \frac{V_L}{m_A}(c_0 - c_{eq}) \tag{3.8}$$

Thus, if the adsorbent dose (m_A/V_L) is known and the concentration difference has been measured, the equilibrium loading corresponding to the equilibrium concentration can be calculated from eq. (3.8). In this way, one isotherm point is found. To get more points of the isotherm, the adsorbent dose or the initial concentration has to be varied.

The procedure described previously can be demonstrated by means of diagrams that show the equilibrium curve together with the operating line of the adsorption process. Given that the material balance equation (eq. (3.8)) is not only valid for the equilibrium state but also valid for each step of the process, it can be formulated in a more general form as

$$q = \frac{V_L}{m_A}(c_0 - c) \tag{3.9}$$

or

$$q = \frac{V_L}{m_A}c_0 - \frac{V_L}{m_A}c \tag{3.10}$$

where c and q are the concentration and the adsorbent loading at a given time, respectively. Equation (3.10) is the equation of the operating line in the q-c diagram (Figure 3.3). The process starts at $c = c_0$, $q = 0$ and ends in the state of equilibrium with $c = c_{eq}$, $q = q_{eq}$. The slope of the operating line is given by $-V_L/m_A$, the negative reciprocal value of the adsorbent dose. As can be seen from the diagrams, different isotherm points can be found by variation of the adsorbent dose at constant c_0 (Figure 3.3a) or by variation of c_0 at constant adsorbent dose (Figure 3.3b).

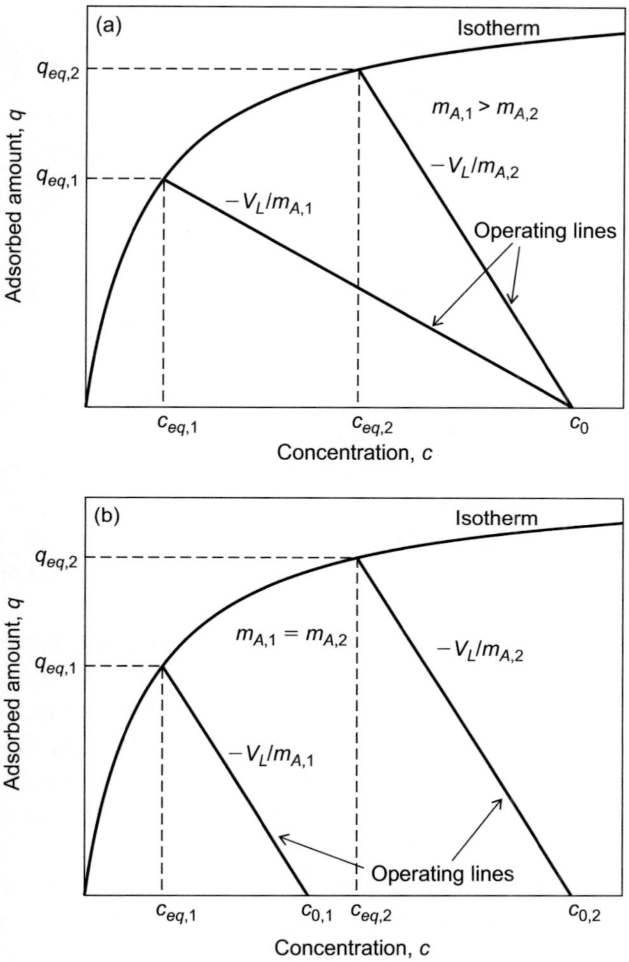

Figure 3.3: Determination of adsorption isotherms by variation of (a) adsorbent dose and (b) initial concentration.

The isotherm points found by one of these methods can be plotted in a diagram, $q_{eq} = \mathrm{f}(c_{eq})$, and can also be fitted by using an isotherm equation (see Section 3.3). Isotherms measured over a broad concentration range are often shown in double logarithmic diagrams. It has to be noted that both methods (variation of adsorbent dose or variation of initial concentration) lead to the same isotherm only in the case of single-solute adsorption. In the case of multisolute adsorption, the different methods do not lead to the same set of equilibrium data.

Generally, the material balance can be expressed in both molar and mass concentrations. Substituting the mass, m, in eqs. (3.3) and (3.4) by the number of moles, n, and applying the definitions of molar concentration ($c = n^l/V_L$) and molar adsorbent loading ($q = n^a/m_A$), the material balance gets the same form as given in eq. (3.8). Thus, there is no restriction concerning the units. The material balance as well as the resulting isotherm equations can be used with mass-related or with mole-related units. Sometimes it may be helpful to express concentration and adsorbent loading as dissolved organic carbon (DOC), for instance, for use in competitive adsorption models (see Chapter 4).

3.2.2 Practical aspects of isotherm determination

To reduce the experimental errors in isotherm determination, the following recommendations should be considered (adapted from Sontheimer et al. 1988):

- A representative adsorbent sample should be taken.
- The adsorbent should be washed prior to use with ultrapure water to remove fine particles.
- After that, the adsorbent has to be dried at ca. 110 °C because, by definition, the adsorbed amount is related to the dry mass of adsorbent. Alternatively, the exact moisture content has to be known.
- The dried adsorbent should be stored in a closed vessel or in a desiccator to avoid the uptake of water vapor.
- Taking into account the unavoidable analytical error in concentration measurement, the adsorbent dose should be chosen in such a way that the difference between the initial and equilibrium concentrations is not too small; otherwise the error in the calculated adsorbed amount may become very high (see eq. (3.8)).
- The applied adsorbent mass should not be too small to reduce errors resulting from particle loss or from heterogeneities in the adsorbent composition (e.g. residuals from the production process). If necessary, the volume has to be increased in parallel to the adsorbent mass to realize a designated adsorbent dose.
- After equilibration, the adsorbent particles have to be removed from the solution by filtration or centrifugation.

A specific problem consists in the choice of the appropriate equilibration time. At a given temperature, the equilibration time depends on the ratio c_{eq}/c_0, the particle radius, and the specific coefficients for the rate-limiting mass transfer. Given that in isotherm experiments, which are carried out under stirring or shaking, the external diffusion is fast and the adsorption rate is only determined by the internal mass transfer processes surface diffusion or pore diffusion, the equilibration time can be calculated from the respective kinetic models (Chapter 5). On the basis of the model solutions for surface diffusion (Suzuki and Kawazoe 1974a; Hand et al. 1983) and for pore diffusion (Suzuki and Kawazoe 1974b), the following equations for the minimum equilibration time, t_{min}, can be derived:

$$t_{min} = \frac{T_{B,min}\, r_P^2}{D_S} \tag{3.11}$$

$$t_{min} = \frac{T_{B,min}\, r_P^2\, \rho_P\, q_0}{D_P\, c_0} \tag{3.12}$$

where r_P is the adsorbent particle radius, D_S is the surface diffusion coefficient, D_P is the pore diffusion coefficient, ρ_P is the particle density (see Chapter 2), q_0 is the adsorbent loading in equilibrium with c_0, and $T_{B,min}$ is the minimum dimensionless time necessary for approaching the equilibrium. For definition of the dimensionless time T_B and for more details of the kinetic models, see Chapter 5.

Crittenden et al. (1987b) have set the minimum dimensionless time to $T_{B,min} = 0.6$ for surface diffusion and $T_{B,min} = 1$ for pore diffusion under the condition that $c_{eq}/c_0 < 0.9$. A closer inspection of the curves presented by Suzuki and Kawazoe shows that under the additional condition $c_{eq}/c_0 > 0.1$, the minimum dimensionless time can be reduced to $T_{B,min} = 0.4$ for surface diffusion and $T_{B,min} = 0.6$ for pore diffusion.

Equations (3.11) and (3.12) give general insights in the factors influencing the rate of equilibration, but their applicability for estimating the equilibration time is limited because in most practical cases the diffusion coefficients are unknown at the time of isotherm measurement. Moreover, in the case of pore diffusion, the equilibrium adsorbent loading for c_0 has to be known; however, equilibrium adsorbent loadings are the results of the isotherm measurement and are not known prior to the measurement. Therefore, the only reliable way to find the appropriate equilibration time and to ensure that no pseudo-equilibrium data are measured is to carry out kinetic tests prior to the equilibrium measurements.

Equations (3.11) and (3.12) show the strong influence of the particle radius on the required equilibration time. This influence will be illustrated by the following example: Assuming a relatively low surface diffusion coefficient of $D_S = 1 \cdot 10^{-13}$ m^2/s, a particle radius of $r_P = 1$ mm, and a minimum dimensionless time of $T_{B,min} = 0.4$, the necessary equilibration time calculated from eq. (3.11) would be 46 days. Reducing the particle radius by a factor of 10 reduces the equilibration time by a factor of 100 to only 11 hours. As an additional effect, an increase of D_S with decreasing particle radius can be expected, which would result in a further decrease of the required equilibration time.

Given that the particle radius has the strongest influence on the required equilibration time, grinding larger adsorbent particles (e.g. granular activated carbon) prior to application in isotherm experiments is often recommended. The resulting shortening of the equilibration time not only saves time but also reduces the risk of experimental errors due to side reactions such as biological or chemical degradation of the adsorbate. However, it was shown in several studies that the achievable adsorbent loading for a given concentration depends on the particle radius of the grinded adsorbent material. Therefore, it has to be taken into account that the isotherms determined with grinded material possibly do not exactly reflect the adsorption on the original particles. Therefore, the grinding of granular adsorbents is not unproblematic. This problem has to be weighed against the negative effects of long equilibration times for larger particles.

3.3 Isotherm equations for single-solute adsorption

3.3.1 Classification of single-solute isotherm equations

Until now, no universal isotherm equation was found that describes all experimental isotherm curves with the same accuracy. At present, a number of isotherm equations exist that have to be tested for applicability as the case arises. Some of the isotherm equations were derived from theoretical considerations; others are empirical. Sometimes, theoretically derived isotherms are applicable to experimental data though the preconditions for the derivation are not fulfilled, so that they are in fact empirical. But this is not really a problem, because for practical application the theoretical background is of secondary relevance. It is much more important to find an appropriate mathematical equation that allows describing the isotherm data as simply as possible. This is especially true if the isotherm equation should be used in adsorber models.

Most of the single-solute isotherms were originally developed for gas or vapor adsorption where the equilibrium loading is typically expressed as a function of gas or vapor pressure. After replacing the equilibrium pressure by the equilibrium concentration, these isotherms can also be applied to adsorption of solutes.

In the following sections, the isotherm equations will be classified by using a criterion that is of practical interest: the number of parameters that have to be determined from the experimental data. Generally, it can be expected that the quality of data fitting increases with increasing number of parameters. On the other hand, a higher number of parameters makes the equations more complex and complicates their application in adsorber models. Therefore, the number of parameters should be as low as possible.

Since isotherm equations always describe equilibrium data, the index *eq* is omitted for simplification in the following sections.

3.3.2 Irreversible isotherm and one-parameter isotherm

For certain limiting cases, very simple isotherm equations can be used. The irreversible (indifferent, horizontal) isotherm

$$q = \text{constant} \tag{3.13}$$

describes a concentration-independent course of the isotherm, which is typical for the saturation range that is often observed at very high concentrations (Figure 3.4a). The stronger the curvature of the isotherm is, the more the validity range of the irreversible isotherm extends to lower concentrations.

In the one-parametric Henry isotherm

$$q = K_H\, c \tag{3.14}$$

a linear relationship between adsorbent loading and concentration is assumed, with K_H as the isotherm parameter (Figure 3.4b). The common unit of K_H is L/g. The Henry equation is the thermodynamically required limiting case of isotherms at very low concentrations ($c \rightarrow 0$).

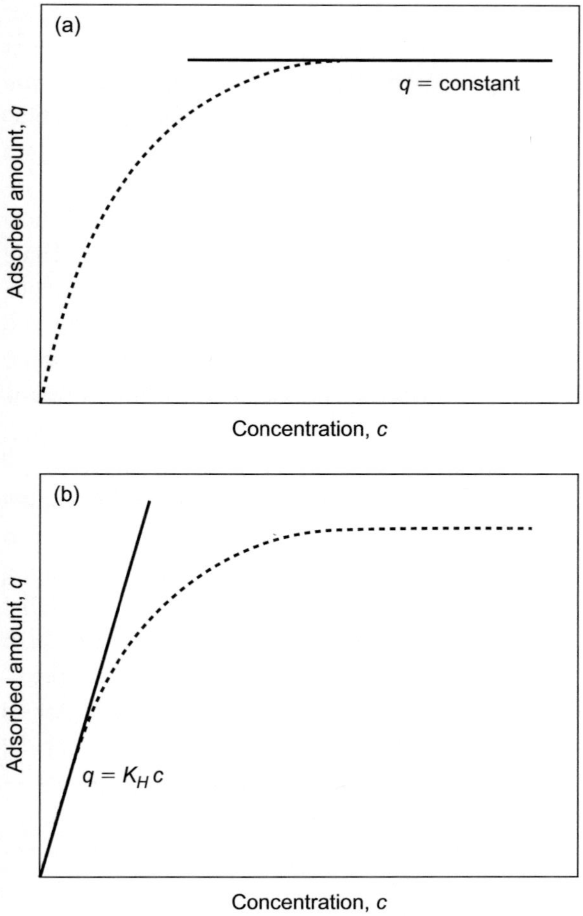

Figure 3.4: Limiting cases of adsorption isotherms: (a) irreversible isotherm and (b) linear isotherm.

Although both equations are not able to describe broader isotherm ranges, they did have relevance for simplifying kinetic and breakthrough curve models in the past. In particular, they enable analytical solutions to these models. With the development of computer technology in recent decades, the importance of these equations as the means to simplify process models has been strongly reduced.

It has to be noted that the linear isotherm is often suitable to describe sorption onto natural adsorbents (Chapter 9) where sorbate-sorbent interactions are typically weaker than in the case of engineered adsorbents such as activated carbon. In geosorption, the isotherm parameter of the linear isotherm is also referred to as the distribution coefficient, K_d.

3.3.3 Two-parameter isotherms

The well-known equations proposed by Langmuir (1918) and Freundlich (1906) are typical representatives of the group of two-parameter isotherms. They belong to the most frequently used isotherm equations.

The Langmuir isotherm has the form

$$q = \frac{q_m\, b\, c}{1 + b\, c} \tag{3.15}$$

where q_m and b are the isotherm parameters. The parameter q_m has the same unit as the adsorbent loading, and the unit of b is the reciprocal of the concentration unit. At low concentrations ($b\,c \ll 1$), eq. (3.15) reduces to the linear Henry isotherm

$$q = q_m\, b\, c = K_H\, c \tag{3.16}$$

whereas at high concentrations ($b\,c \gg 1$) a constant saturation value (maximum loading) results

$$q = q_m = \text{constant} \tag{3.17}$$

While showing plausible limiting cases, the Langmuir isotherm is often not suitable to describe the experimental isotherm data found for aqueous solutions. This might be a consequence of the fact that this theoretically derived isotherm is based on assumptions that are often not fulfilled, in particular monolayer coverage of the adsorbent surface and energetic homogeneity of the adsorption sites. This is in particular true for the most important adsorbent activated carbon. On the other hand, the Langmuir isotherm equation was also found to be applicable in cases where the underlying assumptions were obviously not fulfilled.

The Freundlich isotherm is given by

$$q = K\, c^n \tag{3.18}$$

where K and n are the isotherm parameters. The Freundlich isotherm can describe neither the linear range at very low concentrations nor the saturation effect at very high concentrations. By contrast, the medium concentration range is often very well represented. This isotherm is widely used for describing the adsorption from aqueous solutions, in particular adsorption on activated carbon. It has become a kind of standard equation for characterizing adsorption processes in water treatment. In particular, this equation is applied as the equilibrium relationship in the most kinetic and breakthrough curve models. Moreover, the Freundlich isotherm is often used in prediction models for multisolute adsorption (Chapter 4).

In the Freundlich isotherm, the adsorption coefficient, K, characterizes the strength of adsorption. The higher the value of K, the higher is the adsorbent loading that can be achieved (Figure 3.5a). The exponent n is related to the energetic heterogeneity of the adsorbent surface and determines the curvature of the isotherm. The lower the n value, the more concave (with respect to the concentration axis) is the isotherm shape (Figure 3.5b). If the concentration has a value of 1 in the respective unit, the loading equals the value of K.

In principle, the exponent n can take any value (Figure 3.6). In practice, however, mostly n values lower than 1 are found. With $n = 1$, the isotherm becomes linear. Freundlich isotherms with $n < 1$ show relative high adsorbent loadings at low concentrations. Therefore, they are referred to as favorable isotherms, whereas isotherms with $n > 1$ are characterized as unfavorable.

It has to be noted that the Freundlich isotherm can be considered a composite of Langmuir isotherms with different b values representing patches of adsorption sites with different adsorption energies. It could be shown that summing up a number of Langmuir isotherms leads to Freundlich-type isotherm curves (Weber and DiGiano 1996).

The unit of K ($= q/c^n$) depends on the units used for q and c and includes the exponent n. As discussed before (Section 3.2.1), different liquid- and solid-phase concentrations can be used in isotherm determination (molar concentrations, mass concentrations, carbon-related mass concentrations), which results in different K units. The conversion of these K units is not as simple as for other isotherm parameters due to the included exponent n. Helpful tables for K unit conversions are given in the Appendix (Tables 10.1–10.3).

Sometimes, the Dubinin-Radushkevich (DR) isotherm is also used to describe the adsorption from aqueous solutions. The DR isotherm is based on the theory of volume filling of micropores (TVFM), originally developed for vapor adsorption onto microporous adsorbents (Dubinin et al. 1947). In the modified form for solutes, the equilibrium and saturation pressure of the vapor in the original isotherm equation are substituted

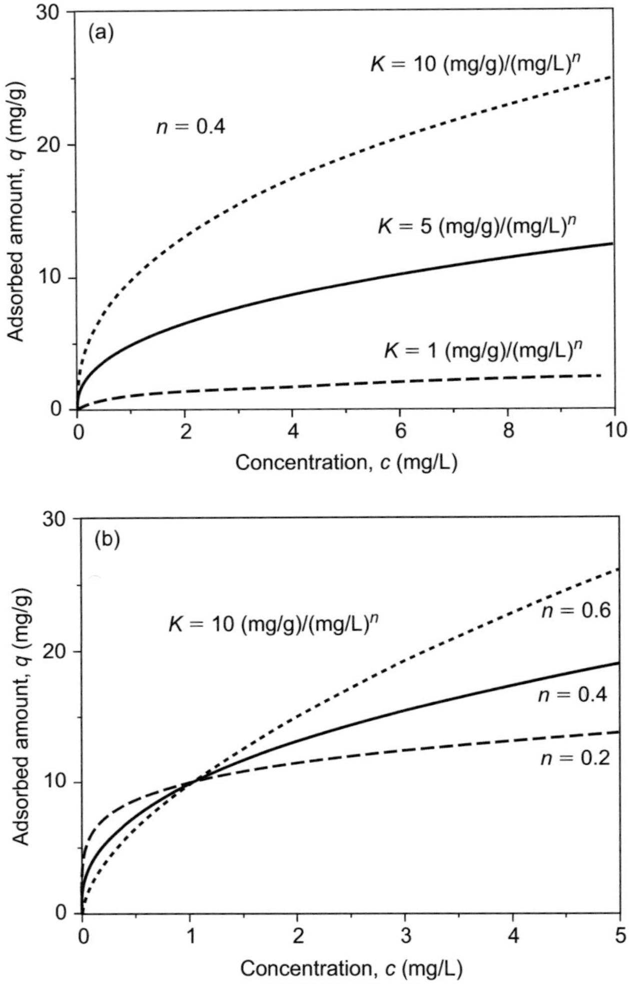

Figure 3.5: Influence of the Freundlich isotherm parameters K (a) and n (b) on the shape of the isotherm.

by the concentration and saturation concentration of the solute. In this form, the DR isotherm reads

$$q = \frac{V_0}{V_m} \exp\left[-\left(\frac{RT \ln \frac{c_{sat}}{c}}{E_C}\right)^2\right] \tag{3.19}$$

where R is the gas constant, T is the absolute temperature, c_{sat} is the saturation concentration (solubility) of the adsorbate, V_0 is the specific micropore volume (isotherm parameter), V_m is the molar volume of the adsorbate, and E_C is a characteristic

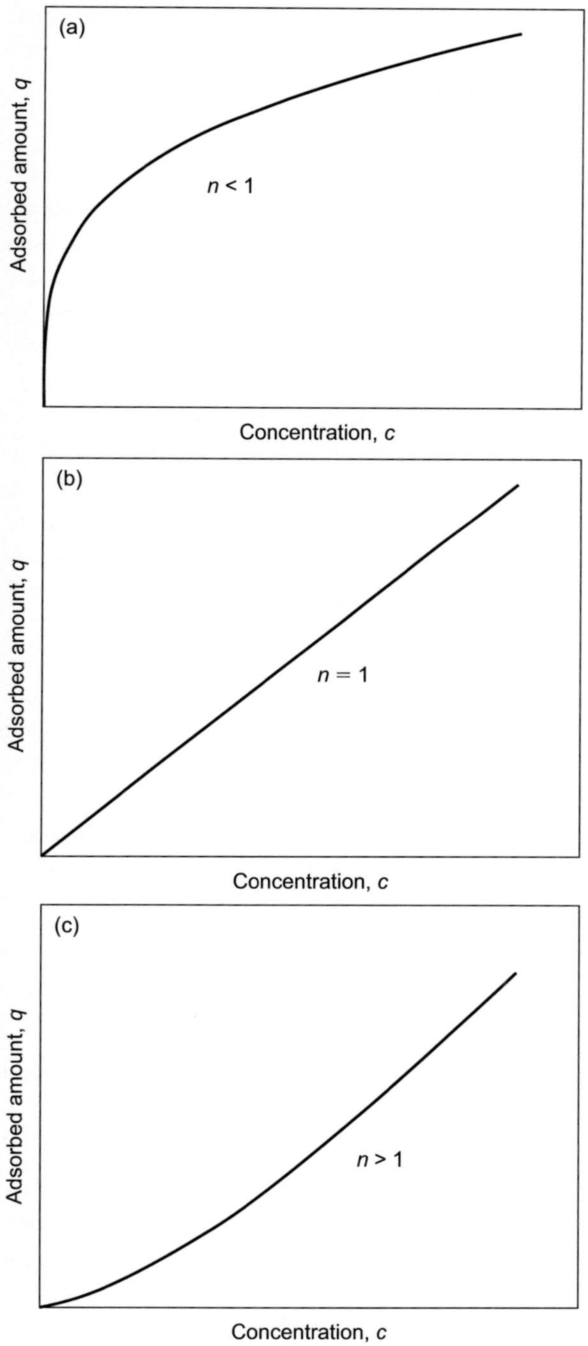

Figure 3.6: Different forms of the Freundlich isotherm: (a) $n < 1$, (b) $n = 1$, (c) $n > 1$.

adsorption energy (isotherm parameter). It has to be noted that here q is the molar adsorbent loading (mol/g) because the unit of V_0 is cm^3/g, and the unit of V_m is cm^3/mol. The term $R\,T\ln(c_{sat}/c)$ is referred to as the adsorption potential ε

$$\varepsilon = R\,T\ln\frac{c_{sat}}{c} \tag{3.20}$$

It is a specific property of the DR isotherm that the adsorption temperature is incorporated in the isotherm equation. The curve $q = f(\varepsilon)$, described by eq. (3.19), is temperature invariant and therefore also referred to as the characteristic curve. Under the assumption that the DR model is valid, the isotherm parameters, V_0 and E_C, determined at a given temperature can be used to predict the equilibrium data for other adsorption temperatures.

It has been further proposed to normalize the DR equation in order to make it independent of the specific adsorbate. For this, an affinity coefficient, β, was introduced, and E_C in eq. (3.19) was replaced by the product $\beta\,E_C$. The affinity coefficient is the scaling factor that has to be applied to let the isotherms of different adsorbates coincide in one curve. Such normalizing would allow for isotherm prediction if β could be estimated independently, for instance, from adsorbate properties. For liquid-phase adsorption, however, up to now no satisfactory method for estimating β has been found. Some aspects of isotherm prediction will be discussed more in detail in Section 3.4.

For the isotherm equations given previously, the estimation of the isotherm parameters can be carried out by nonlinear regression. In this case, an appropriate computer program is necessary. Alternatively, a linear regression is also possible because all two-parameter equations can be linearized.

For the Langmuir isotherm, different types of linearization are possible as follows:

$$\frac{c}{q} = \frac{1}{q_m\,b} + \frac{1}{q_m}c \tag{3.21}$$

$$\frac{1}{q} = \frac{1}{q_m} + \frac{1}{q_m\,b}\frac{1}{c} \tag{3.22}$$

$$q = q_m - \frac{1}{b}\frac{q}{c} \tag{3.23}$$

$$\frac{q}{c} = q_m\,b - q\,b \tag{3.24}$$

The different equations yield slightly different values of the isotherm parameters due to the different weighting of the isotherm sections resulting from the transformation of the variables.

The Freundlich isotherm can be linearized by transforming the equation into the logarithmic form

$$\log q = \log K + n\,\log c \tag{3.25}$$

or

$$\ln q = \ln K + n \ln c \tag{3.26}$$

The linearized DR equation reads

$$\ln q = \ln \frac{V_0}{V_m} - \frac{1}{E_C^2} \left(R\, T \ln \frac{c_{sat}}{c} \right)^2 \tag{3.27}$$

A general problem of all linearized equations consists of the fact that not the original, but the transformed data (e.g. logarithms, reciprocal values, ratios) are used as the basis for regression. Therefore, slightly different results of nonlinear and linear regression can be expected. However, the differences are small and can be neglected in most cases. For illustration, Table 3.1 compares the Freundlich isotherm parameters of two adsorbates estimated by linear and nonlinear regression. The differences in both cases are low, but somewhat higher for the isotherm with the lower n. Only in this case, the differences are visible in the isotherm plot (Figure 3.7).

Table 3.1: Comparison of Freundlich isotherm parameters determined by linear and nonlinear regression (adsorbates: 1,1,1-trichloroethane and 4-nitrophenol; adsorbent: activated carbon F300).

Adsorbate	Linear regression		Nonlinear regression	
	K (mg C/g)/(mg C/L)^n	n	K (mg C/g)/(mg C/L)^n	n
1,1,1-Trichloroethane	21.56	0.73	21.99	0.74
4-Nitrophenol	49.97	0.28	48.68	0.30

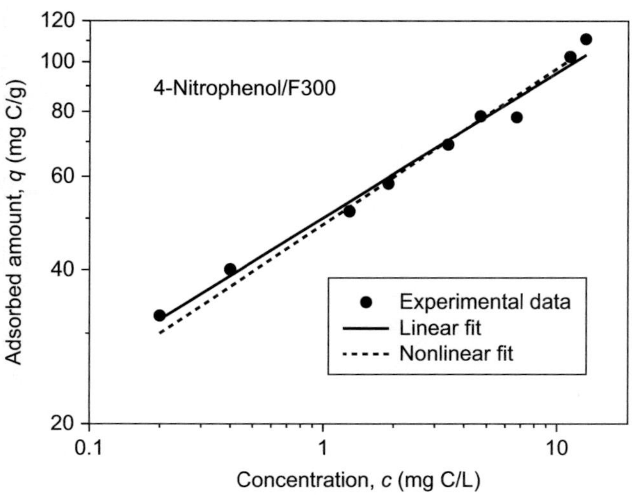

Figure 3.7: Adsorption of 4-nitrophenol onto activated carbon. Comparison of linear and nonlinear isotherm fit based on the Freundlich equation.

Frequently, adsorption isotherms measured over broad concentration ranges cannot be described exactly with only a single set of isotherm parameters. If the simple two-parameter isotherms should be maintained, different parameter sets have to be applied for the different concentration ranges (Figure 3.8); otherwise the application of three-parameter isotherms is recommended.

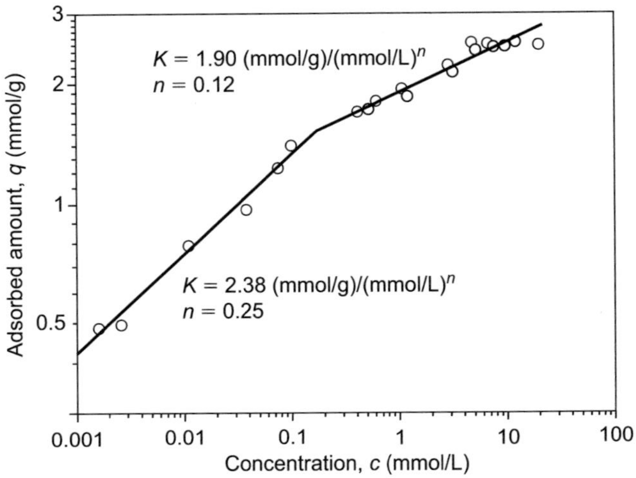

Figure 3.8: Description of an isotherm over a broad concentration range by using different sets of Freundlich parameters (4-chlorophenol – activated carbon DO4).

3.3.4 Three-parameter isotherms

Three-parameter isotherms can be derived from the Langmuir isotherm by introducing an exponent, n, as a third parameter, analogous to the exponent in the Freundlich isotherm. The Langmuir-Freundlich isotherm developed by Sips (1948) reads

$$q = \frac{q_m (b\,c)^n}{1 + (b\,c)^n} \tag{3.28}$$

or

$$q = \frac{q_m b^* c^n}{1 + b^* c^n} \tag{3.29}$$

with

$$b^n = b^* \tag{3.30}$$

This equation describes the saturation phenomenon ($q = q_m$) at higher concentrations ($b^* c^n \gg 1$). At lower concentrations ($b^* c^n \ll 1$), it does not reduce to the linear isotherm but to the Freundlich isotherm

$$q = q_m \, b^* \, c^n = K \, c^n \tag{3.31}$$

The Redlich-Peterson isotherm (Redlich and Peterson 1959) contains an exponent only in the denominator

$$q = \frac{q_m \, b \, c}{1 + (b \, c)^n} \tag{3.32}$$

Alternatively, this isotherm can be written in the form

$$q = \frac{b_1 \, c}{1 + b_2 \, c^n} \tag{3.33}$$

where

$$b_1 = q_m \, b \tag{3.34}$$

and

$$b_2 = b^n \tag{3.35}$$

While this equation reduces to the linear isotherm at lower concentrations ($b_2 \, c^n \ll 1$), it shows no saturation at higher concentrations. Instead of that, with $b_2 \, c^n \gg 1$, it approaches a Freundlich-type isotherm

$$q = \frac{b_1}{b_2} c^{1-n} \tag{3.36}$$

Therefore, except for $n = 1$, the parameter q_m in eq. (3.32) cannot be interpreted as a maximum adsorbent loading, and a formulation according to eq. (3.33) should be preferred.

The Tóth isotherm (Tóth 1971) is a three-parameter isotherm that includes both limiting cases: linear form at low concentrations and maximum adsorbent loading at high concentrations

$$q = \frac{q_m \, b \, c}{\left[1 + (b \, c)^n\right]^{1/n}} \tag{3.37}$$

with

$$q = q_m \, b \, c \quad \text{for} \quad (b \, c)^n \ll 1 \tag{3.38}$$

$$q = q_m \quad \text{for} \quad (b \, c)^n \gg 1 \tag{3.39}$$

The Tóth isotherm is sometimes also written in the form

$$q = \frac{q_m\, c}{(\beta + c^n)^{1/n}} \tag{3.40}$$

where $\beta = 1/b^n$.

The Dubinin-Astakhov (DA) equation, a further three-parameter isotherm, is a generalized Dubinin-Radushkevich isotherm with the exponent m as a third parameter (Dubinin and Astakhov 1971)

$$q = \frac{V_0}{V_m} \exp\left[-\left(\frac{R\,T\,\ln\frac{c_{sat}}{c}}{E_C}\right)^m\right] \tag{3.41}$$

As an example, Figure 3.9 shows the application of the Langmuir-Freundlich equation (eq. (3.29)) to the same experimental data as presented in Figure 3.8. As can be seen from the figure, the three-parameter equation is suitable to describe the isotherm data measured over a broad concentration range.

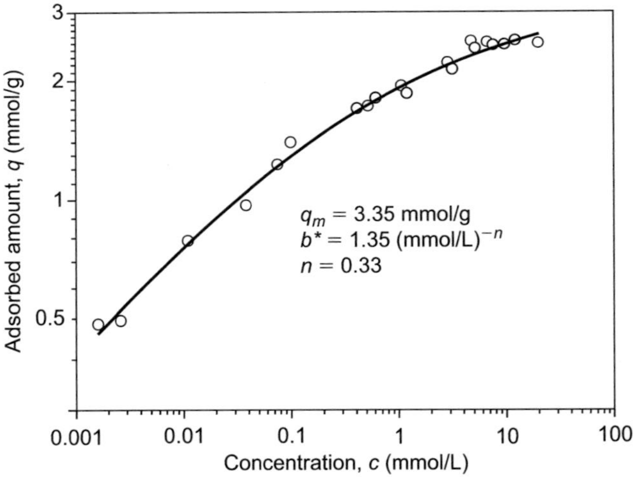

Figure 3.9: Description of an isotherm over a broad concentration range by using the three-parameter Langmuir-Freundlich isotherm (4-chlorophenol – activated carbon DO4).

3.3.5 Isotherm equations with more than three parameters

Because for any regression analysis the number of the experimental data pairs must be much greater than the number of parameters to be estimated, the experimental effort for parameter determination increases with increasing number of parameters. On the other hand, more parameters do not necessarily lead to higher fitting quality because the fitting quality is limited by data scattering resulting from the unavoidable experimental errors. Moreover, isotherm equations with too many parameters are not suitable for application in kinetic or breakthrough curve models because they make the numerical solutions to the models more complicated. For these reasons, isotherm equations with more than three parameters are rarely used in practice. Therefore, only two examples will be shown here.

The generalized Langmuir isotherm proposed by Marczewski and Jaroniec (1983) is a four-parameter isotherm with the additional parameter m

$$q = q_m \left[\frac{(b\,c)^n}{1+(b\,c)^n}\right]^{m/n} \tag{3.42}$$

At high concentrations, q approaches the saturation value, q_m, whereas at low concentrations, the isotherm becomes a Freundlich-type equation

$$q = q_m\,(b\,c)^m = q_m\,b^m\,c^m \tag{3.43}$$

The Fritz-Schlünder isotherm (Fritz and Schlünder 1974) contains the five parameters b_1, b_2, n, m, and d and is an extension of the Langmuir isotherm as well

$$q = \frac{b_1\,c^n}{d + b_2\,c^m} \tag{3.44}$$

At both lower and higher concentrations, eq. (3.44) approaches a Freundlich-type equation, but the Freundlich isotherm parameters are different in the different concentration ranges as can be seen from the following equations:

$$q = \frac{b_1}{d}\,c^n \qquad \text{for} \quad d \gg b_2 c^m \tag{3.45}$$

$$q = \frac{b_1\,c^n}{b_2\,c^m} = \frac{b_1}{b_2}\,c^{n-m} \qquad \text{for} \quad d \ll b_2\,c^m \tag{3.46}$$

It follows from eq. (3.46) that the ratio b_1/b_2 possesses the physical meaning of a constant maximum (saturation) loading in the special case $n = m$.

Both generalized isotherms include, as special cases, a number of isotherms discussed previously (Tables 3.2 and 3.3).

Table 3.2: Special cases of the generalized Langmuir isotherm proposed by Marczewski and Jaroniec (1983).

Condition	Resulting isotherm equation	Isotherm type
$n = m$	$q = \dfrac{q_m (b\,c)^n}{1+(b\,c)^n}$	Langmuir-Freundlich isotherm
$m = 1$	$q = \dfrac{q_m\,b\,c}{\left[1+(b\,c)^n\right]^{1/n}}$	Tóth isotherm
$m = n = 1$	$q = \dfrac{q_m\,b\,c}{1+b\,c}$	Langmuir isotherm

Table 3.3: Special cases of the Fritz-Schlünder isotherm.

Condition	Resulting isotherm equation	Isotherm type
$d = 1,\ m = n$	$q = \dfrac{b_1 c^n}{1 + b_2 c^n}$	Langmuir-Freundlich isotherm
$d = 1,\ n = 1$	$q = \dfrac{b_1 c}{1 + b_2 c^m}$	Redlich-Peterson isotherm
$d = 1,\ m = n = 1$	$q = \dfrac{b_1 c}{1 + b_2 c}$	Langmuir isotherm
$m = 0$	$q = \dfrac{b_1 c^n}{d + b_2} = \dfrac{b_1}{d + b_2} c^n$	Freundlich isotherm
$d = 0$	$q = \dfrac{b_1}{b_2} c^{n-m}$	Freundlich isotherm
$n = 1,\ m = 0$	$q = \dfrac{b_1 c}{d + b_2} = \dfrac{b_1}{d + b_2} c$	Henry isotherm

3.4 Prediction of isotherms

As described in Section 3.2, isotherms can be determined experimentally by batch isotherm tests. To avoid these time-consuming experiments, a number of studies have been carried out to find a model approach that allows predicting isotherms or isotherm parameters. Most of these studies are based on the application of Polanyi's potential theory of adsorption (Polanyi 1914). Below, the potential theory and two other approaches will be discussed in detail. It has to be noted that all these approaches did not allow a completely independent prediction because the respective model equations contain adjustable constants that have to be determined on the basis of experimental data sets. If these constants once have been found, they can be used to predict adsorption data for other substances.

The potential theory of adsorption is based on the assumption that attraction forces of the adsorbent are acting into the adsorption space adjacent to the adsorbent surface. As a consequence, each adsorbate molecule in the neighborhood of the adsorbent surface is subject to a change of its chemical potential in comparison to the state in the bulk liquid. In the framework of the potential theory, this change in the chemical potential is represented by the adsorption potential, ε. Given that the attraction forces decrease with increasing distance from the surface, the adsorption potential must also depend on the proximity to the solid surface. All locations in the adsorption space with the same value of the adsorption potential, ε, form an equipotential surface that together with the solid surface enclose a volume that is filled with adsorbate (Figure 3.10). Therefore, the adsorbed volume must be a function of the adsorption potential

$$V_{ads} = \frac{q}{\rho} = f(\varepsilon) \tag{3.47}$$

where V_{ads} is the adsorbed volume, q is the adsorbent loading (mass/mass), and ρ is the adsorbate density. The function $V_{ads} = f(\varepsilon)$ depends on the structure of the adsorbent and the nature of the adsorbate. For the adsorption of solutes from aqueous solutions, the effective adsorption potential is defined as

$$\varepsilon = R\,T \ln\frac{c_{sat}}{c} \tag{3.48}$$

where R is the gas constant, T is the absolute temperature, c_{sat} is the aqueous solubility of the adsorbate (saturation concentration), and c is the equilibrium concentration related to the adsorbed volume. Since the temperature is included in the adsorption potential, eq. (3.47) describes a temperature-independent function, referred to as the characteristic curve.

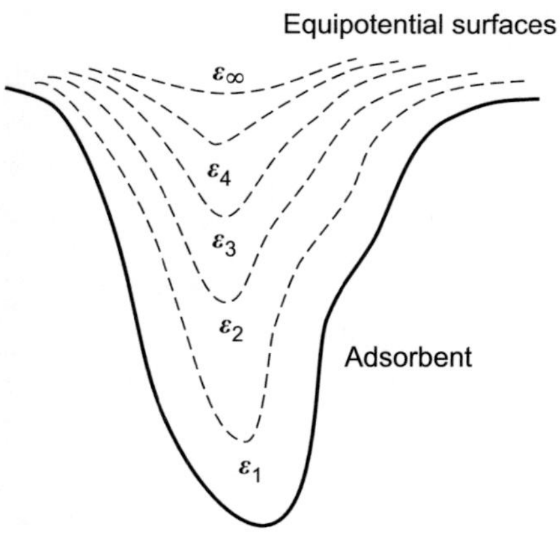

Figure 3.10: Polanyi's potential theory. Equipotential surfaces and pore volume filling.

In the original potential theory, no explicit mathematical equation for the relationship $V_{ads} = f(\varepsilon)$ was given. Later, the potential theory was further developed by Dubinin into the theory of volume filling of micropores (TVFM) with the Dubinin-Radushkevich (DR) isotherm and the Dubinin-Astakhov (DA) isotherm as important outcomes (see Section 3.3).

If Dubinin-type isotherms should be used for prediction purposes, the isotherm equation has to be normalized in such a manner that isotherms of all adsorbates collapse into one curve. As already discussed in Section 3.3.3, the introduction of the affinity coefficient into the Dubinin equation was a first attempt to normalize the characteristic curve. However, normalizing alone is not sufficient for prediction purposes. The normalizing factor must also be available independently from the isotherm experiments. In this case, the knowledge of only one experimental isotherm would be sufficient to predict isotherms for any other adsorbates.

For vapor adsorption, the affinity coefficient of the Dubinin equation was successfully predicted from adsorbate properties such as polarizability, molar volume, or parachor. For adsorption from aqueous solutions, the following Dubinin-type equation was proposed (Crittenden et al. 1987b):

$$\ln V_{ads} = \ln\left(\frac{q}{\rho}\right) = A\left(\frac{\varepsilon}{N}\right)^{B} + \ln V_0 \tag{3.49}$$

where ε is the adsorption potential, V_0 is the maximum volume available for the adsorbate, N is a normalizing factor, and A and B are empirical parameters. Since the adsorbate density in the adsorbed state is unknown, the adsorbate density under normal conditions has to be used to calculate the adsorbed volume. The latter is based on the postulate that liquid and solid solutes will separate out as liquid-like or solid-like adsorbates.

In most cases, the molar volume, V_m, is used as the normalizing factor (e.g. Kuennen et al. 1989; Speth and Adams 1993), but other normalizing factors such as linear solvation energy relationship (LSER) parameters were also proposed (Crittenden et al. 1999):

$$N = k_1 \frac{V_i}{100} + k_2 \pi^* + k_3 \beta + k_4 \alpha + k_5 \tag{3.50}$$

where V_i is the intrinsic molar volume, π^* is the polarity/polarizability parameter, β is the hydrogen-bonding acceptor parameter, α is the hydrogen-bonding donor parameter, and k_1 ... k_5 are empirical constants. The application of eq. (3.50) together with eq. (3.49) is not very comfortable, because five constants have to be found by a regression analysis, which requires a sufficiently large number of experimental data. Furthermore, the LSER parameters for the adsorbates must be known. They have to be taken from the literature or can be calculated according to a procedure developed by Hickey and Passino-Reader (1991).

Equation (3.49) describes an isotherm that theoretically should be characteristic for the adsorbent but independent of the type of adsorbate. That means that for a given adsorbent, the isotherms for all adsorbates collapse into one curve. If the parameters of the characteristic curve are known, it should be possible to calculate isotherms for other adsorbates.

If the molar volume, V_m, which can be calculated from the molecular weight and the density, is used as the normalizing factor, only A, B, and V_0 remain as adjustable parameters. Under this condition, the parameters can be determined from experimental data of only one reference adsorbate.

If furthermore the parameter B has the value 1 as found in many cases, the Freundlich parameters, K and n, are directly related to the parameters of the characteristic curve

$$\ln K = \frac{A\,R\,T}{V_m} \ln c_{sat} + \ln V_0 + \ln \rho \tag{3.51}$$

$$n = -\frac{A\,R\,T}{V_m} \tag{3.52}$$

In most practical cases, the isotherms of different adsorbates can be described satisfactorily by eq. (3.49), but often they do not fall exactly on one single characteristic curve. Frequently, better correlations can be found if only compounds of the same substance class are considered. Figure 3.11 shows the characteristic curves found for the substance groups of phenols and aromatic amines, both adsorbed onto activated carbon F300. While the correlations within the substance groups are acceptable, clear differences exist between the class-specific characteristic curves. It can also be seen that certain deviations of experimental data from the characteristic curve exist even within the same substance group.

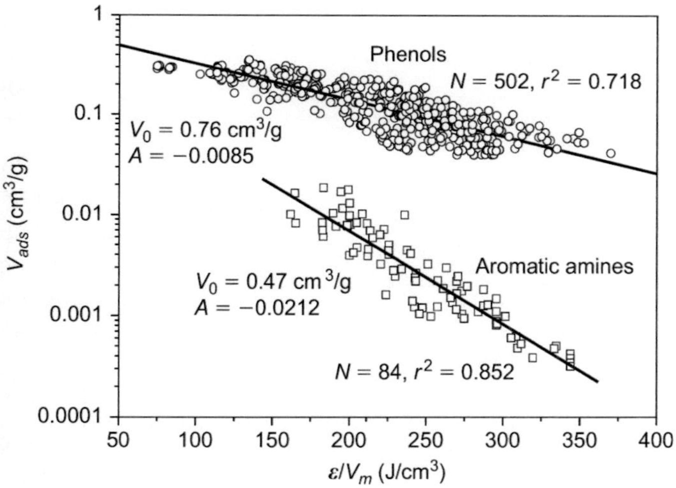

Figure 3.11: Characteristic curves for the substance groups of phenols and aromatic amines adsorbed onto activated carbon F300 (N: number of data points, r^2: coefficient of determination). Experimental data from Slavik (2006) and Eppinger (2000).

Table 3.4 compares experimentally determined isotherm parameters with parameters calculated from the characteristic curves. In some cases, the deviations between the calculated and experimental parameters seem to be unacceptable high, but often the errors in K and n compensate each other, which results in smaller errors in the

Table 3.4: Comparison of experimentally determined Freundlich isotherm parameters with parameters calculated from the Polanyi plots (Figure 3.11) by using eqs. (3.51) and (3.52).

a) Phenols

Adsorbate	Abbreviation	K_{exp} (mg/g)/(mg/L)^n	K_{calc} (mg/g)/(mg/L)^n	n_{exp}	n_{calc}
Phenol	P	75.7	55.0	0.26	0.24
2-Chlorophenol	2-CP	201.9	143.0	0.18	0.20
2-Nitrophenol	2-NP	140.7	195.5	0.31	0.22
2-Methylphenol	2-MP	192.4	101.3	0.13	0.20
3-Chlorophenol	3-CP	113.9	118.1	0.25	0,21
3-Nitrophenol	3-NP	119.1	134.3	0.24	0.22
4-Chlorophenol	4-CP	127.7	115.6	0.22	0.21
4-Nitrophenol	4-NP	125.8	139.5	0.27	0.22
4-Methylphenol	4-MP	192.4	106.7	0.18	0.20
2,4-Dichlorophenol	2,4-DCP	269.3	236.0	0.15	0.18
2,4-Dinitrophenol	2,4-DNP	189.2	280.2	0.27	0.19
2,4,6-Trichlorophenol	2,4,6-TCP	279.4	388.4	0.20	0.18

b) Aromatic amines

Adsorbate	Abbreviation	K_{exp} (mg/g)/(mg/L)^n	K_{calc} (mg/g)/(mg/L)^n	n_{exp}	n_{calc}
2-Nitroaniline	2-NA	24.0	14.1	0.61	0.55
2,5-Dichloroaniline	2,5-DCA	92.7	31.1	0.74	0.50
3,4-Dichloroaniline	3,4-DCA	56.3	36.4	0.47	0.43
2-Chloro-5-nitroaniline	2-C-5-NA	99.0	86.5	0.59	0.46
4-Methyl-2-nitroaniline	4-M-2-NA	73.5	71.9	0.55	0.45

predicted adsorbent loadings. This can be seen from Figure 3.12 where comparisons of calculated and experimental adsorbent loadings for given concentrations are shown. The mean deviations of calculated from experimental adsorbent loadings were found to be 32.2% for amines (at $c = 0.01$ mg/L) and 29.6% for phenols (at $c = 0.5$ mg/L), respectively.

In conclusion, it has to be noted that there are numerous problems connected with the application of the potential theory. The problems result from the model

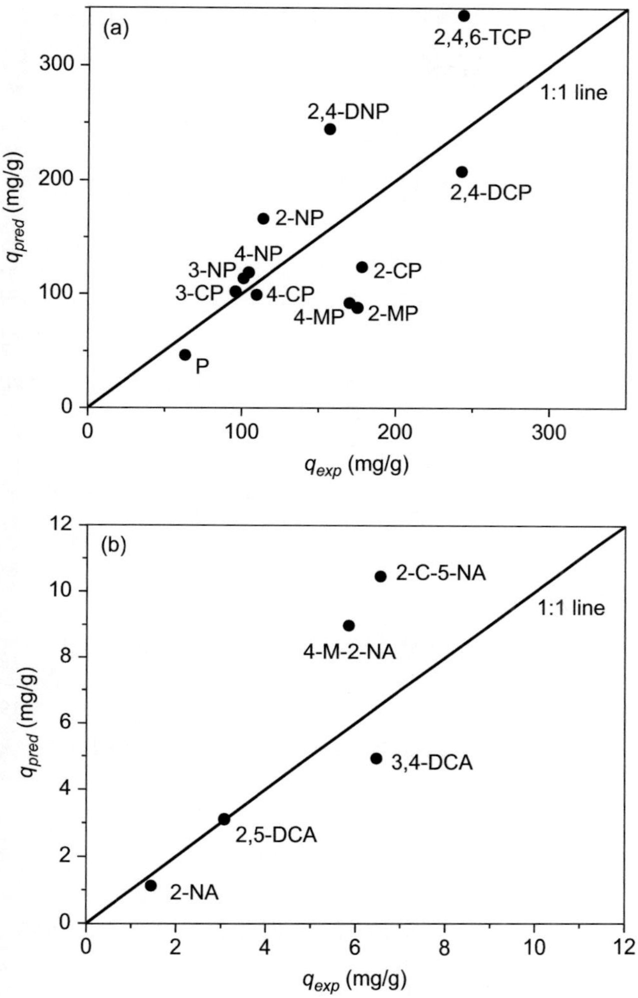

Figure 3.12: Application of the potential theory. Comparison of experimental and predicted adsorbent loadings for (a) phenols (at $c = 0.5$ mg/L) and (b) aromatic amines (at $c = 0.01$ mg/L). For abbreviations, see Table 3.4.

itself (e. g. choice of an appropriate normalizing factor) and from the uncertainties in the substance property data needed for calculation (adsorbate solubilities and densities). In particular, it is not clear if the density of the adsorbate in the pore space under the influence of adsorption forces is really the same as under normal conditions. These uncertainties and the fact that the characteristic curve is only valid for a defined adsorbent are the main factors that limit the applicability of the potential theory as a prediction tool.

Unfortunately, no real alternative is apparent at present. A comprehensive study of the potential theory and two alternative approaches (solubility-normalized Freundlich model, direct correlation of the Freundlich isotherm parameters with LSER parameters) showed that the alternative approaches are also subject to strong limitations (Slavik et al. 2016). As in the case of the potential theory, experimental data are required to find the respective parameters of the model equations by a regression analysis, and a compound grouping is necessary to correlate the data with satisfactory quality.

The solubility-normalized Freundlich model is based on the equation:

$$q = K^* \left(\frac{c}{c_{sat}} \right)^{n^*} \tag{3.53}$$

where K^* and n^* are the modified Freundlich parameters that have to be found from a regression analysis. Combining eq. (3.53) with the Freundlich isotherm (eq. (3.18)), the following relationships are found:

$$K = \frac{K^*}{(c_{sat})^{n^*}} \tag{3.54}$$

$$n = n^* \tag{3.55}$$

The disadvantage of the solubility-normalized Freundlich model is that only one Freundlich exponent can be calculated for a test data set and, consequently, all predicted isotherms have the same exponent.

Direct correlations between the Freundlich isotherm parameters and the LSER parameters can be expressed by

$$\log K = k_{K1} \frac{V_i}{100} + k_{K2}\, \pi^* + k_{K3}\, \beta + k_{K4}\, \alpha + k_{K5} \tag{3.56}$$

$$\log n = k_{n1} \frac{V_i}{100} + k_{n2}\, \pi^* + k_{n3}\, \beta + k_{n4}\, \alpha + k_{n5} \tag{3.57}$$

where the empirical constants, k, with the index K are related to the Freundlich coefficient, K, and the empirical constants with the index n are related to the Freundlich exponent, n. The LSER parameters have the same meaning as in eq. (3.50). Because of the need to know the LSER parameters and the relatively high number of adjustable constants, the LSER approach is not very comfortable for practical purposes.

In summary, it must be noted that there is currently no easy and exact way to predict isotherms and to avoid the experimental determination of equilibrium data.

3.5 Temperature dependence of adsorption

Since physical adsorption is an exothermic process, the adsorbed amount decreases with increasing temperature. Figure 3.13 shows exemplarily the influence of temperature on the adsorption of phenol onto activated carbon. The temperature dependence of adsorption can be described by means of a relationship similar to the well-known Clausius-Clapeyron equation

$$\frac{\mathrm{d}\ln c}{\mathrm{d}T} = -\frac{\Delta H_{ads}^{iso}}{R\,T^2} = \frac{Q_{ads}^{iso}}{R\,T^2} \tag{3.58}$$

where ΔH_{ads}^{iso} is the partial molar adsorption enthalpy at constant adsorbent loading (isosteric conditions, $q = \text{constant}$), Q_{ads}^{iso} is the heat of adsorption at constant adsorbent loading (isosteric heat of adsorption), R is the gas constant, and T is the absolute temperature.

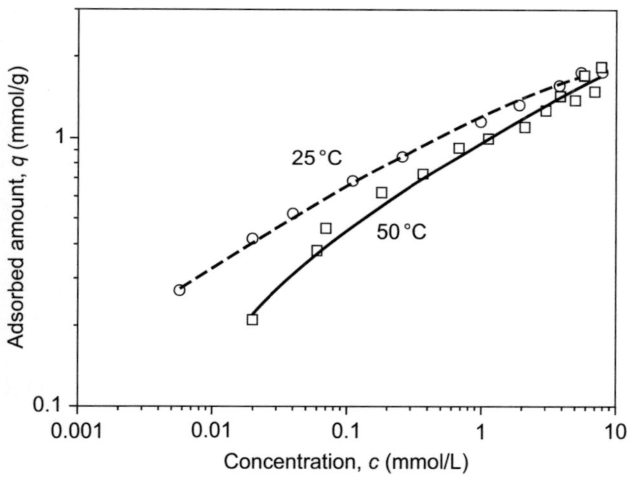

Figure 3.13: Adsorption isotherms of phenol at different temperatures (activated carbon WL2).

The isosteric heat of adsorption can be determined by measuring isotherms at different temperatures and plotting ln c over $1/T$ for given adsorbent loadings. These plots are referred to as adsorption isosters (Figure 3.14). According to

$$Q_{ads}^{iso} = R\left(\frac{\mathrm{d}\ln c}{\mathrm{d}(1/T)}\right)_q \tag{3.59}$$

the isosteric heats of adsorption for different adsorbent loadings can be found from the slopes of the ln $c - 1/T$ plots.

It has to be noted that in the case of adsorption from aqueous phase, the heat of adsorption, measured in the way described previously, includes not only the net heat of adsorption but also the enthalpy of dissolution, ΔH_{sol}, and the heat of adsorption of water, $Q_{ads,w}$,

$$Q_{ads}^{iso} = Q_{ads,\,net}^{iso} - \Delta H_{sol} - n_w\, Q_{ads,\,w} \tag{3.60}$$

where n_w is the number of moles of water that are displaced from the adsorbent by 1 mol adsorbate.

The adsorption isosters shown in Figure 3.14 have different slopes, which indicate different heats of adsorption. Obviously the isosteric heat of adsorption decreases with increasing adsorbent loading. This is typical for adsorbents with energetically heterogeneous surfaces where in the case of low adsorbate concentrations the adsorption sites with higher energy will be preferentially occupied. Later, with increasing concentrations, adsorption sites with lower energy will also be used. By contrast, for energetically homogeneous surfaces, a dependence of the heat of adsorption on the adsorbent loading can only be expected if there are strong adsorbate-adsorbate interactions in the adsorbed phase.

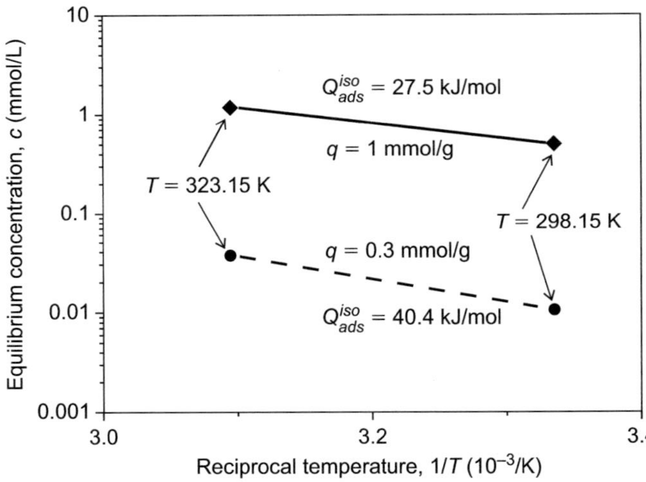

Figure 3.14: Adsorption isosters for phenol adsorbed onto activated carbon WL2.

For some of the isotherm equations presented in Section 3.3, relationships between the energetic quantities and the isotherm parameters can be derived from the fundamentals of the respective isotherm models.

The relationship between the adsorption coefficient of the Langmuir isotherm, b, and the differential heat of adsorption, Q_{ads}^{diff}, reads

$$b = b_0 \exp\left(\frac{Q_{ads}^{diff}}{R\,T}\right) \tag{3.61}$$

where b_0 is a preexponential factor. The differential heat of adsorption is related to the isosteric heat of adsorption by

$$Q_{ads}^{iso} = Q_{ads}^{diff} + R\,T \tag{3.62}$$

Equation (3.61) is in accordance with the basic assumption of the Langmuir isotherm model that postulates an energetically homogeneous surface of the adsorbent. Since the adsorption coefficient, b, of the Langmuir isotherm equation is constant over the whole concentration range of the isotherm, the heat of adsorption must also be constant.

The Freundlich isotherm – valid for heterogeneous surfaces – is related to a logarithmic decrease of the adsorption heat with increasing loading. Between the isosteric heat of adsorption extrapolated to zero loading, $Q_{ads,0}^{iso}$, and the Freundlich exponent, n, the following relationship holds:

$$\frac{Q_{ads,0}^{iso}}{RT} = \frac{1}{n} \tag{3.63}$$

Analogous relationships can also be derived for other isotherm equations. Though such relationships are interesting in view of interpreting the physical meaning of the isotherm parameters, from a practical point of view, they are of minor relevance. They do not allow for predicting the temperature dependence of adsorption equilibria, because their application requires the knowledge of the heat of adsorption, which itself can only be evaluated from isotherm measurements at different temperatures.

The DR isotherm as well as the DA isotherm is based on the potential theory, which postulates the existence of a temperature-invariant characteristic curve (Section 3.3). These isotherm equations should therefore be able to describe the temperature dependence of the adsorption. By using these equations to predict isotherms for other temperatures, it has to be considered that some of the parameters in the equations depend on the temperature as well, in particular the molar volume and the solubility of the adsorbate. Because the molar volume in the adsorbed state is unknown, it has to be calculated from the molecular weight, M, and the density in the normal state, ρ,

$$V_m = \frac{M}{\rho} \tag{3.64}$$

Therefore, in addition to the solubility, the density of the adsorbate at the respective temperature must also be known.

In Figure 3.15, the phenol isotherms at 25 °C and 50 °C, already shown in Figure 3.13, are plotted in terms of Dubinin-type isotherms (adsorbed volume vs. adsorption potential). Obviously, the isotherms actually collapse into a single characteristic curve.

In general, problems in predicting temperature-dependent isotherms by using these equations can arise from the uncertainty or nonavailability of the required density and/or solubility data.

It has to be noted that in the practice of water treatment only minor temperature fluctuations occur, and the temperature effect can therefore be neglected in most cases.

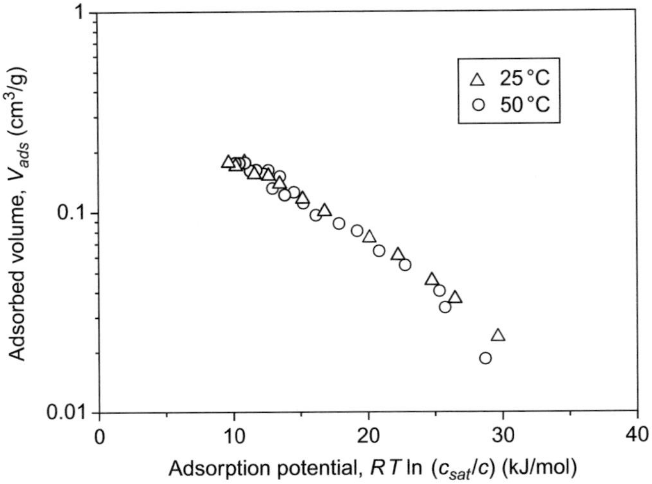

Figure 3.15: Dubinin-type plot of the phenol adsorption isotherms shown in Figure 3.13.

3.6 Slurry adsorber design

3.6.1 General aspects

In adsorption practice, different process variants and reactor types are in use. The technological options are strongly related to the particle size of the applied adsorbents. As a rule, slurry reactors are used for powdered adsorbents, whereas fixed-bed reactors are applied for granular adsorbents.

From the kinetic point of view, powdered adsorbents such as powdered activated carbon (PAC) have the advantage that the adsorption rate is very high and the equilibrium is established within a short contact time. However, powdered adsorbents cannot be used in fixed-bed adsorbers, because the flow resistance, which increases with decreasing particle size, would be too high. Therefore, slurry reactors are applied for powdered adsorbents with the consequence that an additional separation step is necessary to remove the loaded adsorbent particles from the water. Furthermore, the adsorbent consumption to achieve a given treatment goal is higher for slurry adsorbers than for fixed-bed adsorbers. This aspect will be discussed more detailed in Chapter 6.

Granular adsorbents, such as granular activated carbon (GAC), are generally used in fixed-bed adsorbers. Here, the adsorbent is fixed in the reactor, and therefore no additional separation step is necessary. The adsorbent consumption is lower in comparison to slurry reactors, but the adsorption process is slower due to the larger particle size. In the case of activated carbon, there is a further important aspect that argues for GAC application in fixed-bed adsorbers. In contrast to PAC, which cannot be efficiently regenerated and has to be burned or deposited, GAC can be regenerated (reactivated) without difficulties and used repeatedly.

While fixed-bed adsorbers provide some advantages and are increasingly applied, there may also be some situations in which the application of powdered adsorbents in slurry adsorbers is advantageous. Slurry adsorbers are often used in cases where an adsorption step within the treatment train is not continuously needed – for instance, when the raw water quality varies over time. The seasonal occurrence of taste and odor compounds in raw waters from reservoirs is a typical example of such a situation.

The design methods for the different reactor types are quite different. Fixed-bed adsorber design typically requires consideration of the adsorption kinetics besides equilibrium relationships and material balances. Fixed-bed adsorber models will be the subject of Chapters 6 and 7. For slurry reactors, by contrast, under certain conditions only equilibrium data in combination with the material balance equation are needed to describe the adsorption process, as will be demonstrated below.

In principle, slurry adsorbers can be operated either discontinuously as batch adsorbers or continuously as continuous-flow slurry adsorbers. In practice, the continuous process is favored.

In batch adsorbers, the adsorbent is in contact with the adsorbate solution until the equilibrium is reached. Batch adsorber design for single-solute adsorption is therefore very simple and requires only combining the material balance with the isotherm equation. The material balance equation for the batch reactor is the same as used for the determination of isotherms (Section 3.2)

$$m_A(q_{eq} - q_0) = V_L(c_0 - c_{eq}) \tag{3.65}$$

For continuous-flow slurry adsorbers (tanks or tubes), a comparable material balance equation can be used under the assumption that the contact time in the adsorber is longer than the time needed for establishing the equilibrium. Under this condition, the balance equation reads

$$\dot{m}_A\,(q_{eq} - q_0) = \dot{V}(c_0 - c_{eq}) \tag{3.66}$$

where $\dot{m}_A$ is the adsorbent mass added to the aqueous solution per time unit (e.g. kg/h), and $\dot{V}$ is the volumetric flow rate (e.g. m^3/h). Given that the ratio $\dot{m}_A/\dot{V}$ equals the ratio m_A/V_L according to

$$\frac{\dot{m}_A}{\dot{V}} = \frac{m_A}{t}\frac{t}{V_L} = \frac{m_A}{V_L} \tag{3.67}$$

there is no fundamental difference in the design of batch or continuous-flow adsorbers as long as establishment of equilibrium can be assumed.

Although adsorption kinetics is rapid due to the small particle size, equilibrium may not be achieved within the contact time provided. Generally, the rate of adsorption depends not only on the particle size but also on further factors such as adsorbent type, adsorbate properties, and process conditions (e.g. adsorbate concentration, adsorbent dose, mixing conditions). On the other hand, often the contact time in practical treatment processes cannot be extended arbitrarily due to technical or economic restrictions. In the case of short contact times without establishment of equilibrium, the simplified model approach only based on material balance and isotherm equation leads to an overestimation of the removal efficiency. This can be compensated by an empirical safety margin to the estimated adsorbent dose. Alternatively, short-term isotherms, with the same contact time as in the practical treatment process, can be determined. These short-term isotherms then have to be used in the adsorber design instead of the equilibrium isotherms. As an alternative, an exact modeling under consideration of adsorption kinetics can be carried out (see Chapter 5).

For the following discussion, the establishment of equilibrium will be assumed. Thus, the adsorber design requires only considering the material balance and the isotherm. Depending on the kind of adsorbent addition (all at once into one reactor or consecutively in different portions into different reactors), it can be distinguished between single-stage and multistage adsorption processes. In water treatment plants, adsorption in slurry adsorbers is most frequently carried out as a single-stage process. However, the application of a multistage reactor, in the simplest case a two-stage reactor, can reduce the adsorbent consumption.

3.6.2 Single-stage adsorption

Figure 3.16 shows the process scheme for single-stage adsorption under continuous-flow conditions. For simplification, the downstream separation step (sedimentation or filtration) necessary for the removal of the loaded adsorbent particles is not shown in the scheme.

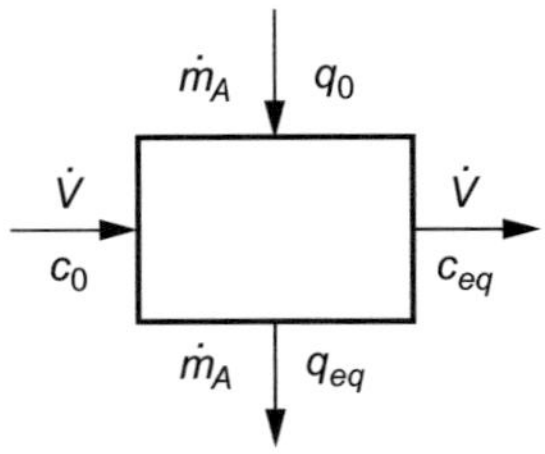

Figure 3.16: Scheme of a single-stage adsorption process.

According to eq. (3.10) in Section 3.2.1 and taking into consideration eq. (3.67), the operating line for a single-solute batch or continuous-flow adsorption process is given by

$$q(t) = \frac{V_L}{m_A} c_0 - \frac{V_L}{m_A} c(t) \tag{3.68}$$

During the process, the actual concentration $c(t)$ decreases from $c(t) = c_0$ to $c(t) = c_{eq}$ (Figure 3.17). The slope of the operating line is given by $-V_L/m_A$, the negative inverse of the adsorbent dose.

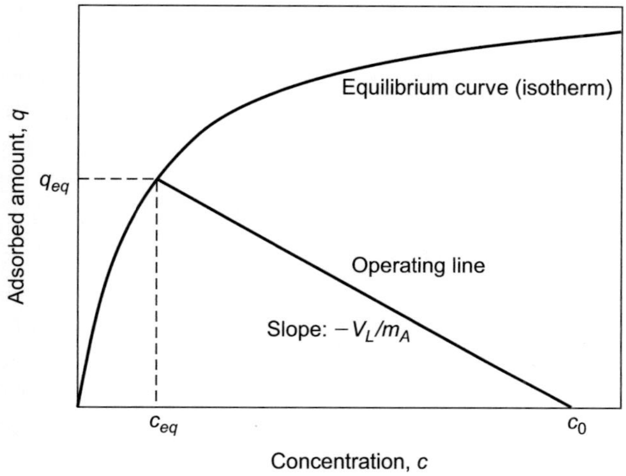

Figure 3.17: Operating line for single-stage adsorption.

For the practical implementation of the adsorption process in a slurry adsorber, it is necessary to know the optimal adsorbent dose for the given treatment goal. Under the assumptions that the adsorbent is free of adsorbate at the beginning of the process and that the contact time is long enough to achieve the equilibrium state, the adsorbent dose needed to reach a given residual concentration (= equilibrium concentration) can be found from the balance equation

$$\frac{m_A}{V_L} = \frac{c_0 - c_{eq}}{q_{eq}} \tag{3.69}$$

As can be derived from eq. (3.69), the adsorbent dose depends on the initial concentration, the equilibrium (residual) concentration, and the equilibrium adsorbent loading. The latter is given by the isotherm $q_{eq} = f(c_{eq})$. Figure 3.18 illustrates the effects of these influence factors.

Figure 3.18a shows the influence of the strength of adsorption on the required adsorbent dose. In comparison to isotherm 2, isotherm 1 reflects a higher adsorbability (e.g. a compound that is better adsorbable at the given adsorbent or another adsorbent with higher capacity for the same adsorbate). Given that the slope of the operating line represents the negative inverse of the adsorbent dose, it can be derived from a comparison of the slopes that in the case of isotherm 1, a lower adsorbent dose is needed to reach the defined residual concentration, c_{eq}.

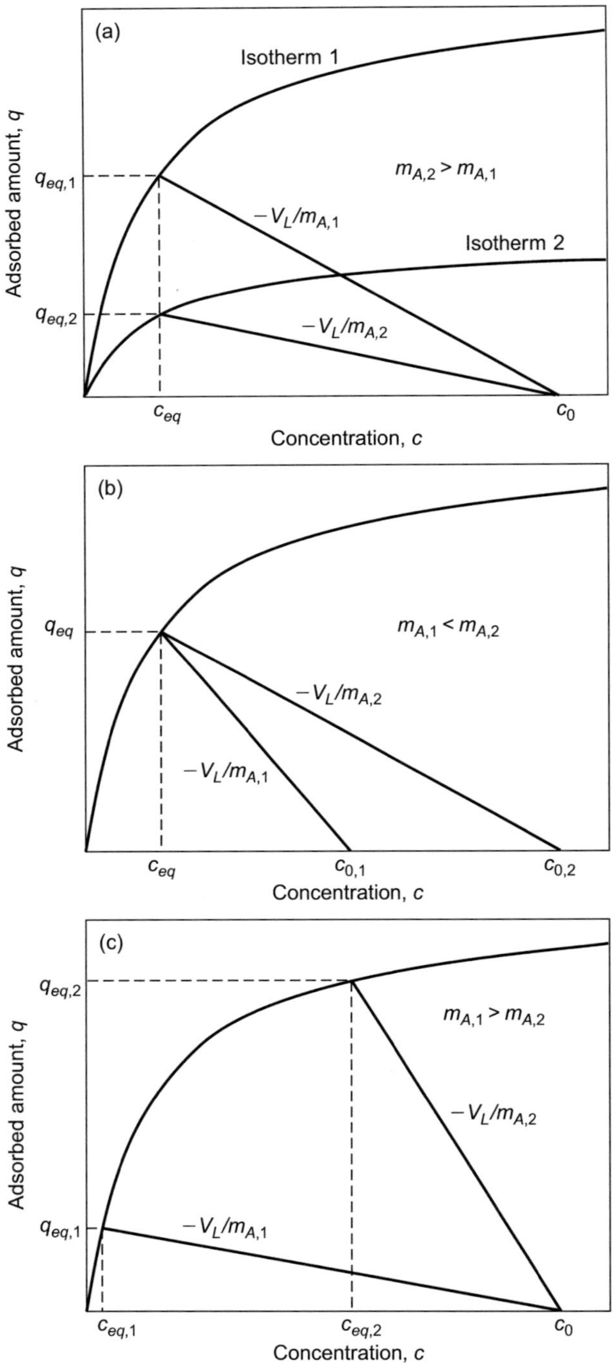

Figure 3.18: Factors influencing the adsorbent demand: (a) adsorbability, (b) initial concentration, and (c) desired residual concentration.

The influence of the initial concentration on the required adsorbent dose is demonstrated in Figure 3.18b. For the same adsorbate and the same desired residual concentration, the necessary adsorbent dose to be applied increases with increasing initial concentration.

At last, Figure 3.18c shows that very high adsorbent doses are necessary to achieve very low residual concentrations. Furthermore, it can also be seen that a residual concentration of $c_{eq} = 0$ cannot be realized in practice, because this would require that the slope of the operating line becomes zero, which means that the adsorbent mass must be infinitely large ($c_{eq} \rightarrow 0$, $-V_L/m_A \rightarrow 0$, $m_A \rightarrow \infty$). Nevertheless, for strongly adsorbable substances, very low concentrations (often under the limit of detection) can be achieved with an acceptable adsorbent dose.

A mathematical relationship that can be used to compute the required adsorbent dose for a given target concentration, c_{eq}, can be derived from substituting q_{eq} in eq. (3.69) by the respective isotherm equation. For the frequently used Freundlich isotherm, the following equation results:

$$\frac{m_A}{V_L} = \frac{c_0 - c_{eq}}{K\,c_{eq}^n} \tag{3.70}$$

Due to its nonlinear character, eq. (3.70) has to be solved by numerical methods. Design equations for other isotherms can be found in an analogous manner.

In Figure 3.18b, the influence of the initial concentration on the adsorbent dose required to reach a given residual concentration was shown. If instead of c_{eq} the adsorbent dose m_A/V_L is held constant, the initial concentration influences the achievable residual concentration, c_{eq} (see also Figure 3.3b). The influence of c_0 on the relative residual concentration, c_{eq}/c_0, at constant adsorbent dose depends on the isotherm shape. In the case of a favorable isotherm ($n < 1$), the relative residual concentration, c_{eq}/c_0, decreases with decreasing c_0, whereas in the special case of a linear isotherm ($n = 1$), the relative removal remains constant. The latter can be derived from eq. (3.70) by setting $n = 1$ and rearranging the equation:

$$\frac{c_{eq}}{c_0} = \frac{1}{K\frac{m_A}{V_L} + 1} \tag{3.71}$$

3.6.3 Two-stage adsorption

The adsorption in slurry adsorbers can also be carried out as two-stage process. The main advantage of this technological option consists in a lower adsorbent demand for the same removal effect (or stronger adsorbate removal at the same adsorbent dose). On the other hand, the higher complexity compared to a single-stage process has to be considered a drawback.

Figure 3.19 shows the process scheme for the two-stage adsorption process. For simplification, again the separation steps necessary for the removal of the loaded adsorbent particles are not shown.

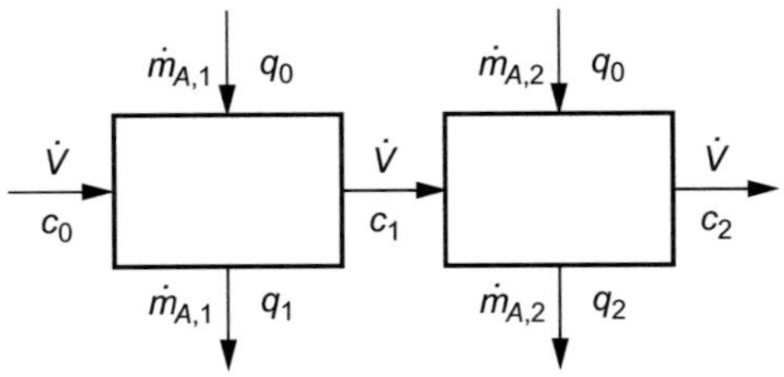

Figure 3.19: Scheme of a two-stage adsorption process.

Figure 3.20 presents the operating lines for the two-stage process. The value of the residual concentration of the first stage is determined by the adsorbent mass applied in this stage. The residual concentration of the first stage is then the initial concentration for the second stage.

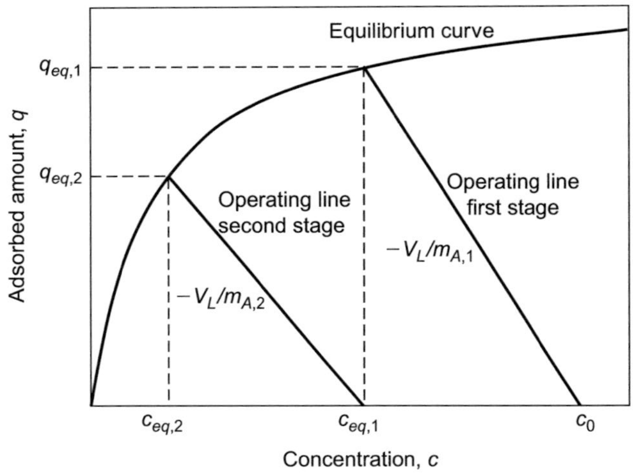

Figure 3.20: Operating lines for two-stage adsorption.

The splitting of the total adsorbent mass into two doses for the first and the second stage can be optimized by using the material balance equations for both process steps

$$\frac{m_{A,1}}{V_L} = \frac{c_0 - c_1}{q_1} \tag{3.72}$$

$$\frac{m_{A,2}}{V_L} = \frac{c_1 - c_2}{q_2} \tag{3.73}$$

Here, the subscripts 1 and 2 indicate the equilibrium state achieved in the respective adsorption stage; the index *eq* is omitted for clarity. The volume, V_L, is the total volume to be treated and is therefore the same for both adsorption steps. For the total process, the material balance equations have to be added

$$\frac{m_{A,T}}{V_L} = \frac{m_{A,1}}{V_L} + \frac{m_{A,2}}{V_L} = \frac{c_0 - c_1}{q_1} + \frac{c_1 - c_2}{q_2} \tag{3.74}$$

where $m_{A,T}$ is the total adsorbent mass applied.

The adsorbent loadings, q_1 and q_2, have to be substituted by the isotherm equation. For the Freundlich equation, the total material balance reads

$$\frac{m_{A,T}}{V_L} = \frac{m_{A,1}}{V_L} + \frac{m_{A,2}}{V_L} = \frac{c_0 - c_1}{K\,c_1^n} + \frac{c_1 - c_2}{K\,c_2^n} \tag{3.75}$$

Equation (3.75) can be used to find the concentration at the outlet of the first reactor, c_1, and the related adsorbent doses, $m_{A,1}/V_L$ and $m_{A,2}/V_L$, for which the total adsorbent mass becomes a minimum. As an example, Figure 3.21 shows the dependence of the total adsorbent demand on the adsorbent dose used in the first stage. The diagram shows that there is an optimum adsorbent splitting that leads to a minimum total adsorbent demand. Furthermore, the total adsorbent demand in a two-stage process is lower than in a comparable single-stage process with the same final concentration. The adsorbent demand for the single-stage process estimated from eq. (3.70) is given in the diagram as the lower and upper endpoint of the curve, representing the addition of the total mass either to the first or to the second adsorber.

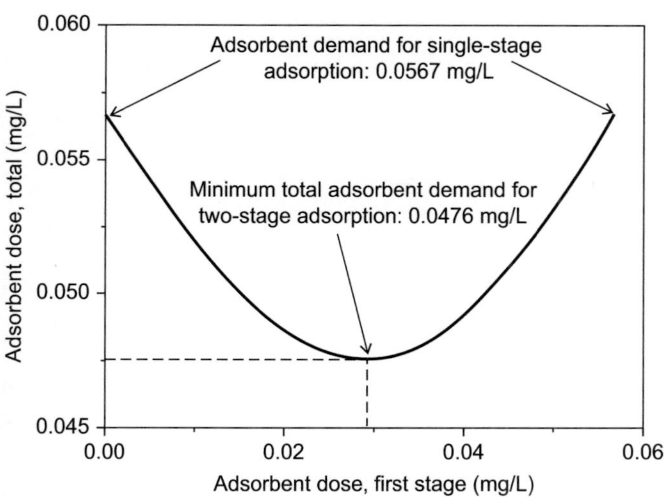

Figure 3.21: Dependence of the total adsorbent demand on the adsorbent splitting ratio in a two-stage adsorption process; model calculation with $c_0 = 1$ µg/L, $c_2 = 0.1$ µg/L, $K = 100$ (mg/g)/(mg/L)^n, and $n = 0.2$.

3.7 Application of isotherm data in kinetic or breakthrough curve models

Models for describing the adsorption kinetics in slurry adsorbers or the breakthrough behavior in fixed-bed adsorbers generally include equilibrium relationships as an essential component. These models are complex and often require numerical solution methods (Chapters 5–7). To simplify the solution algorithms, the basic equations are typically formulated in dimensionless form. Below, it will be demonstrated how the frequently used Freundlich and Langmuir isotherms can be transformed into their dimensionless forms for application in kinetic or breakthrough curve models.

Dimensionless concentrations, X, and adsorbent loadings, Y, can be defined by using the initial (or inlet) concentration, c_0, and the related equilibrium loading, q_0, as normalizing parameters

$$X = \frac{c}{c_0} \tag{3.76}$$

$$Y = \frac{q}{q_0} \tag{3.77}$$

Dividing the Freundlich isotherm expressed for c

$$q = K\, c^n \tag{3.78}$$

by the Freundlich isotherm expressed for c_0

$$q_0 = K\, c_0^n \tag{3.79}$$

gives the dimensionless Freundlich isotherm

$$Y = X^n \tag{3.80}$$

It has to be noted that the dimensionless isotherm contains only one parameter, the exponent n.

A dimensionless Langmuir isotherm can be found after introducing a separation factor, R^*, according to

$$R^* = \frac{X(1-Y)}{Y(1-X)} \tag{3.81}$$

Rearranging leads to the dimensionless isotherm

$$Y = \frac{X}{R^* + (1-R^*)X} \tag{3.82}$$

Equation (3.82) is the dimensionless form of the Langmuir isotherm as can be proved by dividing the Langmuir isotherm equations for c and c_0

$$q = \frac{q_m\, b\, c}{1 + b\, c} \tag{3.83}$$

$$q_0 = \frac{q_m\, b\, c_0}{1 + b\, c_0} \tag{3.84}$$

and setting

$$R^* = \frac{1}{1 + b\, c_0} \tag{3.85}$$

Thus, the dimensionless Langmuir isotherm contains only one parameter, the constant separation factor R^*, which is related to the adsorption coefficient b and the initial concentration c_0.

The definition of the separation factor given in eq. (3.81) can also be formally applied to other isotherms. Inserting isotherm data in the interval between $c = 0$ ($q = 0$) and $c = c_0$ ($q = q_0$) into eq. (3.81) gives separation factors, which are, in contrast to the special case of Langmuir isotherm, not constant over the considered concentration range. Therefore, in order to get a constant R^* for application in kinetic or breakthrough curve models, a mean value has to be estimated. This can be done in an appropriate manner by estimating R^* at the isotherm point $X = 1 - Y$ (Figure 3.22). In this way, any isotherm can be formally reduced to a Langmuir isotherm with a constant separation factor, R^*. Such an isotherm transformation can be advantageous because for the case R^* = constant, several analytical solutions for breakthrough curve models exist. It has to be noted that the formal application of eq. (3.81) to other isotherms can

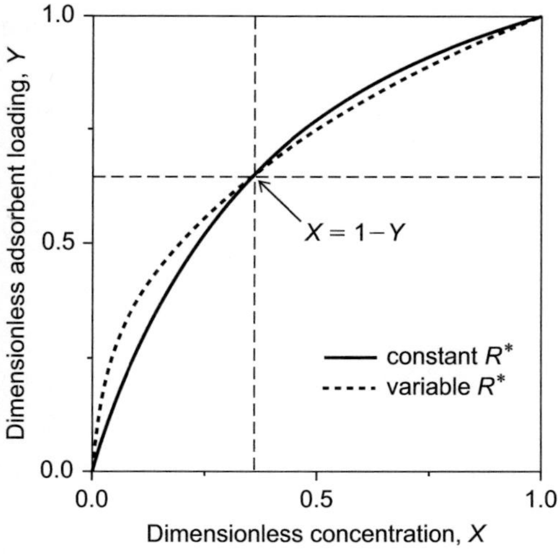

Figure 3.22: Approximation of an isotherm with variable separation factor R^* by a Langmuir-type isotherm with constant R^*.

lead to separation factors of $R^* > 1$, a range that is originally not covered by the Langmuir isotherm. For the Langmuir isotherm, R^* is always lower than 1 as can be seen from eq. (3.85).

A relationship between the Freundlich isotherm and the separation factor R^* can be found by inserting eq. (3.80) into eq. (3.81)

$$R^* = \frac{X^{1-n} - X}{1 - X} \tag{3.86}$$

Thus, it is in principle possible to characterize Freundlich isotherms by the separation factor R^*. Since in this case R^* depends on the concentration as can be seen from eq. (3.86), a mean value has to be used for further application in such models, which require a constant R^*.

As already shown in Section 3.3.3, the isotherm shape (e.g. favorable, linear, unfavorable) can be attributed to typical values of the Freundlich exponent ($n < 1$, $n = 1$, $n > 1$). According to eq. (3.86), the different isotherm shapes are also related to characteristic values of R^* (Table 3.5).

Table 3.5: Isotherm parameters and isotherm shape.

Freundlich exponent, n	Separation factor, R^*	Isotherm shape
$n = 0$	$R^* = 0$	Horizontal (irreversible)
$n < 1$	$R^* < 1$	Concave (favorable)
$n = 1$	$R^* = 1$	Linear
$n > 1$	$R^* > 1$	Convex (unfavorable)

4 Adsorption equilibrium II: Multisolute adsorption

4.1 Introduction

As shown in Chapter 3, the adsorption equilibrium of a single adsorbate can be described by the adsorption isotherm

$$q_{eq} = \mathrm{f}(c_{eq}), \quad T = \text{constant} \tag{4.1}$$

where c_{eq} is the adsorbate concentration in the state of equilibrium, q_{eq} is the adsorbed amount (adsorbent loading) in the state of equilibrium, and T is the temperature. In practice, however, raw waters or wastewaters to be treated by adsorption processes typically contain more than only a single adsorbate. With regard to the composition, a distinction has to be made between two types of multisolute systems, which exhibit different levels of complexity. In the simpler case, the aqueous solution contains only a limited number of components, and the concentrations of all constituents are known. In particular, specific industrial process wastewaters belong to this type of defined multisolute system. By contrast, the composition of raw waters in drinking water treatment is much more complex. These raw waters contain not only defined micropollutants but also ubiquitously occurring natural organic matter (NOM), which is a mixture of different natural compounds (e.g. humic substances). The exact qualitative and quantitative composition of NOM is unknown; only the total concentration can be measured by means of collective parameters such as dissolved organic carbon (DOC). A comparable situation is found for effluents from municipal wastewater treatment plants, which also contain background dissolved organic matter (DOM) besides micropollutants. The effluent organic matter (EfOM) consists of a heterogeneous mixture of refractory organic compounds with diverse structures and varying origin. Such complex multisolute adsorption systems require specific model approaches and will be discussed later in this chapter.

If a solution contains more than one adsorbable component, the adsorbates compete for the available adsorption sites on the adsorbent surface. In this case, the equilibrium loading, $q_{eq,i}$, of a considered component depends not only on the concentration of this component, $c_{eq,i}$, but also on the equilibrium concentrations of all other components. Therefore, for an N-component mixture, the following set of equilibrium relationships has to be formulated:

$$\begin{aligned} q_{eq,1} &= \mathrm{f}(c_{eq,1}, c_{eq,2}, c_{eq,3}, \ldots, c_{eq,N}) \\ q_{eq,2} &= \mathrm{f}(c_{eq,1}, c_{eq,2}, c_{eq,3}, \ldots, c_{eq,N}) \\ q_{eq,3} &= \mathrm{f}(c_{eq,1}, c_{eq,2}, c_{eq,3}, \ldots, c_{eq,N}) \\ &\vdots \\ q_{eq,N} &= \mathrm{f}(c_{eq,1}, c_{eq,2}, c_{eq,3}, \ldots, c_{eq,N}) \end{aligned} \tag{4.2}$$

https://doi.org/10.1515/9783110715507-004

Of course, the experimental effort for determining equilibrium data in such multisolute systems increases very strongly with the increasing number of components. Because in an *N*-component mixture *N* adsorbent loadings are related to *N* concentrations, the equilibrium relationships constitute a 2*N*-dimensional system. Therefore, in contrast to single-solute adsorption, the complete experimental determination of equilibrium data taking into consideration all dependencies is not feasible. Experimental data can only be determined under certain restrictive conditions, and their applicability is therefore limited to these conditions. As an alternative, prediction models for mixture equilibrium data on the basis of the more easily accessible single-solute data have been developed.

In principle, the competitive adsorption can be described by multicomponent isotherms or by thermodynamic models. In comparison to multicomponent isotherm equations, thermodynamic models possess a more general character and allow a broader application. Among the thermodynamic models, the ideal adsorbed solution theory (IAST) takes a dominant position because it allows predicting the multicomponent adsorption from single-solute isotherm parameters. At present, it can be considered the standard method to describe and predict multisolute adsorption. Based on the fundamental equations of the IAST, calculation methods for different tasks and boundary conditions can be derived.

The applicability of the original IAST as a prediction tool is restricted to multisolute systems of known composition, which means that the concentrations as well as the single-solute isotherm parameters of all components must be known. Since the composition of background DOM in drinking water or wastewater (NOM or EfOM) is unknown, the IAST cannot be directly applied to describe the DOM adsorption. A well-known solution to this problem is to apply a special fictive component approach, referred to as adsorption analysis. Combining the IAST with the concept of the adsorption analysis provides the opportunity to characterize adsorption processes in real multicomponent systems with unknown composition as typically found in drinking water treatment or wastewater treatment.

Besides background DOM adsorption, the competitive adsorption of micropollutants and DOM is of special interest in drinking water treatment as well as in wastewater treatment, because the removal of micropollutants is a main objective for application of adsorption technology. Because the original IAST is not able to predict the adsorption in micropollutant/DOM systems, specific model approaches were developed which are modifications of the IAST and allow characterizing the adsorption of micropollutants in the presence of DOM.

4.2 Experimental determination of equilibrium data

In principle, multisolute isotherms can be determined in the same way as described for single-solute adsorption (Section 3.2). For the partial adsorbent loadings of each component in an N-component system as well as for the total adsorbent loading, balance equations can be written analogously to eq. (3.8)

$$q_{eq,i} = \frac{V_L}{m_A}(c_{0,i} - c_{eq,i}) \qquad i = 1 \dots N \tag{4.3}$$

$$q_{eq,T} = \sum_{i=1}^{N} q_{eq,i} = \frac{V_L}{m_A}\left(\sum_{i=1}^{N} c_{0,i} - \sum_{i=1}^{N} c_{eq,i}\right) = \frac{V_L}{m_A}(c_{0,T} - c_{eq,T}) \tag{4.4}$$

As already discussed in Section 4.1, an extensive experimental determination of isotherm data is not possible due to the high complexity of multicomponent systems and the resulting experimental effort. In particular, it is not possible to determine partial isotherms – for example, $q_{eq,1} = f(c_{eq,1})$ – at constant equilibrium concentrations of the other components, because during the multisolute adsorption process all concentrations change simultaneously and decrease from their initial value to an equilibrium value that depends on the adsorbent dose (Figure 4.1). Given that the resulting equilibrium concentrations cannot be predicted, the concentrations cannot be fixed to special values. Therefore, mixture adsorption data are generally determined in such a manner that for a given initial composition the equilibrium concentrations of all components are measured as a function of adsorbent dose as shown in Figure 4.1. This is in accordance with the practical conditions in batch reactors. The related adsorbent loadings can be found from eq. (4.3).

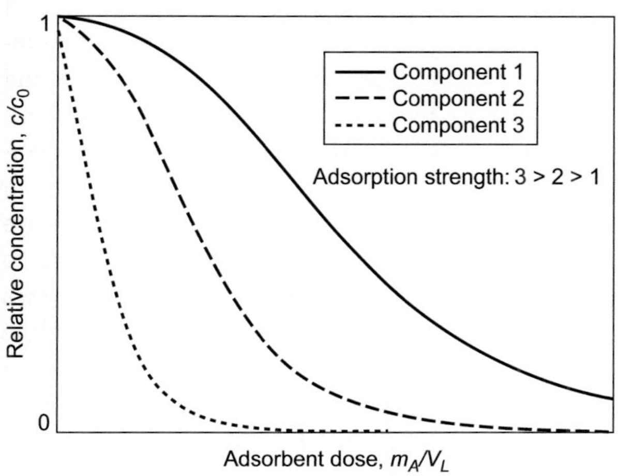

Figure 4.1: Adsorption of a three-component adsorbate mixture. Decrease of the component concentrations as a function of adsorbent dose.

In the case of complex mixtures of unknown composition (e.g. DOM-containing waters), adsorption isotherms can only be determined in terms of collective parameters such as DOC. In this case, the obtained isotherms are total isotherms (according to eq. (4.4)) with total concentrations and adsorbent loadings given as mg DOC/L and mg DOC/g, respectively.

4.3 Overview of existing multisolute adsorption models

The mathematical models for multisolute adsorption equilibria can be divided into two main groups. In the first group, mixture adsorption isotherms are classified, which are extensions of known single-solute isotherm equations. The second group comprises thermodynamic calculation methods.

Multisolute adsorption isotherm equations contain the single-solute isotherm parameters of all adsorbates present in the solution. Their application is therefore always linked to the condition that the single-solute adsorption equilibria of all components can be described by that single-solute isotherm equation, which provides the basis for the considered multisolute adsorption isotherm. Frequently, multisolute isotherm equations contain additional mixture parameters that have to be determined by competitive adsorption experiments. In these cases, consequently, only a mathematical description but not a prediction of the mixture adsorption data is possible. Furthermore, some of the proposed extended isotherm equations are restricted to bisolute adsorption.

The best-known thermodynamic models are the ideal adsorbed solution theory (IAST), the vacancy solution theory (VST), and the potential theory for multisolute adsorption.

Of these models, the IAST (Myers and Prausnitz 1965; Radke and Prausnitz 1972) is most frequently applied. It allows for predicting mixture adsorption equilibria on the basis of single-solute data without being linked to a particular single-solute isotherm model. In principle, it is even possible to apply the IAST to mixture components whose single-solute adsorption behavior is described by different isotherm equations.

Although the IAST provides significant advantages in comparison to the extended isotherm equations, it is also subject to certain restrictions. For example, for both the liquid and the adsorbed phase, ideal behavior of the adsorbates is assumed. For the liquid phase, which is usually a dilute adsorbate solution, this condition can be regarded as fulfilled. In the adsorbed phase, however, interactions between the adsorbates cannot be excluded. These interactions have to be considered in the model by an additional introduction of activity coefficients. Since these activity coefficients are only available from mixture data, the possibility of predicting the equilibrium mixture data gets lost in this case. Even neglecting nonideal behavior of the adsorbates, the solution of the IAST generally requires a higher computational effort than the application of mixture adsorption isotherms.

The VST (Suwanayuen and Danner 1980a, 1980b; Fukuchi et al. 1982) is bound to the use of a special four-parameter single-solute isotherm equation that contains parameters describing the interactions between the adsorbate and the solvent water. The parameters for the adsorbate-adsorbate interactions, which are required for an exact calculation of multisolute adsorption, however, can only be determined from multisolute adsorption data. A prediction solely on the basis of the single-solute adsorption data is, therefore, as with the IAST, only possible under the assumption of ideal behavior, which means neglecting all adsorbate-adsorbate interactions. The binding to a very specific single-solute isotherm equation, which is typically not used in practice, is an additional constraint. Moreover, the mathematical effort for applying the VST is even higher than for solving the IAST.

The potential theory for mixtures (Rosene and Manes 1976, 1977) is an extension of Polanyi's potential theory for single-solute systems (Chapter 3). As shown in Section 3.4, the potential theory requires the knowledge of physical parameters, which are difficult to access or are afflicted with uncertainties such as solubilities and densities of the adsorbates at the given adsorption temperature. On the strength of past experience, no significant advantages over the IAST, which does not require such adsorbate properties, are obvious.

To summarize, all currently known calculation methods for multisolute adsorption are subject to more or less severe restrictions. After weighing all advantages and disadvantages, at present, the IAST seems to be the most favorable method for predicting multisolute adsorption equilibria. This view is supported by a series of studies in which the IAST has been successfully applied. Moreover, the IAST provides a good basis for special model approaches that allow describing competitive adsorption of mixtures with unknown composition (e.g. DOM) and for predicting competitive adsorption of DOM and defined micropollutants.

For the sake of completeness, some extended isotherm equations will be presented in the next section. The other sections of the chapter are reserved for the IAST and its derivatives according to their great practical relevance. As in Chapter 3, for the sake of simplification, the index *eq*, which indicates the equilibrium state, is omitted in the following equations.

4.4 Multisolute isotherm equations

One of the most famous isotherm equations for describing competitive adsorption, the extended Langmuir isotherm, was developed by Butler and Ockrent (1930). Here, the partial isotherm of component *i* in an *N*-component adsorbate solution reads

$$q_i = \frac{q_{m,i}\, b_i\, c_i}{1 + \sum\limits_{j=1}^{N} b_j\, c_j} \tag{4.5}$$

The parameters q_m and b for each component are the same as in the respective single-solute isotherm equations (Chapter 3, Section 3.3.3). However, it could be shown that eq. (4.5) is thermodynamically consistent only under the condition that the maximum adsorbent loadings of all components have the same value (Broughton 1948). If this condition is not fulfilled, the extended Langmuir isotherm can indeed be used but has then an empirical character.

Equation (4.5) was modified by Schay et al. (1957) for the special case of bisolute solutions

$$q_1 = \frac{q_{m,1}\, b_1\, c_1}{E_1 + b_1\, c_1 + \frac{E_1}{E_2} b_2\, c_2} \tag{4.6}$$

$$q_2 = \frac{q_{m,2}\, b_2\, c_2}{E_2 + b_2\, c_2 + \frac{E_2}{E_1} b_1\, c_1} \tag{4.7}$$

The advantage in comparison to the Butler-Ockrent equation consists of the fact that through the introduction of the correction parameters E_1 and E_2, the case $q_{m,1} \neq q_{m,2}$ is consistently included. However, while q_m and b can be determined from the Langmuir isotherms of the individual components, the additional parameters E_1 and E_2 are only available from multisolute adsorption data. Consequently, these equations cannot be used to predict competitive adsorption equilibria.

Jain and Snoeyink (1973) have proposed an extension of the Langmuir equation for binary mixtures that is based on the assumption that only a fraction of the adsorption sites that are available for component 1 can also be occupied by component 2. They obtained the following equations:

$$q_1 = \frac{(q_{m,1} - q_{m,2})\, b_1\, c_1}{1 + b_1\, c_1} + \frac{q_{m,2}\, b_1\, c_1}{1 + b_1\, c_1 + b_2\, c_2} \tag{4.8}$$

$$q_2 = \frac{q_{m,2}\, b_2\, c_2}{1 + b_1\, c_1 + b_2\, c_2} \tag{4.9}$$

If $q_{m,2} = q_{m,1}$, eqs. (4.8) and (4.9) become identical to eq. (4.5) of the Butler-Ockrent model.

DiGiano et al. (1978) have developed a Freundlich isotherm (Chapter 3, Section 3.3.3) for N components based on the assumption that all components have the same value of the Freundlich exponent, n, and differ only in their Freundlich coefficients, K:

$$q_i = \frac{K_i^{1/n}\, c_i}{\left(\sum\limits_{j=1}^{N} K_j^{1/n}\, c_j\right)^{1-n}} \tag{4.10}$$

with

$$n_1 = n_2 = \ldots = n_N = n \tag{4.11}$$

Another extension of the Freundlich isotherm was proposed by Sheindorf et al. (1981) for the special case of bisolute adsorbate systems

$$q_1 = \frac{K_1\, c_1}{\left(c_1 + K_{1,2}\, c_2\right)^{1-n_1}} \tag{4.12}$$

$$q_2 = \frac{K_2\, c_2}{\left(K_{1,2}^{-1}\, c_1 + c_2\right)^{1-n_2}} \tag{4.13}$$

Herein, K_i and n_i ($i = 1, 2$) are the single-solute Freundlich isotherm parameters, whereas the competition coefficient $K_{1,2}$ has to be determined from mixture adsorption data.

Jaroniec and Tóth (1976) have extended the Tóth isotherm (Chapter 3, Section 3.3.4) to binary adsorbate systems. As in the case of eq. (4.10), it is assumed that the exponent n has the same value for both adsorbates

$$q_i = \frac{q_m\, c_i}{\left[\beta_i + \left(c_i + \beta_{i,j}\, c_j\right)^n\right]^{1/n}} \tag{4.14}$$

with

$$\beta_{i,j} = \frac{\beta_i}{\beta_j} \tag{4.15}$$

$$n_1 = n_2 = n \tag{4.16}$$

Regarding the parameter q_m, only the general condition

$$\min\left(q_{m,1}, q_{m,2}\right) \le q_m \le \max\left(q_{m,1}, q_{m,2}\right) \tag{4.17}$$

was given by the authors. As can be derived from eq. (4.17), the value of the multisolute parameter q_m lies somewhere between the values of the single-solute parameters $q_{m,1}$ and $q_{m,2}$ and is, therefore, uniquely defined only under the restrictive condition $q_{m,1} = q_{m,2}$.

An extension of the Redlich-Peterson isotherm (Chapter 3, Section 3.3.4) to N-component systems was developed by Mathews and Weber (1980)

$$q_i = \frac{b_{1,i}\, \dfrac{c_i}{\eta_i}}{1 + \sum_{j=1}^{N} b_{2,j} \left(\dfrac{c_j}{\eta_j}\right)^{n_j}} \tag{4.18}$$

Besides the single-solute isotherm parameters b_1, b_2, and n, eq. (4.18) contains interaction parameters, η, which have to be determined from multisolute adsorption data. Often, these interaction parameters are not constant in a given adsorbate system but depend additionally on the mixture composition. Thus, a large number of multisolute equilibrium data have to be determined to find representative mean values. For a bisolute system and under the condition $n = 1$, eq. (4.18) gets the same form as eqs. (4.6) and (4.7), with $E_i = \eta_i$, $q_{m,i}\, b_i = b_{1,i}$, and $b_i = b_{2,i}$.

Fritz and Schlünder (1974) have extended their five-parameter isotherm equation (Chapter 3, Section 3.3.5) to multisolute adsorption. The respective isotherm equation reads

$$q_i = \frac{b_{1,i}\, c_i^{n_i}}{d_i + \sum\limits_{j=1}^{N} b_{2,i,j}\, c_j^{m_{i,j}}} \tag{4.19}$$

Herein, the parameters $b_{1,i}$, n_i, and d_i are the single-solute isotherm parameters of component i. The parameters after the summation sign in the denominator are single-solute isotherm parameters if the indices are identical ($b_{2,i,i} = b_{2,i}$, $m_{i,i} = m_i$). In contrast, the cross coefficients with $i \neq j$ have to be determined from multisolute adsorption data.

A closer examination of the discussed bisolute and multisolute isotherm equations makes clear that prediction of multisolute equilibria from single-solute isotherm data is only possible by using the extended Langmuir isotherms given by eqs. (4.5), (4.8), and (4.9); the extended Freundlich isotherm given by eq. (4.10); or the extended Tóth isotherm given by eq. (4.14). All other equations contain multisolute adsorption parameters. Given that for aqueous solutions, in most cases, the single-solute isotherms can be better described by the Freundlich isotherm than by the Langmuir isotherm and that furthermore eqs. (4.8) and (4.9) are restricted to the case of bisolute systems, the extended Langmuir isotherms are of limited practical significance. From this point of view, the extended Freundlich isotherm (eq. (4.10)) should be more feasible, but this equation is subject to the restriction that the single-solute isotherm exponent, n, must have the same value for all components. However, this condition is only rarely fulfilled. The same limitation holds for the extended Tóth isotherm. Summing up, it has to be stated that extended isotherm equations are unsuitable for most practical cases.

4.5 The ideal adsorbed solution theory (IAST)

4.5.1 Basics of the IAST

The IAST was originally developed by Myers and Prausnitz (1965) on the basis of the interfacial thermodynamics to describe competitive adsorption in the gas phase. Later, this theoretical approach was extended by Radke and Prausnitz (1972) to the competitive adsorption from dilute solutions. Hereinafter, the IAST will be explained only to the extent that is necessary for practical application. The basic thermodynamic equations and more details can be found in the original literature.

The IAST is based on the following conditions:

- The adsorbed phase is considered a two-dimensional layer, which is in equilibrium with the liquid phase.
- Both the liquid phase and the adsorbed phase show ideal behavior (i.e. no interactions occur between the adsorbate molecules).
- The adsorbent surface is accessible to all adsorbates in equal measure.
- To consider the adsorbed phase, the Gibbs fundamental equation is extended by the product of surface area as extensive variable, and spreading pressure as intensive variable.

The spreading pressure, π, is defined as the difference of the surface tensions at the interfaces water-solid, σ_{ws}, and adsorbate solution-solid, σ_{as} (Chapter 1),

$$\pi = \sigma_{ws} - \sigma_{as} \tag{4.20}$$

It can be concluded from this equation that π depends on the kind of adsorbent and on the strength of the adsorbate-adsorbent interactions. Consequently, π must be correlated to the adsorption isotherm. For a single-solute system, this relation is given by the so-called Gibbs adsorption isotherm, which is a special case of the Gibbs fundamental equation. In its integrated form, this relation reads

$$\frac{\pi_i\, A_m}{R\, T} = \int\limits_0^{c_i^0} \frac{q_i^0}{c_i^0}\, \mathrm{d}\, c_i^0 \tag{4.21}$$

where A_m is the specific surface area (surface area per mass), R is the gas constant, T is the absolute temperature, c_i^0 is the equilibrium concentration of the component i in single-solute adsorption, and q_i^0 is the equilibrium loading at c_i^0. Since q_i^0 and c_i^0 are related by the respective single-solute isotherm equation, eq. (4.21) can be used to find a mathematical relationship between the spreading pressure, π_i, and the single-solute isotherm data c_i^0 and q_i^0 of the component i.

The spreading pressure plays a key role in the IAST because it determines the distribution of the adsorbates between the liquid and adsorbed phase. For a fixed π of the adsorbate mixture, the following relationship, analogous to Raoult's law, holds:

$$c_i = c_i^0(\pi)\, z_i \tag{4.22}$$

where c_i is the equilibrium concentration of the component i in the multiadsorbate solution; $c_i^0(\pi)$ is the equilibrium concentration of component i that causes in single-solute adsorption the same spreading pressure, π, as the multisolute system ($\pi = \pi_i$ with $i = 1 \ldots N$); and z_i is the mole fraction of component i in the adsorbed phase. It has to be noted that the definition of the mole fraction is here restricted to the adsorbates; the solvent is not considered.

Further, the total adsorbed amount in multisolute adsorption is related to the adsorbed amounts of the mixture components during single-solute adsorption at the given spreading pressure, π, by

$$q_T = \left[\sum_{i=1}^{N} \frac{z_i}{q_i^0(\pi)} \right]^{-1} \tag{4.23}$$

where q_T is the total adsorbed amount during multisolute adsorption, and $q_i^0(\pi)$ is the adsorbed amount of component i during its single-solute adsorption at the considered spreading pressure, π, of the multisolute system.

The partial adsorbent loading of each component, q_i, in multisolute adsorption can be found from

$$q_i = z_i \, q_T \tag{4.24}$$

and, by definition, the sum of the mole fractions must be 1

$$\sum_{i=1}^{N} z_i = 1 \tag{4.25}$$

In order to predict multisolute adsorption data from single-solute isotherms, the set of equations from (4.21) to (4.25) has to be solved after introducing the single-solute isotherm equation into eq. (4.21) and solving the resulting integral.

For the sake of simplification, the left-hand side of eq. (4.21) can be summarized to a spreading pressure term, φ. The spreading pressure term includes, besides the spreading pressure, the constant parameters A_m (specific surface area), R (gas constant), and T (temperature)

$$\varphi_i = \frac{\pi_i \, A_m}{R \, T} \tag{4.26}$$

According to this, the standard state for the IAST equations is now defined as $\varphi = \varphi_i =$ constant instead of $\pi = \pi_i =$ constant. Since φ is proportional to π, both conditions are equivalent. Accordingly, $c_i^0(\pi)$ and $q_i^0(\pi)$ in eqs. (4.22) and (4.23) can be replaced by $c_i^0(\varphi)$ and $q_i^0(\varphi)$, respectively.

Analytical solutions to the spreading pressure term integral (eq. (4.21)) are listed in Table 4.1 for the most important isotherm equations. These solutions provide the relationships between c_i^0 und φ that are necessary for the application of eq. (4.22). The relationships between q_i^0 and φ, as required in eq. (4.23), can be found easily by inserting the terms for c_i^0 into the respective isotherm equations. These relationships are also given in Table 4.1.

Table 4.1: Solutions to the spreading pressure integral.

Isotherm equation	φ_i	c_i^0	q_i^0
$q_i^0 = K_i\left(c_i^0\right)^{n_i}$ (Freundlich)	$\varphi_i = \frac{K_i}{n_i}\left(c_i^0\right)^{n_i}$	$c_i^0 = \left(\frac{\varphi_i\, n_i}{K_i}\right)^{1/n_i}$	$q_i^0 = \varphi_i\, n_i$
$q_i^0 = \frac{q_{m,i}\, b_i\, c_i^0}{1 + b_i\, c_i^0}$ (Langmuir)	$\varphi_i = q_{m,i}\ln\left(1 + b_i\, c_i^0\right)$	$c_i^0 = \frac{\exp\left(\frac{\varphi_i}{q_{m,i}}\right) - 1}{b_i}$	$q_i^0 = q_{m,i}\left[1 - \exp\left(\frac{-\varphi_i}{q_{m,i}}\right)\right]$
$q_i^0 = \frac{q_{m,i}\, b_i\left(c_i^0\right)^{n_i}}{1 + b_i\left(c_i^0\right)^{n_i}}$ (Langmuir-Freundlich)	$\varphi_i = \frac{q_{m,i}}{n_i}\ln\left[1 + b_i\left(c_i^0\right)^{n_i}\right]$	$c_i^0 = \frac{\left[\exp\left(\frac{n_i\,\varphi_i}{q_{m,i}}\right) - 1\right]^{1/n_i}}{b_i^{1/n_i}}$	$q_i^0 = q_{m,i}\left[1 - \exp\left(\frac{-n_i\,\varphi_i}{q_{m,i}}\right)\right]$

In view of the practical application of the IAST, a distinction has to be made between two different cases, which differ in the kind of the available initial data. In the simpler case, the equilibrium concentrations are known. This is the typical situation in fixed-bed adsorption where the inlet concentrations are in equilibrium with the adsorbent loading (see Chapters 6 and 7) and therefore possess the character of equilibrium concentrations. In the case of batch adsorption systems, the initial concentrations are different from the equilibrium concentrations, and both are connected by the material balance equation. Typically, only the initial concentrations are given in this case, and therefore the IAST has to be combined with material balance equations. In this case, the solution is more complex. The general solution approaches for both cases are shown schematically in Figure 4.2; the mathematical details are given in the following sections.

It has to be noted that different methods were proposed to solve the IAST equations. These methods differ in the form of the solution equations that are derived from the basic IAST equations. All procedures have in common that numerical methods are necessary to solve the derived equations. In the following sections, general solution methods are presented that can be used for different types of isotherm equations (e.g. Freundlich, Langmuir, Langmuir-Freundlich). An alternative approach, especially for Freundlich isotherms, can be found in the Appendix (Section 10.2).

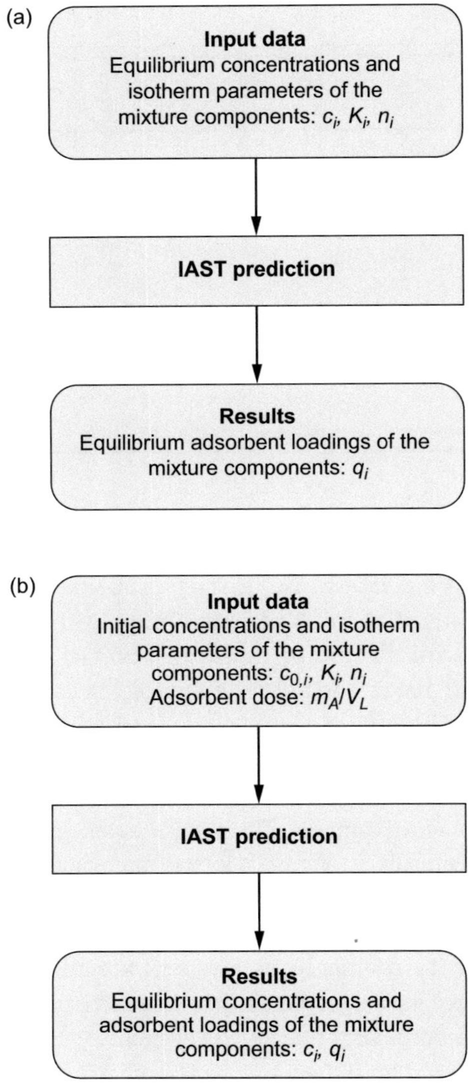

Figure 4.2: IAST solution schemes for given equilibrium concentrations (a) and given initial concentrations (b) of the multisolute adsorbate system.

4.5.2 Solution to the IAST for given equilibrium concentrations

From eqs. (4.22) and (4.25), the following relationship can be derived:

$$\sum_{i=1}^{N} z_i = \sum_{i=1}^{N} \frac{c_i}{c_i^0(\varphi)} = 1 \tag{4.27}$$

Introducing the expression $c_i^0(\varphi)$ for the Freundlich isotherm, as given in Table 4.1, eq. (4.27) becomes

$$\sum_{i=1}^{N} z_i = \sum_{i=1}^{N} \frac{c_i}{\left(\frac{\varphi\, n_i}{K_i}\right)^{1/n_i}} = 1 \tag{4.28}$$

Equation (4.28) has to be solved by an iterative method under variation of φ. Appropriate start values for the iteration can be found from

$$\varphi_{start} = \max(\varphi_i(c_T)) \qquad i = 1 \dots N \tag{4.29}$$

$$\varphi_{start} = \min(\varphi_i(c_T)) \qquad i = 1 \dots N \tag{4.30}$$

where c_T is the sum of the equilibrium concentrations of all components. The value of $\varphi_i(c_T)$ for each component can be found from the respective equation given in Table 4.1 by using the single-solute isotherm parameters, K_i, and n_i, of the component together with the total concentration, c_T.

If that spreading pressure term is found that fulfills eq. (4.28), then the adsorbed-phase mole fractions, z_i, are also fixed (see left-hand side of the equation). With φ and z_i, the total adsorbed amount, q_T, can be calculated from eq. (4.23) after substituting $q_i^0(\varphi)$ by the respective relationship given in Table 4.1. For the Freundlich isotherm, the total loading, q_T, is given by

$$q_T = \left[\sum_{i=1}^{N} \frac{z_i}{\varphi\, n_i}\right]^{-1} \tag{4.31}$$

Finally, the partial adsorbent loadings can be calculated by using eq. (4.24).

The solution approach demonstrated previously can be used not only for the Freundlich isotherm but also for other isotherm equations as will be shown exemplarily for the Langmuir and the Langmuir-Freundlich isotherms. For the Langmuir isotherm, the respective solution equations are

$$\sum_{i=1}^{N} z_i = \sum_{i=1}^{N} \frac{c_i\, b_i}{\exp\left(\frac{\varphi}{q_{m,i}}\right) - 1} = 1 \tag{4.32}$$

$$q_T = \left[\sum_{i=1}^{N} \frac{z_i}{q_{m,i}\left[1 - \exp\left(\frac{-\varphi}{q_{m,i}}\right)\right]}\right]^{-1} \tag{4.33}$$

For the Langmuir-Freundlich isotherm, the following set of equations can be derived:

$$\sum_{i=1}^{N} z_i = \sum_{i=1}^{N} \frac{c_i \, b_i^{1/n_i}}{\left[\exp\left(\frac{n_i \, \varphi}{q_{m,i}}\right) - 1\right]^{1/n_i}} = 1 \tag{4.34}$$

$$q_T = \left[\sum_{i=1}^{N} \frac{z_i}{q_{m,i}\left[1 - \exp\left(\frac{-n_i \, \varphi}{q_{m,i}}\right)\right]}\right]^{-1} \tag{4.35}$$

For a binary adsorbate system, the basic relationships of the IAST can be depicted in a diagram as shown in Figure 4.3. The diagram exhibits the spreading pressure curves for both components of the binary adsorbate mixture as can be found from eq. (4.21). The solid lines between the curves are the graphical presentation of eq. (4.22) where c_i is substituted by the product of mole fraction and total concentration, $x_i \, c_T$. It can be seen that for a given spreading pressure term and a given total concentration, the mole fractions in the liquid phase (x_i) and in the solid phase (z_i) are represented by the distances between the intersection of the lines φ = constant and c_T = constant and the related points on the spreading pressure curves. As can further be seen, each liquid-phase composition is related to a specific value of the spreading pressure term and to a specific adsorbed-phase composition. The diagram also explains the conditions for the start values of φ to be used for the iterative solution of eqs. (4.28), (4.32), or (4.34) (minimum and maximum, according to eqs. (4.29) and (4.30)).

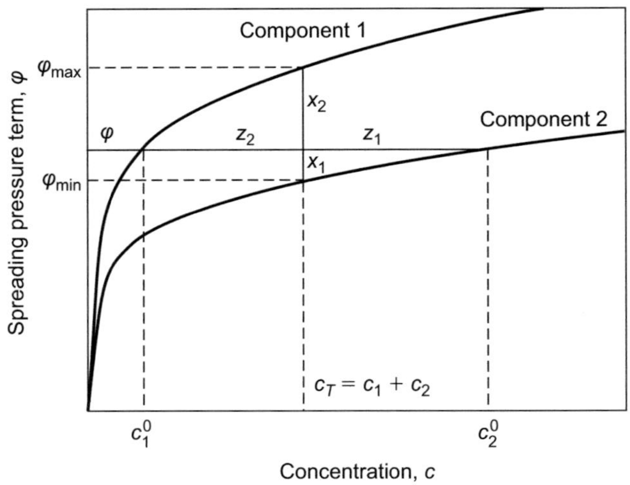

Figure 4.3: Spreading pressure curves in a bisolute adsorbate system and graphical presentation of the basic IAST relationships.

4.5.3 Solution to the IAST for given initial concentrations

In a batch adsorption system, the initial concentrations are typically known, whereas the equilibrium concentrations that are reached during the adsorption process depend on the adsorbent dose and are unknown at the beginning of the process. Initial and equilibrium concentrations are related by the material balance equation. For IAST predictions, the basic equations discussed in the previous sections have to be combined with a material balance for each component. According to eq. (4.3) and under omitting the index *eq* for simplification, the balance equation for component *i* reads

$$q_i = \frac{V_L}{m_A}(c_{0,i} - c_i) \tag{4.36}$$

The combination of the balance equation with the basic IAST equations will be demonstrated at first for the Freundlich isotherm. By using the expression for c_i^0 from Table 4.1, eq. (4.22) can be written for a given constant spreading pressure term, φ, as

$$c_i = z_i \left(\frac{\varphi\, n_i}{K_i}\right)^{1/n_i} \tag{4.37}$$

The partial adsorbent loading, q_i, in equilibrium is related to the total equilibrium loading by

$$q_i = z_i\, q_T \tag{4.38}$$

Introducing eqs. (4.37) and (4.38) into the material balance and rearranging the resulting equation gives

$$z_i = \frac{c_{0,i}}{\dfrac{m_A}{V_L} q_T + \left(\dfrac{\varphi\, n_i}{K_i}\right)^{1/n_i}} \tag{4.39}$$

and, together with eq. (4.25), we obtain

$$\sum_{i=1}^{N} z_i = \sum_{i=1}^{N} \frac{c_{0,i}}{\dfrac{m_A}{V_L} q_T + \left(\dfrac{\varphi\, n_i}{K_i}\right)^{1/n_i}} = 1 \tag{4.40}$$

Equation (4.40) contains two unknowns, the total adsorbent loading, q_T, and the spreading pressure term, φ. To determine these unknowns, a second equation is needed. This second equation can be found by inserting eq. (4.39) into eq. (4.31)

$$\sum_{i=1}^{N} \frac{1}{\varphi\, n_i} \cdot \frac{c_{0,i}}{\dfrac{m_A}{V_L} q_T + \left(\dfrac{\varphi\, n_i}{K_i}\right)^{1/n_i}} = \frac{1}{q_T} \tag{4.41}$$

Solving eq. (4.40) together with eq. (4.41) by means of a numerical method provides the total loading, q_T, and the spreading pressure term, φ, for the given adsorbent dose, m_A/V_L. Moreover, the mole fractions, z_i, are fixed according to eq. (4.39). With z_i and q_T, the partial adsorbent loadings, q_i, can be calculated. Finally, the equilibrium concentrations can be found from eqs. (4.36) or (4.37).

Analogous sets of equations can be derived for other single-solute isotherms. For the Langmuir isotherm, the solution equations read

$$\sum_{i=1}^{N} z_i = \sum_{i=1}^{N} \frac{c_{0,i}}{\frac{m_A}{V_L} q_T + \frac{E_i - 1}{b_i}} = 1 \tag{4.42}$$

and

$$\sum_{i=1}^{N} \frac{E_i}{q_{m,i}(E_i - 1)} \cdot \frac{c_{0,i}}{\frac{m_A}{V_L} q_T + \frac{E_i - 1}{b_i}} = \frac{1}{q_T} \tag{4.43}$$

with

$$E_i = \exp\left(\frac{\varphi}{q_{m,i}}\right) \tag{4.44}$$

In the case of the Langmuir-Freundlich isotherm, we obtain

$$\sum_{i=1}^{N} z_i = \sum_{i=1}^{N} \frac{c_{0,i}}{\frac{m_A}{V_L} q_T + \left(\frac{E_i - 1}{b_i}\right)^{1/n_i}} = 1 \tag{4.45}$$

and

$$\sum_{i=1}^{N} \frac{E_i}{q_{m,i}(E_i - 1)} \cdot \frac{c_{0,i}}{\frac{m_A}{V_L} q_T + \left(\frac{E_i - 1}{b_i}\right)^{1/n_i}} = \frac{1}{q_T} \tag{4.46}$$

with

$$E_i = \exp\left(\frac{n_i \varphi}{q_{m,i}}\right) \tag{4.47}$$

As an example, Figure 4.4 shows a comparison of experimental and predicted equilibrium data for a three-component model solution. The calculation was carried out on the basis of the single-solute Freundlich isotherms of the components. As can be seen from this diagram, the IAST prediction fits the experimental data well.

Taking the same three-component system as an example, the typical course of equilibrium adsorbent loadings as a function of equilibrium concentrations can be shown (Figure 4.5). In comparison to single-solute adsorption isotherms, these

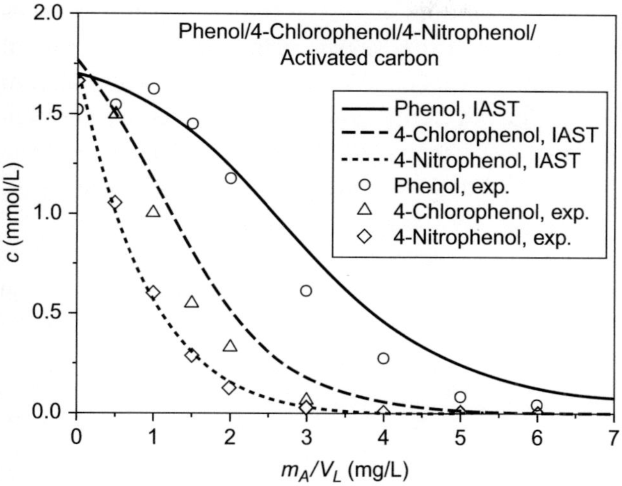

Figure 4.4: Comparison of experimental and predicted multisolute equilibrium data for the three-component adsorbate system phenol/4-chlorophenol/4-nitrophenol (adsorbent: activated carbon).

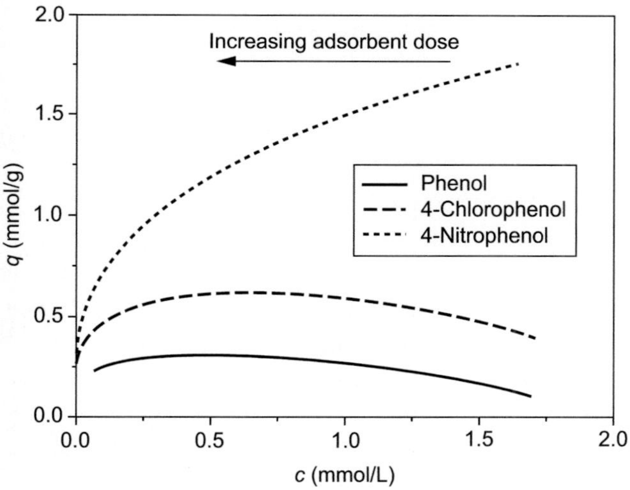

Figure 4.5: Equilibrium adsorbent loadings as a function of equilibrium concentrations for the three-component system phenol/4-chlorophenol/4-nitrophenol calculated by the IAST, according to the data shown in Figure 4.4.

equilibrium curves show quite different characteristics. As a consequence of the kind of determination of these equilibrium data (starting with a given solution composition and varying the adsorbent dose), the concentrations of all components change in parallel. This concentration change takes place not in the same measure for all components but depends on their adsorption strength. Consequently, a different liquid and solid phase concentration distribution results for each adsorbent dose. The decrease of the adsorbed amounts of the weaker adsorbable components in the higher concentration range can be explained by the stronger competition at low adsorbent doses (see also Figure 4.4). Furthermore, the concentration distribution depends on the starting concentrations. In Figure 4.6, the change of the equilibrium curves after reducing all initial concentrations by half is shown.

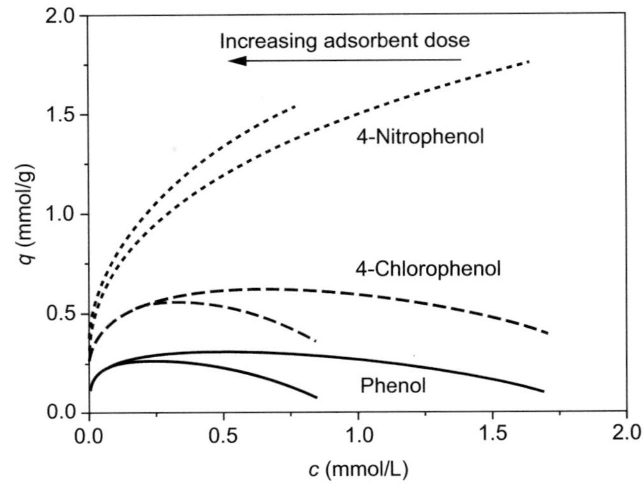

Figure 4.6: Influence of the initial concentrations on the course of the isotherms in the three-component system phenol/4-chlorophenol/4-nitrophenol. The shorter curves represent the data for initial concentrations reduced by half.

The model calculations shown in Figures 4.5 and 4.6 demonstrate the complexity of multisolute adsorption and underline the impossibility of determining complete mixture equilibria by experimental measurements as already discussed in Section 4.1.

4.5.4 Solution to the IAST for the special case of linear single-solute isotherms

The solution to the IAST for the special case of linear single-solute isotherms can be derived from the equations valid for the Freundlich isotherm and setting the Freundlich exponents $n_i = 1$. Under this condition, the expressions for c_i^0 and q_i^0 given in Table 4.1 read

$$c_i^0 = \frac{\varphi_i}{K_i} \tag{4.48}$$

and

$$q_i^0 = \varphi_i \tag{4.49}$$

For a constant $\varphi = \varphi_i$, eqs. (4.27) and (4.31) can then be written as

$$\sum_{i=1}^{N} z_i = \sum_{i=1}^{N} \frac{K_i\, c_i}{\varphi} = 1 \tag{4.50}$$

and

$$\frac{1}{q_T} = \sum_{i=1}^{N} \frac{z_i}{q_i^0} = \sum_{i=1}^{N} \frac{z_i}{\varphi} = \frac{1}{\varphi} \sum_{i=1}^{N} z_i = \frac{1}{\varphi} \tag{4.51}$$

Equation (4.51) can be used to substitute φ by q_T in eq. (4.50)

$$\sum_{i=1}^{N} z_i = \sum_{i=1}^{N} \frac{K_i\, c_i}{q_T} = 1 \tag{4.52}$$

Multiplying eq. (4.52) by q_T and considering eq. (4.24) finally gives

$$q_T = \sum_{i=1}^{N} q_i = \sum_{i=1}^{N} K_i\, c_i \tag{4.53}$$

It can be derived from eq. (4.53) that the total adsorbed amount is made up additively by the adsorbed amounts that result from the (linear) single-solute adsorption isotherms. That means that the mixture compounds are adsorbed as with single-solute adsorption and competition does not occur.

It has to be noted that this special case is seldom found in engineered adsorption processes, where the Freundlich exponents are typically much lower than 1. In geosorption processes, however, the isotherms found for the sorption of individual sorbates on soil or sediment material are often linear, in particular at low concentrations. The modeling of mixture sorption in such natural processes is therefore considerably simplified (Chapter 9).

4.6 The pH dependence of adsorption: A special case of competitive adsorption

A number of water constituents are weak organic acids or bases. The adsorption of such adsorbates is strongly influenced by the proton activity in the aqueous solution, commonly expressed as pH. A detailed theoretical description of this phenomenon is complicated because it is influenced by the adsorbent as well as by the adsorbate properties in a complex manner. The main pH-dependent effects are the protonation/deprotonation of the adsorbate and the change of the surface charge of the adsorbent. Under certain conditions, in particular if the surface charge density is low, only the influence of the pH value on the adsorbate properties has to be considered, and the pH-dependent adsorption can be described by a simplified multisolute adsorption approach.

Weak acids (HA) and bases (B), dissolved in aqueous solutions, are subject to protolysis according to

$$HA \rightleftharpoons H^+ + A^-$$

$$B + H_2O \rightleftharpoons BH^+ + OH^-$$

where A^- is the acid anion and BH^+ is the protonated base. It follows from these reaction equations that the neutral adsorbates are transformed to charged species depending on the pH of the solution. The fractions of the charged species in relation to the total adsorbate concentration can be estimated from the respective law of mass action

$$K_a = \frac{a(H^+)\,a(A^-)}{a(HA)} \tag{4.54}$$

or

$$K_b = \frac{a(BH^+)\,a(OH^-)}{a(B)} \tag{4.55}$$

and the respective material balance equation

$$c_T(HA) = c(A^-) + c(HA) \tag{4.56}$$

or

$$c_T(B) = c(B) + c(BH^+) \tag{4.57}$$

where K_a is the acidity constant, K_b is the basicity constant, a is the equilibrium activity, c is the equilibrium concentration, and c_T is the total concentration of the acidic or basic adsorbate.

By using the logarithmic parameters $pH = -\log a(H^+)$, $pOH = -\log a(OH^-)$, $pK_a = -\log K_a$, and $pK_b = -\log K_b$, and assuming an ideal dilute solution ($a \approx c$), the following equations can be derived:

$$\alpha = \frac{c(A^-)}{c_T(HA)} = \frac{1}{1 + 10^{pK_a - pH}} \tag{4.58}$$

$$\alpha = \frac{c(BH^+)}{c_T(B)} = \frac{1}{1 + 10^{pK_b - pOH}} \tag{4.59}$$

where α is the degree of protolysis, representing the portion of the charged species in the adsorbate solution.

Since pH and pOH are related by the dissociation constant of water (pH + pOH = pK_w, with $pK_w = 14$ at 25 °C), eq. (4.59) can also be written as

$$\alpha = \frac{c(BH^+)}{c_T(B)} = \frac{1}{1 + 10^{pK_b - pK_w + pH}} \tag{4.60}$$

For basic solutes, often the pK_a of the protonated base instead of pK_b is given in tables. In this case, the portion of the protonated species can be found from

$$\alpha = \frac{c(BH^+)}{c_T(B)} = \frac{1}{1 + 10^{pH - pK_a}} \tag{4.61}$$

because pK_a and pK_b of the conjugate acid/base pair (BH^+/B) are related by

$$pK_a + pK_b = pK_w \tag{4.62}$$

Using the equations given previously, it is possible to find the fractions, f, of charged ($f = \alpha$) and uncharged ($f = 1 - \alpha$) species for any given pH. This is schematically shown in Figures 4.7a and 4.7b for an acid and a base, respectively. As can be seen from the diagrams, there are three ranges, which differ in the solution composition. The boundaries between these ranges can be defined as pH = $pK_a - 2$ and pH = $pK_a + 2$.

In the case of acids (Figure 4.7a), in the range pH < $pK_a - 2$, the solution contains practically only neutral species ($\alpha < 0.01$), whereas in the range pH > $pK_a + 2$, the adsorbate exists nearly totally in the charged (anionic) form ($\alpha > 0.99$). The adsorbate solutions in both ranges can be considered single-solute systems, and the adsorption equilibria can be described by single-solute adsorption isotherms. In the medium range, $pK_a - 2 < pH < pK_a + 2$, the adsorbate solution is a two-component system consisting of neutral and charged species.

Bases (Figure 4.7b) show, in principle, a comparable behavior, but the sequence of the three ranges is reversed in comparison to the acids. The charged species (protonated base) dominate at lower pH values, whereas at higher pH values mainly neutral species occur. As in the case of acids, the medium range is characterized by the coexistence of neutral and charged species.

Generally, the pH-dependent protonation/deprotonation changes the polarity of the adsorbate and therefore also its adsorbability. Additionally, the pH may affect the surface charge of the adsorbent. As described in Chapter 2 (Section 2.5.6), in particular oxidic and, to a lesser extent, carbonaceous adsorbents possess functional groups on their surface that can be protonated or deprotonated. As a consequence, the surface of such adsorbents is typically positively charged at low pH values and negatively charged at high pH values. The pH value at which the sum of positive charges equals the sum of negative charges is referred to as point of zero charge, pH_{pzc}. Whereas it can be assumed that the surface charge has no influence on the adsorption of neutral adsorbates, charged species (acid anions and protonated bases) can be subject to additional attraction or repulsion forces, depending on the signs of the species and surface charges.

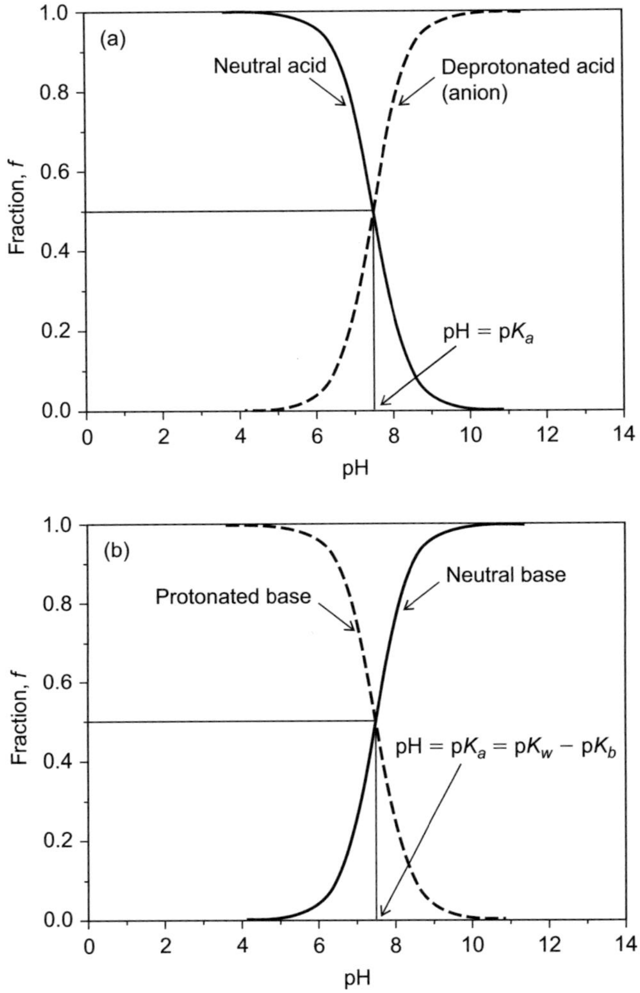

Figure 4.7: Fractions of neutral (–) and charged (- -) species as a function of pH in the case of an acid (a) and a base (b). The assumed pK_a in both examples is 7.5.

In pH ranges where the adsorbate species have the same charge as the surface, the adsorption is relatively weak, not only as a result of the higher polarity of the charged adsorbate species but also due to additional repulsion forces. Under certain conditions ($pH_{pzc} > pK_a$ for acids; $pH_{pzc} < pK_a$ for bases), pH ranges exist where the adsorbate species and the adsorbent surface show opposite charges and, consequently, attraction forces occur. Since these attraction forces act in addition to the van der Waals forces, an adsorption maximum can often be observed in these pH ranges. Table 4.2 summarizes the conditions under which attraction and repulsion forces can be expected.

It follows from Table 4.2 and the previous discussion that an adsorption maximum can be expected only under special conditions ($pH_{pzc} > pK_a$ for acids; $pH_{pzc} < pK_a$

Table 4.2: Conditions for electrostatic interactions during adsorption of weak acids and bases onto charged surfaces.

Adsorbate character	Relative position of pH_{pzc} and pK_a	pH range	Dominating adsorbate charge	Dominating adsorbent surface charge	Resulting electrostatic interaction
Acidic	$pH_{pzc} < pK_a$	$pH > pK_a$	Negative	Negative	Repulsion
	$pH_{pzc} > pK_a$	$pK_a < pH < pH_{pzc}$	Negative	Positive	Attraction
		$pH > pH_{pzc}$	Negative	Negative	Repulsion
Basic	$pH_{pzc} > pK_a$	$pH < pK_a$	Positive	Positive	Repulsion
	$pH_{pzc} < pK_a$	$pH_{pzc} < pH < pK_a$	Positive	Negative	Attraction
		$pH < pH_{pzc}$	Positive	Positive	Repulsion

for bases). In all other cases, the adsorption will decrease continuously with increasing degree of deprotonation (acids) or protonation (bases). If there are no (or negligible) attraction forces, the pH-dependent adsorption can be described by using multisolute adsorptions models such as the IAST.

As shown before, the adsorbate solution in the range $pK_a - 2 < pH < pK_a + 2$ can be considered a bisolute adsorbate system consisting of neutral and charged species. The single-solute adsorption of these components can be found from isotherm measurements under pH conditions where only the neutral or only the charged species exist. Subsequently, the pH-depending adsorption in the medium pH range where both species exist can be predicted by the IAST if the boundary isotherms (single-solute isotherms of neutral and charged species) as well as the composition of the solution are known. The latter is available from eqs. (4.58) and (4.59) for given pK_a, pH, and total concentration. With the component concentrations and the parameters of the boundary isotherms, the IAST, in the specific form described in Section 4.5.2 (solution for given equilibrium concentrations), can be applied.

As an example, the pH-dependent adsorption of phenol ($pK_a = 10$) on activated carbon is shown in Figure 4.8. The dashed curves are the isotherms of the neutral phenol and the phenolate anion, respectively. For the sake of clarity, the experimental data of the boundary isotherms are not shown. These isotherms could be best described by the Langmuir-Freundlich isotherm with $q_m = 3.01$ mmol/g, $b = 0.673$ (L/mmol)^n, $n = 0.37$ for the neutral phenol and $q_m = 1.09$, $b = 0.586$ (L/mmol)^n, $n = 0.77$ for the phenolate anion, respectively. The isotherms for the pH values, where both species coexist, were calculated by the IAST, in particular by using eqs. (4.34) and (4.35). As can be seen from Figure 4.8, the predicted isotherms are in good agreement with the experimental data.

Finally, it has to be stated that the pH dependence of the adsorption of weak acids or bases is of practical relevance only if the pK_a is near the pH of the water to

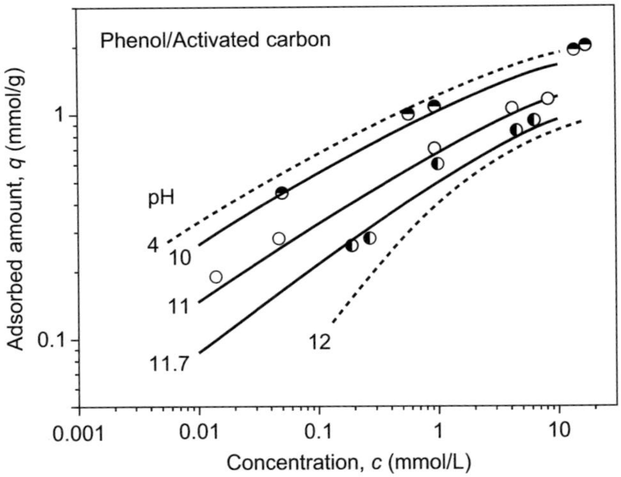

Figure 4.8: pH dependence of the phenol adsorption onto activated carbon.

be treated because only under this condition do both species occur in comparable concentrations.

4.7 Adsorption of background dissolved organic matter (DOM)

4.7.1 The significance of DOM in activated carbon adsorption

Background DOM plays an important role as competitor in adsorption processes that are used in drinking water treatment or wastewater treatment to remove micropollutants. It occurs as natural organic matter (NOM) in drinking water treatment or as effluent organic matter (EfOM) in wastewater treatment.

NOM is a constituent of all natural waters. Therefore, raw waters to be treated with activated carbon always contain NOM besides other compounds. NOM consists of humic substances and other naturally occurring organics. It is a mixture of molecules of different sizes and structures, but neither the qualitative nor the quantitative composition can be specified. In practice, only the total concentration is measurable by using collective parameters. In particular, the sum parameter dissolved organic carbon (DOC) is frequently used to characterize the NOM content in raw waters. Typical DOC concentrations are 0.1–1.5 mg/L in groundwater and 1–10 mg/L in surface water; sometimes higher values are also found, in particular in peatland lakes.

During activated carbon adsorption in drinking water treatment, NOM is relatively well adsorbed and competes with other compounds for the adsorption sites, in particular with micropollutants that occur in raw water in much lower concentrations (ng/L or µg/L). As a consequence, the adsorption capacity for micropollutants will be reduced (Figure 4.9). Since the concentration of NOM is much higher in comparison to the micropollutants, the effect of competitive adsorption can be expected even for stronger adsorbable micropollutants. Because of the relatively strong capacity reduction for micropollutants, the NOM adsorption onto activated carbon has to be considered a drawback in view of micropollutant removal. On the other hand, the reduction of the NOM concentration by activated carbon adsorption prior to the following disinfection with chlorine leads to a reduction of the disinfection by-product formation, which is a positive aspect in view of the water quality.

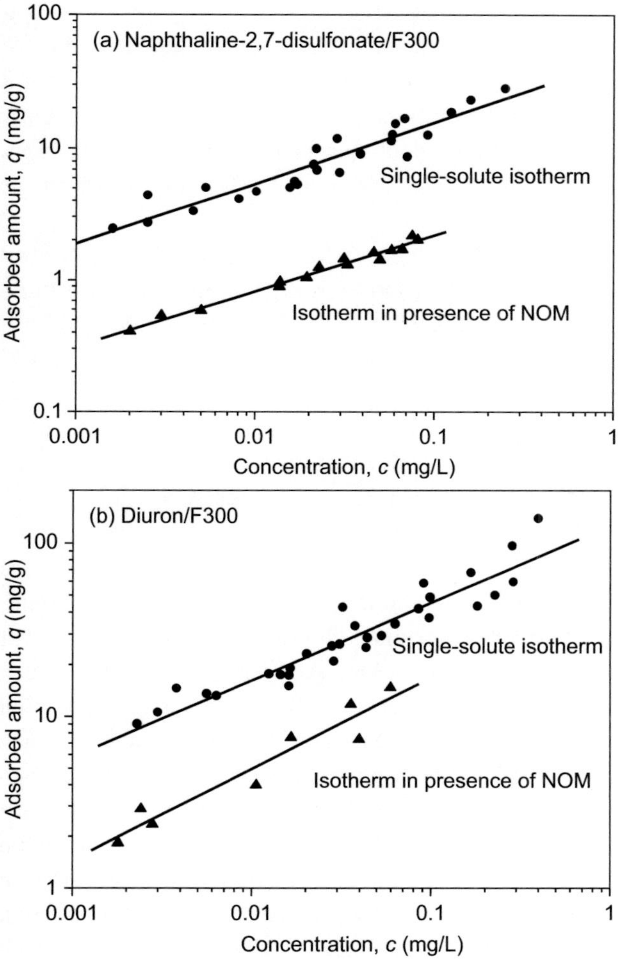

Figure 4.9: Adsorption of naphthaline-2,7-disulfonate (a) and diuron (b) onto activated carbon F300 from pure water and from NOM-containing river water (c_0[DOC] = 5.6 mg/L). Experimental data from Rabolt (1998).

From the viewpoint of process modeling, two main problems result from the occurrence of NOM in raw waters. The first problem consists of how to describe the adsorption of NOM. NOM is a mixture of components, but the composition of this mixture is unknown. Therefore, the IAST in its original form cannot be applied to predict the adsorption equilibrium, because IAST requires the knowledge of concentrations and isotherm parameters of all mixture components. The second problem is how to describe the competitive adsorption of NOM and micropollutants and is strongly connected to the first problem. Knowledge about the impact of NOM on micropollutant adsorption is of particular importance in view of evaluating the micropollutant removal efficiency during water treatment.

A comparable situation is found for wastewater treatment. In recent years, adsorption processes have gained increasing importance as an additional treatment step to remove micropollutants from the effluents of wastewater treatment plants. Here, the EfOM, which is also a multicomponent system of unknown composition, acts as competing background organic matter. The typical DOC concentrations are ranging from 5 to 15 mg/L.

From the modeling point of view, there is no fundamental difference between NOM and EfOM adsorption and their competition with micropollutants. Therefore, the model approaches discussed in the following sections can be applied for both cases. To clarify the general validity of the models, the more general term "DOM" will be used in the next sections instead of NOM or EfOM.

4.7.2 Modeling of DOM adsorption: The fictive component approach (adsorption analysis)

The fact that DOM is an unknown mixture has consequences not only for the experimental determination but also for the mathematical description of its adsorption behavior. Since the constituents of DOM cannot be identified, only total isotherms can be measured experimentally, commonly by using the collective parameter DOC. The measured isotherms are then given as adsorbed amount of DOC as a function of DOC concentration. Figure 4.10 shows a typical DOC isotherm. The shape of such DOC isotherms differs significantly from the shape of single-solute isotherms. At high adsorbent doses (low equilibrium concentrations), DOC isotherms typically approach a value on the concentration axis that is different from zero and corresponds to a nonadsorbable DOC fraction. Furthermore, there is typically no uniform change in the slope of the isotherms as can be found for nonlinear single-solute isotherms such as Freundlich or Langmuir isotherms. Instead of that, the DOC isotherms show step-like characteristics, which indicate that DOM consists of a number of fractions with different adsorbabilities. Consequently, DOM cannot be considered a pseudo single-solute system and cannot be described by simple single-solute isotherm equations. However, the IAST in its original form (Section 4.5) also cannot be applied to

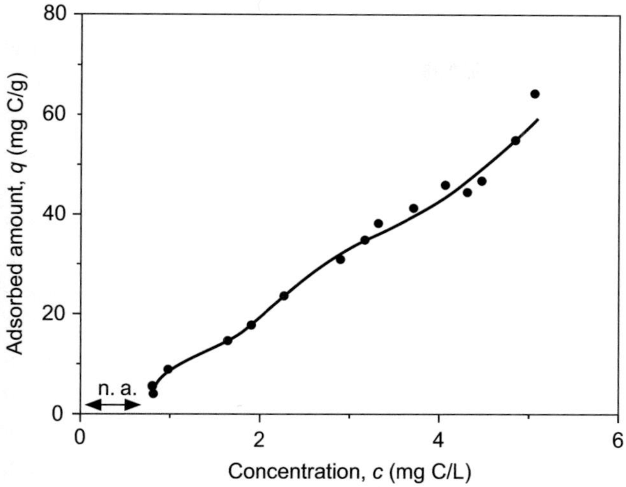

Figure 4.10: Typical DOC isotherm (n. a. = nonadsorbable fraction).

this specific multisolute system, because neither the concentrations nor the isotherm parameters of the different DOM fractions are known.

To overcome the difficulties in the description of DOM adsorption, a fictive component approach was developed by Frick (1980), which allows describing DOC isotherms and characterizing the DOC composition from the practical standpoint of adsorption. This approach, also referred to as adsorption analysis, was further developed by several authors and comprehensively documented in the monograph of Sontheimer et al. (1988). An efficient mathematical method for accomplishing the adsorption analysis was presented by Johannsen and Worch (1994).

The basic principle of the adsorption analysis consists in a formal transformation of the unknown multicomponent system DOM into a defined mixture of a limited number of fictive components. Each fictive component of this mixture system stands for a DOC fraction with a characteristic adsorbability. Within the framework of the adsorption analysis, the different adsorbabilities of the DOC fractions are defined by assigning characteristic Freundlich isotherm parameters. For simplification, the exponent n is normally held constant (often set to 0.2 or 0.25), and only different coefficients K are used to characterize the graduation of the adsorption strength. The nonadsorbable fraction is characterized by setting the K value to zero. For the other fractions, typically K values in the range between 0 and 100 (mg C/g)/(mg C/L)^n are chosen. Sometimes, if very strongly adsorbable fractions are present, the choice of higher K values up to 150 (mg C/g)/(mg C/L)^n can also be useful. Generally, the choice of three to five fictive components has been proved to be reasonable for most practical cases. Table 4.3 shows an example of the definition of fictive components. It has to be noted that there is no general rule for the choice of the number of components and their specific K values. The appropriate K values as

Table 4.3: Definition of fictive components (example).

Component	Adsorbability	K ((mg C/g)/(mg C/L)^n)	n
1	No	0	–
2	Poor	10	0.2
3	Medium	50	0.2
4	Strong	80	0.2

well as the optimum number of fictive components have to be found in each individual case by trial and error.

After defining the number and the isotherm parameters of the fictive components, an IAST-based search routine is used to find the concentration distribution of the DOC fractions that best describe the measured DOC isotherm. The principle of the adsorption analysis is schematically shown in Figure 4.11.

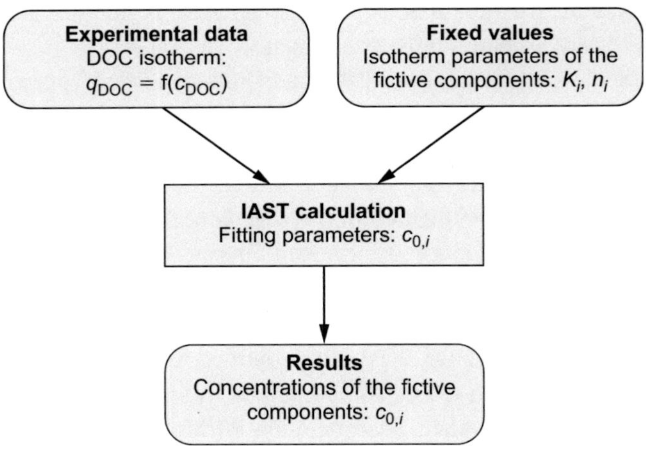

Figure 4.11: Principle of the fictive-component approach (adsorption analysis).

After running the adsorption analysis, the DOC is characterized in the same manner as a defined adsorbate mixture; that is, the isotherm parameters as well as the initial concentrations of the DOC fractions are now known and can be used for further process modeling purposes, for instance, to design slurry or fixed-bed adsorbers.

As an example, Figure 4.12 shows the adsorption analyses of two water samples collected at different times from the Elbe River in Dresden, Germany. Table 4.4 shows the related concentration distributions found from the adsorption analyses. As can be seen from Figure 4.12 and Table 4.4, the definition of only three components is already sufficient to describe the DOC isotherms in satisfactory quality.

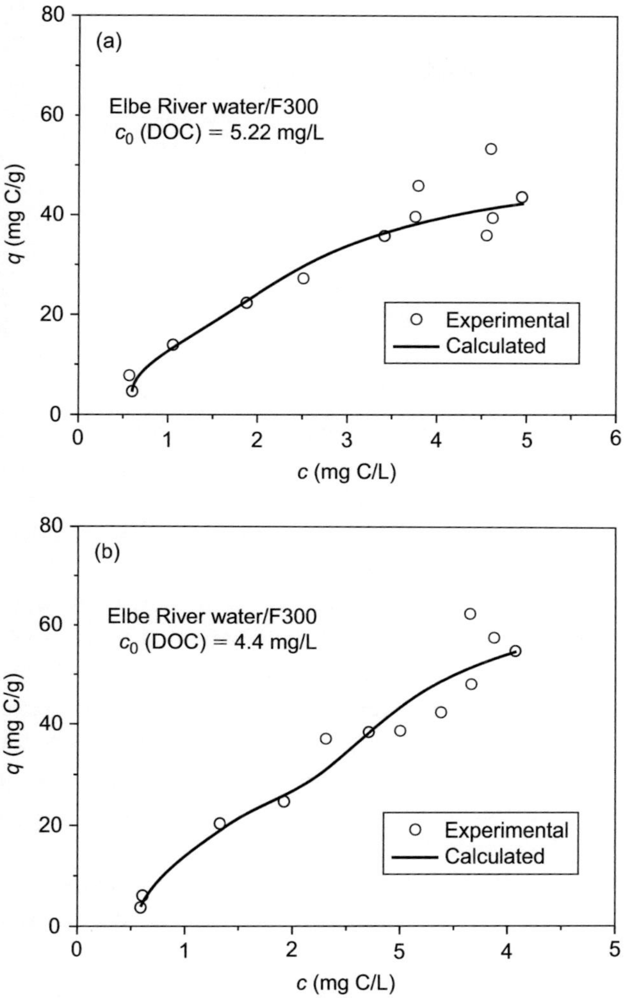

Figure 4.12: Adsorption analyses of Elbe River water: (a) c_0(DOC) = 5.22 mg/L and (b) c_0(DOC) = 4.4 mg/L (adsorbent: activated carbon F300). Experimental data from Rabolt (1998).

At this point, it has to be noted that there is a problem concerning the units in the adsorption analysis. As mentioned in Section 4.5, the IAST is based on fundamental thermodynamic equations and consequently requires the use of molar concentrations. However, from the operational definition of the sum parameter DOC, it follows that it can only be measured as mass concentration. Basically, mass concentrations could be used in the IAST only under the restrictive condition that all components have the same molecular weight, which is unrealistic for the considered DOM system. Therefore, if using the adsorption analysis on the basis of DOC, an error, resulting from the unavoidable use of the unit mg/L instead of mmol/L, has to be accepted.

Table 4.4: Adsorption analyses of Elbe River water.

Sample 1, $c_{0,total}$(DOC) = 5.22 mg/L		Sample 2, $c_{0,total}$(DOC) = 4.4 mg/L	
K ((mg/g)/(mg/L)^n)	c_0 (mg/L)	K ((mg/g)/(mg/L)^n)	c_0 (mg/L)
0	0.60	0	0.59
12	1.34	18	1.94
34	3.29	50	1.88

Mean percentage errors: 4.38% (sample 1), 4.23% (sample 2); Freundlich exponent of the adsorbable fractions: $n = 0.2$.

It follows from the theory that the use of mass concentration instead of molar concentration leads to different adsorbed-phase mole fractions and consequently to a different concentration distribution. On the other hand, the adsorption analysis approach is based on the arbitrary definition of fictive components. Therefore, no "true" concentration distribution exists. Moreover, a comparison of Figures 4.11 and 4.2b shows that the adsorption analysis is the reversal of an IAST prediction. That means a shift in the concentration distribution as a consequence of the use of the wrong unit in the adsorption analysis will be compensated in subsequent IAST applications (e.g. in process modeling) for the same DOM system due to the reverse direction of the calculation. Consequently, as long as only fictive components are considered, the use of mass concentration instead of molar concentration has no negative effects on adsorption modeling.

By contrast, difficulties in process modeling can be expected for adsorbate systems that contain defined solutes additionally to the fictive DOM fractions, for instance, micropollutant/DOM systems. For such systems, special model approaches are necessary to overcome this problem.

4.7.3 Competitive adsorption of micropollutants and DOM

Problems connected with the application of the IAST to competitive adsorption of micropollutants and DOM

In both drinking water and wastewater treatment, it is of practical interest to quantify the influence of the background dissolved organic matter (either NOM or EfOM) on the micropollutant adsorption. Examples for this influence were exemplarily shown in Figure 4.9 (Section 4.7.1). From the modeling point of view, it would seem at first consideration a simple matter to add the trace pollutant to the fictive-component mixture (DOM) as a further component and to apply the IAST to this extended system in order to characterize the micropollutant adsorption in the presence of background organic matter. In principle, if the concentration and the single-solute isotherm parameters of the micropollutant are known and the result of an adsorption analysis of the DOM-containing water is also given, all required data for an IAST calculation are available. However, this kind of prediction fails in many cases. Typically, the IAST overestimates the competition effect of DOM, as shown exemplarily in Figure 4.13 for the adsorption of two pesticides in the presence of DOM.

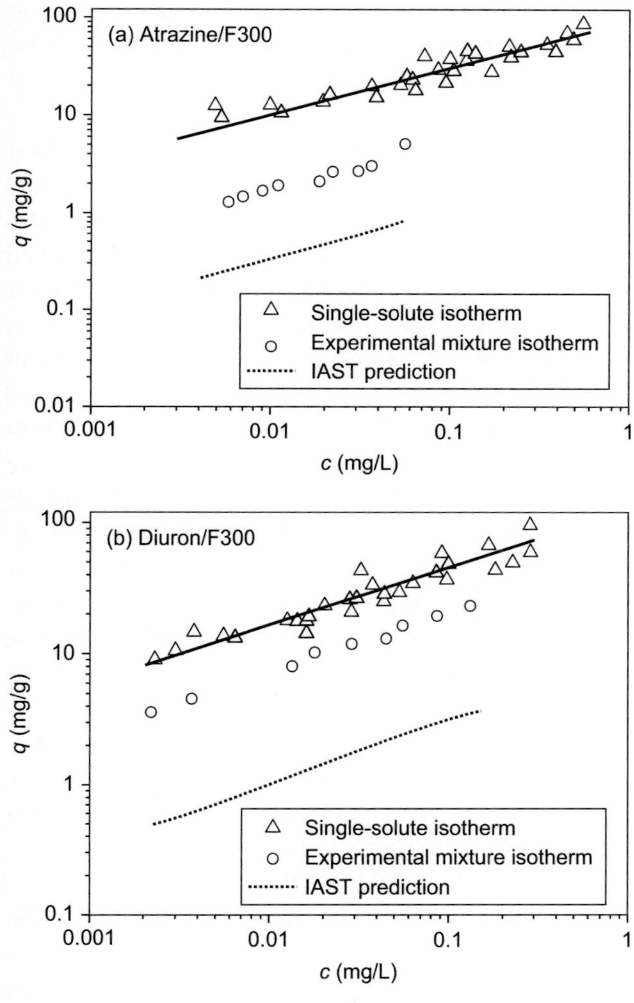

Figure 4.13: Isotherms of atrazine (a) and diuron (b) on activated carbon F300 measured in pure water and in Elbe River water. DOC concentrations: (a) 5.22 mg/L and (b) 4.4 mg/L. Comparison of experimental data with IAST predictions. Experimental data from Rabolt (1998).

There are manifold reasons that may be responsible for the failure of the IAST in micropollutant/DOM systems:

- The equations of the IAST are only valid for ideal conditions. This means that all possible interactions between the adsorbates in the liquid phase as well as in the adsorbed phase are neglected. Experiences from studies with defined model mixtures show that in most cases the differences between experiment and prediction are relatively small, indicating no strong deviation from the ideal behavior. On the other hand, stronger interactions and therefore larger deviations from the ideal behavior can be expected for mixtures of micropollutants and DOM due to possible specific interactions, in particular complex formation. Such complex formation processes between micropollutants and DOM have been reported in several studies.
- As discussed in the previous section, an unavoidable error results from the need of using the unit mg/L for DOC, whereas the exact IAST requires the use of molar concentrations. There are only two possible ways to deal with micropollutant/DOM mixtures: (a) either the micropollutant concentration must be transformed to DOC to have at least the same unit for all components or (b) different units for DOM and micropollutant have to be applied in the calculation. However, in both cases the requirement of using molar concentrations for all components is not fulfilled.
- Different accessibility to the micropores for DOM and trace compounds has to be considered a further error source. Within the IAST, it is assumed that during the competitive adsorption all adsorption sites are similarly accessible to all adsorbates. However, it cannot be excluded that in the case of microporous adsorbents – for instance, activated carbons – the larger DOM molecules cannot enter the fine micropores (Pelekani and Snoeyink 1999). Thus, mainly the low molecular weight fraction of the DOM would be able to compete with the micropollutant in the sense of the IAST.
- As a further effect, a possible blockage of the micropore entries by larger DOM molecules has to be considered (Carter et al. 1992; Weber 2004) with the consequence that the micropollutant adsorption is reduced independently of the direct competitive adsorption. On the other hand, an uptake of micropollutants by the bound DOM due to partitioning is also possible (Weber 2004).

Possible effects that may influence the competitive adsorption of micropollutants and DOM are schematically shown in Figure 4.14.

To overcome the problems connected with the application of the IAST to systems consisting of DOM and trace organic compounds, two different approaches have been developed, the tracer model (TRM) and the equivalent background compound model (EBCM). Both models are based on the IAST (with Freundlich single-solute isotherms) but modify the original theory in order to make it applicable to micropollutant/DOM systems. In contrast to the original IAST, these models require the

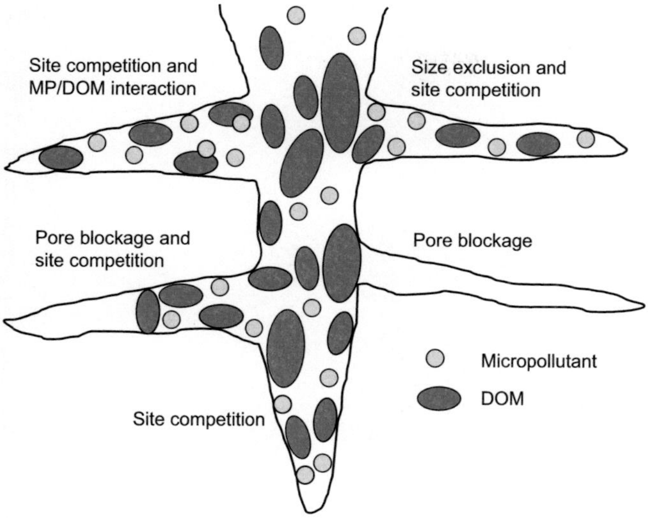

Figure 4.14: Possible effects that may influence the simultaneous adsorption of micropollutants and DOM.

measurement of a micropollutant isotherm in the presence of DOM. This micropollutant isotherm is then used to find parameters that are able to compensate for the errors resulting from the unfulfilled preconditions of the IAST application. The basic difference between the approaches consists in the fact that in the first case (TRM), the isotherm parameters of the micropollutant are corrected for further application in the IAST, whereas in the second case (EBCM), the mathematical description of the DOM adsorption is modified. In the following sections, these models will be discussed in more detail.

The tracer model

The TRM (Burwig et al. 1995; Rabolt et al. 1998) is based on the results of the adsorption analysis, which are used unmodified as input data for calculating the micropollutant/DOM competitive adsorption. The correction, necessary for the correct description of the competitive adsorption by the IAST, is done by modifying the single-solute isotherm parameters of the micropollutant. The corrected isotherm parameters have to be determined from experimental competitive adsorption data (i.e. from the micropollutant isotherm in the presence of DOM) by using a fitting procedure based on the IAST (Figure 4.15). The results of this fitting procedure are modified (corrected) isotherm parameters of the micropollutant, which allows, together with the adsorption analysis data, a more precise description of the competitive adsorption data by the IAST.

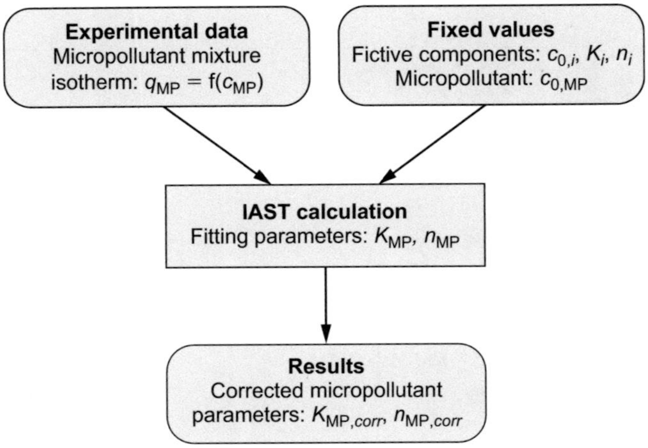

Figure 4.15: Calculation scheme of the tracer model (TRM).

Through the correction of the isotherm parameters, all possible errors discussed in the previous section will be compensated. The corrected parameters are fictive (empirical) single-solute isotherm parameters, which improve the quality of the IAST calculations in comparison to the use of the original single-solute parameters. Often, the corrected isotherm parameters differ considerably from the original single-solute isotherm parameters. It has to be noted that the modified parameters are only valid for the respective DOM-containing water and allow only IAST calculations in the system under consideration, for instance, for adsorber design purposes. A further theoretical interpretation of the parameter correction is not possible. The benefit of this model approach consists of the fact that it is based on the results of the adsorption analysis, and therefore both micropollutant and DOM adsorption can be described in parallel with the same set of parameters.

The equivalent background compound model

The EBCM (Najm et al. 1991) is based on a different approach. It leaves unchanged all micropollutant data (concentration, single-solute isotherm parameters) and makes the corrections needed for the IAST application by means of a simplified description of the DOM. Instead of a number of fictive components, only a single fictive DOM background component (EBC = equivalent background compound) is defined, which represents the DOM fraction that is able to compete with the micropollutant within the interior of the adsorbent particle. In principle, all parameters of the EBC (isotherm parameters, concentration) are unknown and have to be determined from the competitive adsorption isotherm of the micropollutant by means of a fitting procedure based on the IAST. However, it was found that fitting of all three parameters (c_0, n, and K for EBC) often does not lead to unique results; that means that different sets of parameters result in a comparable fitting quality (Najm et al.

1991; Graham et al. 2000). For this reason, the EBCM was further simplified. It was proposed to use the same Freundlich exponent, n, for the EBC as for the micropollutant and to determine only K and the concentration, c_0, by fitting (Newcombe et al. 2002a, 2002b). It was also suggested to use the two micropollutant isotherm parameters, K and n, also for the EBC and to fit only the EBC concentration (Schideman et al. 2006a, 2006b). The latter approach is based on the postulate that a DOM component that is able to compete with the micropollutant should have more or less the same adsorption properties as the micropollutant. It has to be noted that this assumption is a very strong simplification because in practice competitive adsorption can also be found for components with different adsorption behavior. Figure 4.16 shows the calculation scheme of the EBCM.

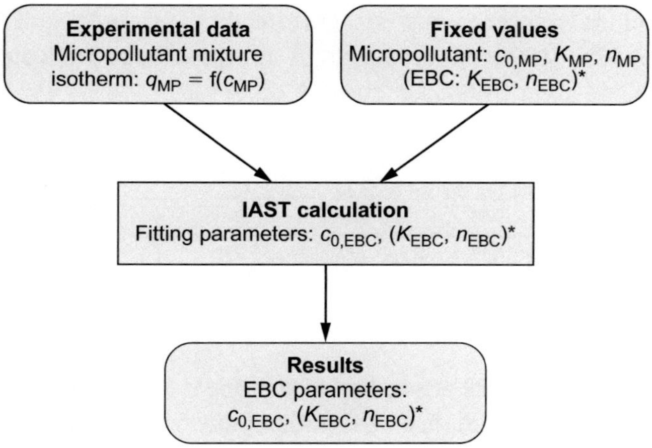

* Optionally, K and n of the EBC can be set to the values of the micropollutant.

Figure 4.16: Calculation scheme of the equivalent background compound model (EBCM).

The EBCM avoids the problems concerning the concentration units in IAST calculations because only micropollutant data, which can be expressed in molar quantities, are required as input parameters. If mmol/L is used as the unit for the initial micropollutant concentration, K is expressed in a related unit (e.g. [mmol/g]/[mmol/L]^n), and the isotherm data are also given in molar units, then the fitting result will be the molar EBC concentration. Therefore, the EBCM is consistent with the thermodynamic fundamentals of the IAST. On the other hand, a direct comparison of the calculated EBC concentration (in mmol/L) with the measurable DOC concentration (in mg/L) is not possible because both the molecular weight and the carbon content of the EBC are unknown.

In principle, the assumption that only a fraction of DOM is able to compete with the micropollutant is in accordance with the reported findings that the IAST overestimates the competition if all DOM fractions are considered in the calculation. On the other hand, the EBC approach cannot explain the typical form of DOC isotherms. According to the model approach, EBC is a single solute, whereas DOC isotherms show a typical multicomponent curve shape (Figure 4.10). This contradiction is a limitation of the EBCM, in particular in view of theoretical interpretation.

Especially for slurry adsorber design, a simplified EBCM was developed. This approach will be presented in Section 4.8.3.

Example: Atrazine adsorption in the presence of DOM

The application of the models discussed previously will be demonstrated by taking the adsorption of the pesticide atrazine from DOM-containing river water onto activated carbon F300 as an example. The experimental data were taken from the PhD thesis of Rabolt (1998). The DOC concentration of the water used for the experiments was 5.22 mg/L. The result of the adsorption analysis was already shown in Section 4.7.2 (Table 4.4). The Freundlich isotherm parameters of atrazine, found from single-solute isotherm by nonlinear regression, are given in Table 4.5.

Table 4.5: Single-solute isotherm parameters of atrazine (activated carbon F300). For conversion of K values, see Table 10.1 in the Appendix (Chapter10).

K (mg/g)/(mg/L)^n	K (mg C/g)/(mg C/L)^n	K (mmol/g)/(mmol/L)^n	n
92.4	60.6	5.6	0.48

As already shown in Figure 4.13a, the adsorption of atrazine is strongly reduced in the presence of DOM, but the conventional IAST without any data correction overestimates the DOM influence. Atrazine in the presence of DOM is much better adsorbed than predicted by the IAST. As mentioned previously, better results can be expected if the TRM or the EBCM are used to describe the micropollutant adsorption in the presence of DOM.

In Figure 4.17, the results of the application of the TRM are shown. In TRM calculations, the same concentration units (mass concentrations related to C) for the micropollutant and the fictive components were used. Therefore, the Freundlich K value of the micropollutant is also related to the C content. From the fitting procedure based on the TRM, the corrected single-solute isotherm parameters $K = 31.2$ (mg C/g)/(mg C/L)^n und $n = 0.13$ were found. For comparison, the original isotherm parameters estimated from single-solute isotherm are $K = 60.6$ (mg C/g)/(mg C/L)^n and $n = 0.48$ (see Table 4.5).

By using these corrected parameters together with the results of the adsorption analysis, a good description of the experimental data can be achieved. The mean percentage error of the equilibrium concentrations and loadings is about 5%.

When using the EBCM, the number of fitting parameters can vary between one and three. One-parameter fitting means that n and K of the EBC are set equal to the

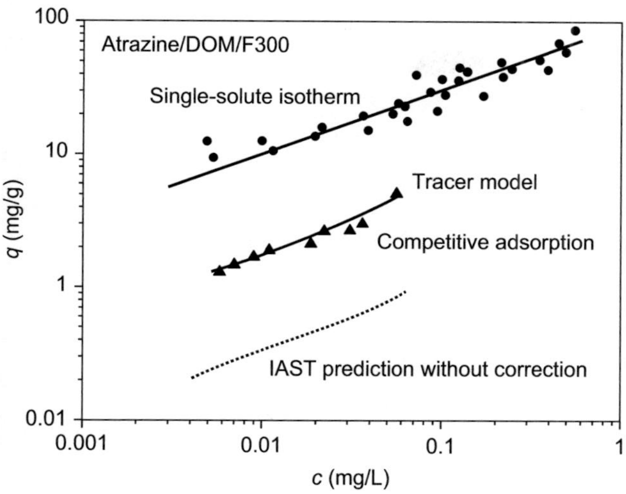

Figure 4.17: Atrazine adsorption in the presence of DOM (5.22 mg/L DOC). Comparison of the IAST and TRM results with experimental data.

values of the respective micropollutant and the EBC concentration is the only remaining fitting parameter, whereas in two-parameter fitting, only n is set equal to the micropollutant value and K is fitted together with the EBC concentration. In the case of three-parameter fitting, the concentration and both isotherm parameters of the EBC are determined by curve fitting. If both isotherm parameters as well as the initial concentration of the equivalent background component are used as fitting parameters, problems concerning the uniqueness of the results may arise. Reducing the number of fitting parameters leads to unique results, but the fitting quality slightly decreases (Table 4.6). In the example discussed here, however, the assumption of an EBC that has the same isotherm parameters as the micropollutant (atrazine: $K = 5.6$ (mmol/g)/(mmol/L)^n, $n = 0.48$) would be absolutely sufficient to describe the competitive adsorption of micropollutant and DOM (Figure 4.18). In this case, the only fitting parameter, the EBC concentration, was found to be 0.009 mmol/L for the considered system atrazine/DOM.

Table 4.6: EBC parameters found from the competitive adsorption isotherm of atrazine by one-, two-, and three-parameter fitting. (A): EBC parameter set equal to the value of atrazine (fixed value).

Fitted EBC parameter	c_0 mmol/L	K (mmol/g)/(mmol/L)^n	n	Mean percentage error, %
1 (c_0)	0.009	5.60 (A)	0.48 (A)	5.13
2 (c_0, K)	0.009	4.97	0.48 (A)	5.04
3 (c_0, K, n)	0.003	0.67	0.26	4.08

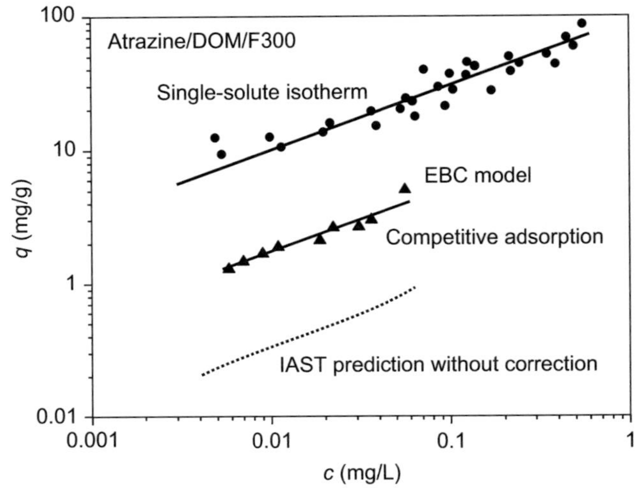

Figure 4.18: Atrazine adsorption in the presence of DOM (5.22 mg/L DOC). Comparison of the IAST and EBCM results with experimental data.

As already discussed, a direct comparison of the EBC concentration and the measured DOC concentration is not possible because the molecular weight and the carbon content of the EBC are unknown. However, to get an impression of the order of magnitude of the EBC concentration in terms of DOC, the DOC concentration was calculated for an assumed DOM molecular weight of 500 g/mol and an assumed carbon content of 50%. Under these conditions, the DOC concentration of the EBC would be 2.25 mg/L. This value is lower than the total DOC concentrations of 5.22 mg/L and therefore in accordance with the assumption of the EBCM that only a fraction of the DOC is able to compete with the micropollutant.

4.8 Slurry adsorber design for multisolute adsorption

4.8.1 Basics

Generally, the slurry adsorber design for multicomponent solutions is based on the same principles as discussed in Section 3.6 for single-solute adsorption. Consequently, the effects of adsorbability, adsorbent dose, and initial concentration on the treatment efficiency are comparable to that of the single-solute adsorption. The main difference consists in the fact that additionally the competitive effects have to be considered.

The objective in slurry adsorber design is to find the adsorbent dose necessary to achieve a defined treatment goal. The IAST provides the required design equations. Given that the initial composition of the multisolute system is known and the single-solute adsorption of the mixture components can be described by the Freundlich isotherm, the following set of equations has to be applied (see Section 4.5.3):

$$\sum_{i=1}^{N} z_i = \sum_{i=1}^{N} \frac{c_{0,i}}{\frac{m_A}{V_L} q_T + \left(\frac{\varphi\, n_i}{K_i}\right)^{1/n_i}} = 1 \tag{4.63}$$

$$\sum_{i=1}^{N} \frac{1}{\varphi\, n_i} \cdot \frac{c_{0,i}}{\frac{m_A}{V_L} q_T + \left(\frac{\varphi\, n_i}{K_i}\right)^{1/n_i}} = \frac{1}{q_T} \tag{4.64}$$

$$c_i = z_i \left(\frac{\varphi\, n_i}{K_i}\right)^{1/n_i} \tag{4.65}$$

Solving this set of equations gives the equilibrium concentrations of all adsorbates as a function of the adsorbent dose. A precondition for the solution is that the initial concentrations and the single-solute isotherm parameters of all components are known. In the case of DOM or micropollutant/DOM systems, specific model modifications or extensions have to be applied as already discussed in Sections 4.7.2 and 4.7.3. Some application examples will be given in the following sections.

4.8.2 DOM adsorption

In order to predict the adsorption of DOM in a slurry reactor, at first DOM has to be characterized by the fictive component approach (adsorption analysis). After that, eqs. (4.63) to (4.65) can be applied to find the concentrations of the fictive components as well as the total DOC concentration for a given adsorbent dose. As an example, the results of a model calculation for Elbe River water will be shown. The required adsorption analysis is given in Table 4.4 and shows that this water (total DOC: 5.22 mg/L) can be described as a mixture of three fictive components: nonadsorbable fraction (0.6 mg/L DOC), weakly adsorbable fraction (1.34 mg/L DOC), and strongly adsorbable fraction (3.29 mg/L DOC). Figure 4.19 depicts the calculated DOC as a function of the adsorbent dose in comparison with experimental data. Obviously, the fictive component approach reflects the adsorption behavior of the DOM very well. The different adsorption behavior of the fictive components is shown in Figure 4.20.

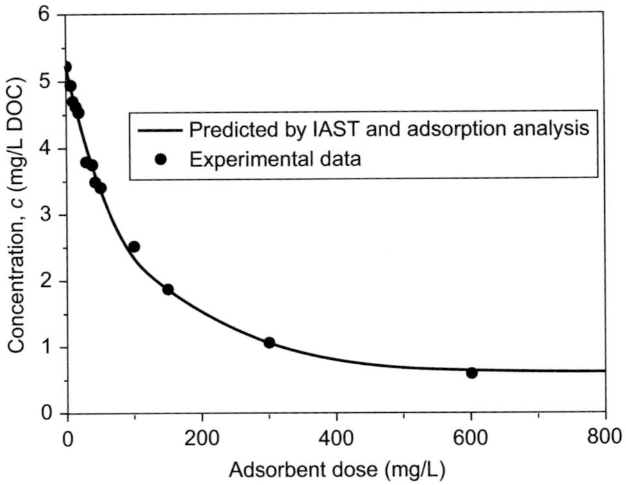

Figure 4.19: Prediction of DOM removal by using the IAST and the fictive component approach.

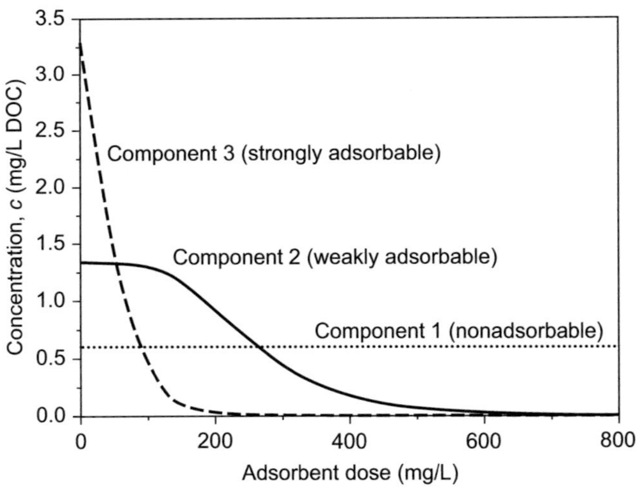

Figure 4.20: Removal of the different DOM fractions as calculated by the IAST and the fictive component approach.

4.8.3 Competitive adsorption of micropollutants and DOM

The TRM and the EBCM described in Section 4.7.3 are model approaches that can be used to find the characteristic input data needed for subsequent slurry adsorber modeling. Figure 4.21 shows the atrazine removal from DOM-containing river water in comparison with predictions based on the EBCM and the TRM in combination with eqs. (4.63) to (4.65). Both models reflect the experimental data very well. On the

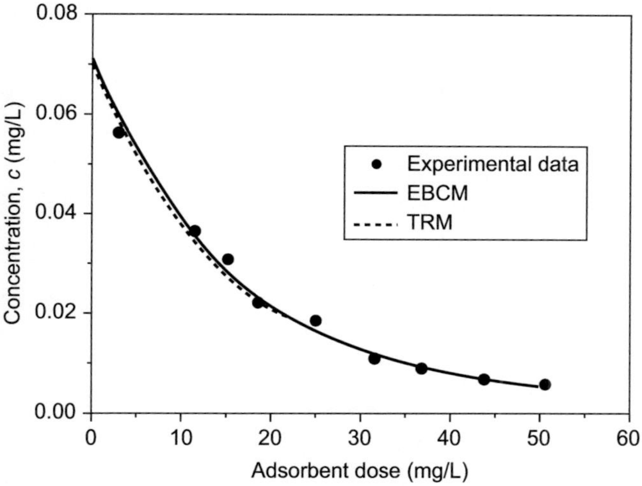

Figure 4.21: Removal of atrazine from DOM-containing river water (5.22 mg/L DOC) as predicted by the TRM and the EBCM.

other hand, the application of these prediction models is laborious because they require parameters that have to be determined in time-consuming experiments. In the case of the TRM, prior to the prediction, the DOC isotherm (required for the adsorption analysis) and the isotherm of the micropollutant in the presence of DOM (to find the corrected micropollutant isotherm parameters) have to be determined. The EBCM requires the knowledge of the single-solute isotherm of the micropollutant as well as its isotherm in the presence of DOM in order to find the EBC parameters.

As an alternative, for the special case of slurry reactors, a simplified model approach can also be applied. Under certain conditions, in particular if the concentration of the micropollutant is low in comparison to the DOM concentration, the relative remaining concentration, c/c_0, for the micropollutant at a given adsorbent dose becomes independent of the initial micropollutant concentration. This can be easily demonstrated by IAST model calculations. As an example, Figure 4.22 shows the relative concentration of diuron adsorbed from Elbe River water as a function of the adsorbent dose for different initial concentrations. It can be seen that, with exception of the curve for the highest initial concentration, all other curves have nearly the same shape. Obviously, the differences between the curves vanish with decreasing initial concentration.

This effect can also be demonstrated by plotting the relative residual concentration, c/c_0, of the micropollutant as a function of the initial concentration, c_0 (Figure 4.23). The plot reveals that there is a maximum concentration below which the fractional removal of the micropollutant remains independent from its initial concentration. In a recent study it was shown that this concentration threshold depends on the adsorbent dose, the adsorbate type, the adsorbent type, and the concentration

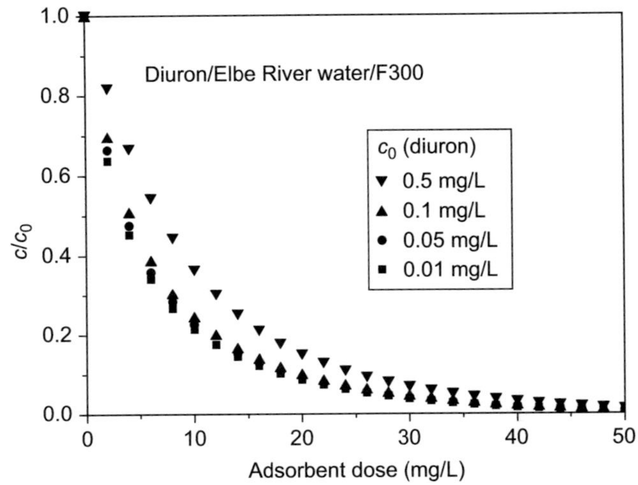

Figure 4.22: Diuron adsorption in a slurry reactor in the presence of DOM (4.4 mg/L DOC). Influence of the micropollutant initial concentration. The model calculation is based on the IAST with $K = 72.8$ (mg C/L)/(mg C/L)^n and $n = 0.19$ for diuron and with the adsorption analysis data for DOM given in Table 4.4.

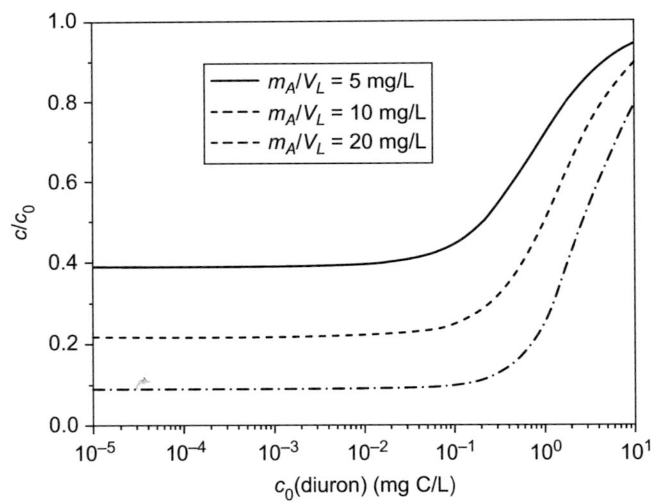

Figure 4.23: Relative residual concentration of diuron as a function of its initial concentration for different adsorbent dosages calculated with the IAST using the data given in the caption of Figure 4.22.

of the background organic matter (Shimabuku et al. 2017). However, the dependence on the adsorbent dose is not always very pronounced as can be seen in Figure 4.23. In this context, it should be pointed out that an exact value of the concentration threshold is difficult to determine because of the asymptotic curve shape. To overcome this difficulty, a limit value of c/c_0 relative to the constant end value can be applied to define the concentration threshold (e.g. $(c/c_0)_{const} + x$ %).

Unfortunately, there is no independent way to find the concentration threshold without IAST calculations (TRM or EBCM). However, it can be assumed that in most practical cases (drinking water treatment, wastewater treatment), the concentration of the micropollutant in comparison with the DOM is low enough to fall below the concentration threshold, where c/c_0 becomes independent of the initial concentration.

The effect of constant c/c_0 at low concentrations can be explained by the fact that the micropollutant in the presence of strongly competing background compounds shows a linear adsorption behavior (see Appendix, Section 10.3), and linear isotherms yield to constant relative residual concentrations in batch processes as shown in Chapter 3, Section 3.6.2.

From the previously discussed findings, it can be concluded that for a micropollutant that occurs in the water to be treated in very low concentrations, only one curve, $c/c_0 = f(m_A/V_L)$, has to be determined experimentally. This characteristic curve can then be used to estimate the micropollutant removal for any initial concentrations and adsorbent doses. Knappe et al. (1998) have demonstrated the applicability of this simple approach for a number of micropollutant/DOM systems.

Based on the IAST and the two-component EBC approach, Qi et al. (2007) have derived a simplified equation that describes the micropollutant (component 1) removal in the presence of the equivalent background compound (EBC, component 2) as a function of the adsorbent dose. Inspecting their simplified model, the same effect as discussed before (independence of percentage removal from the initial concentration) can be recognized. Furthermore, this approach provides the opportunity to describe mathematically the characteristic micropollutant removal curve. In the following discussion, this approach will be referred to as the simplified equivalent background model (SEBCM).

Under the assumption that the EBC, which represents the competing DOM fraction, is much better adsorbed than the micropollutant ($q_2 \gg q_1$) and that furthermore the values of the Freundlich exponents of micropollutant and EBC are similar ($n_1 \approx n_2$), the following relationship can be found from the basic IAST equations:

$$\frac{c_{0,1}}{c_1} = \frac{1}{A}\left(\frac{m_A}{V_L}\right)^{1/n_1} + 1 \tag{4.66}$$

with

$$A = c_{0,2}^{(1/n_1)-1}\left(\frac{n_1}{n_2\, K_1}\right)^{1/n_1} \tag{4.67}$$

Herein, $c_{0,1}$ is the initial concentration of the micropollutant, c_1 is the residual concentration of the micropollutant, $c_{0,2}$ is the initial concentration of the EBC, n_1 and n_2 are the Freundlich exponents of the respective components, and K_1 is the Freundlich coefficient of the micropollutant. Details of the derivation of eq. (4.66) are given in the Appendix (Section 10.3).

As can be seen from the right-hand sides of eqs. (4.66) and (4.67), there is no influence of the initial micropollutant concentration, $c_{0,1}$, on the relative removal, $c_1/c_{0,1}$. The logarithmic form of eq. (4.66) can be used to find the parameters A and n_1 from an experimentally determined curve, $c_1/c_{0,1} = \mathrm{f}(m_A/V_L)$, by linear regression

$$\ln\left(\frac{c_{0,1}}{c_1} - 1\right) = \frac{1}{n_1}\ln\left(\frac{m_A}{V_L}\right) - \ln A \tag{4.68}$$

If using eq. (4.68), it has to be considered that the unit of m_A/V_L (mg/L or g/L) influences the numerical value of ln A that will be found from linear regression. Furthermore, it has to be noted that, due to the simplifications made in the model derivation, the fitted n_1 does not necessarily have exactly the same value as the Freundlich exponent obtained from a single-solute isotherm measurement.

Once the two parameters A and n_1 are known, they can be used to calculate the relative removal of the micropollutant from DOM-containing water for any adsorbent dose. For this type of calculation, it is not necessary to know the exact values of the parameters included in A.

After rearranging, eq. (4.66) can be written in the form

$$\frac{m_A}{V_L} = A^{n_1}\left(\frac{c_{1,0}}{c_1} - 1\right)^{n_1} \tag{4.69}$$

By using this equation, the adsorbent dose needed for a given treatment objective can be calculated.

The SEBCM is a simple, user-friendly model that can be recommended for slurry (batch) adsorber design. It allows estimating the removal efficiency for micropollutants in dependence on the applied adsorbent dose for the case where the micropollutant is adsorbed in the presence of background organic matter. In contrast to the other models discussed previously (TRM, EBCM), only one experimental curve has to be determined as a precondition for subsequent predictions of the micropollutant adsorption behavior.

However, it has to be noted that this simplified model is also subject to some limitations. The approach is restricted to batch adsorption processes, whereas the results of TRM and EBCM can be used not only for batch processes but also as input data for calculating fixed-bed adsorber breakthrough curves (see Chapter 7). Furthermore, it is not easy to verify the fulfillment of the preconditions for the simplified approach. The simplified IAST presented by Qi et al. (2007) is based on the assumption that $q_2 \gg q_1$ and $n_1 \approx n_2$. In their paper, the authors have shown that the deviation of the simplified IAST from the original IAST is less than 10% in the full range of expected

n_1 and n_2 values if the EBC dominates over the trace compound by a factor of 130 or more on the adsorbent surface. It can be expected that this condition is fulfilled in most practical cases, but it cannot be proved exactly, because typically only the concentrations of micropollutant and DOM (measured as DOC) are available, whereas the EBC parameters are unknown. Since the simplified approach is a special case of the EBCM, it underlies the same restrictions. In particular, both models are not able to describe simultaneously the adsorption of micropollutants and DOM. A comparison of TRM, EBCM, and SEBCM is given in Table 4.7.

Table 4.7: Comparison of models suitable for describing competitive adsorption of micropollutants and DOM.

	TRM	EBCM	SEBCM
Number of isotherms to be measured	2 (DOM and micropollutant in the presence of DOM)	2 (Micropollutant in the presence and absence of DOM)	1 (Micropollutant in the presence of DOM)
Applicability to batch or fixed-bed adsorbers	Batch Fixed-bed	Batch Fixed-bed	Batch
Simultaneous modeling of micropollutant and DOM adsorption	Yes	No	No

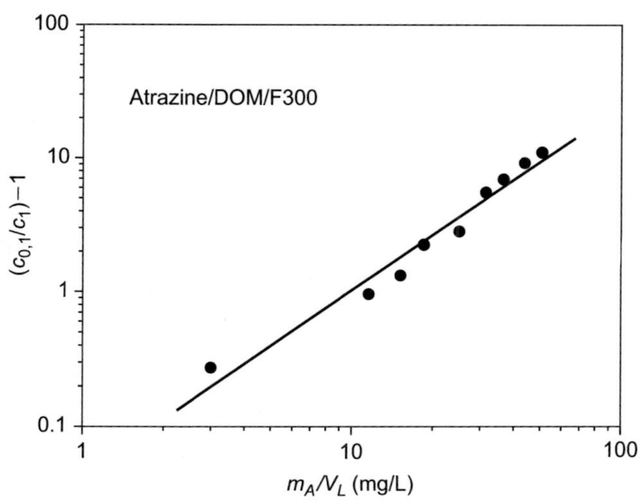

Figure 4.24: Atrazine adsorption in the presence of DOM (Elbe River water, 5.22 mg/L DOC, activated carbon F300). Experimental data plotted according to eq. (4.68).

The application of the SEBCM will be demonstrated below, using the adsorption of atrazine in the presence of DOM (Elbe River water, 5.22 mg/L DOC) as an example. Figure 4.24 shows the experimental data plotted according to eq. (4.68). The parameters found by linear regression are $n_1 = 0.73$ and $\ln A = -3.38$ (adsorbent dose in mg/L). These parameters can be used to calculate the characteristic removal curve $c/c_0 = f(m_A/V_L)$ for the micropollutant (Figure 4.25). As discussed before, this removal curve is independent from the initial concentration. In the given example, the curve calculated with the parameters found from the experiment with c_0(atrazine) = 0.07 mg/L is able to describe not only the experimental data for c_0(atrazine) = 0.07 mg/L but also for c_0(atrazine) = 0.1 mg/L.

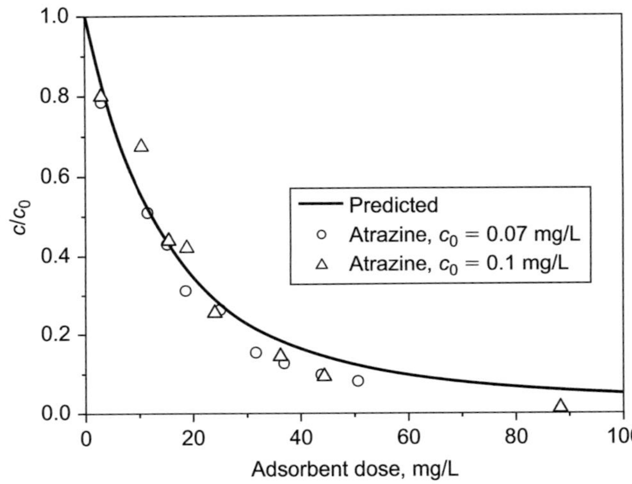

Figure 4.25: Atrazine adsorption in the presence of DOM, predicted by the SEBCM (Elbe River water, 5.22 mg/L DOC, activated carbon F300).

4.8.4 Nonequilibrium adsorption in slurry reactors

In practice, the contact time between the adsorbent and the water to be treated in slurry reactors is often too short to reach the adsorption equilibrium. Consequently, the removal efficiency for a considered adsorbate is lower than that which is predicted from the equilibrium models discussed before. In principle, there are two different ways to overcome this problem. The more exact but also more complicated way is to apply kinetic models, which allow predicting the time-depending adsorption process. To consider adsorption kinetics, however, not only equilibrium data but also kinetic parameters have to be determined experimentally. Kinetic models are discussed in detail in Chapter 5. A simpler approach is to apply short-term (nonequilibrium) isotherms as the basis for the prediction of the adsorption behavior in slurry reactors. In this case, the isotherms have to be determined for the same contact time as occurs in the considered reactor. Consequently, these isotherms are only valid for the given contact time, and predictions for other contact times are not possible. This practice-oriented method can be used in particular to describe the micropollutant adsorption in the presence of DOM by the TRM, the EBCM, or the SEBCM. In the case of the TRM and the EBCM, the micropollutant isotherm in the presence of DOM has to be determined as a short-term isotherm, whereas the DOC isotherm (in TRM) or the single-solute isotherm (in EBCM) can be determined either as a short-term isotherm or as an equilibrium isotherm. The different methods lead to different values of the empirical fitting parameters. For the SEBCM, only the short-term isotherm of the micropollutant in the presence of DOM is required.

The prediction of the adsorption behavior for short contact times will be demonstrated by means of a practical example (Zoschke et al. 2011). In the considered waterworks, powdered activated carbon is applied to remove geosmin, an odor compound that occurs seasonally in the raw water from a reservoir. Figure 4.26 shows the adsorption kinetics of geosmin. Whereas the equilibration in the system geosmin/DOM requires at least 10 hours, the contact time in the considered waterworks is limited to 30 minutes due to technical restrictions.

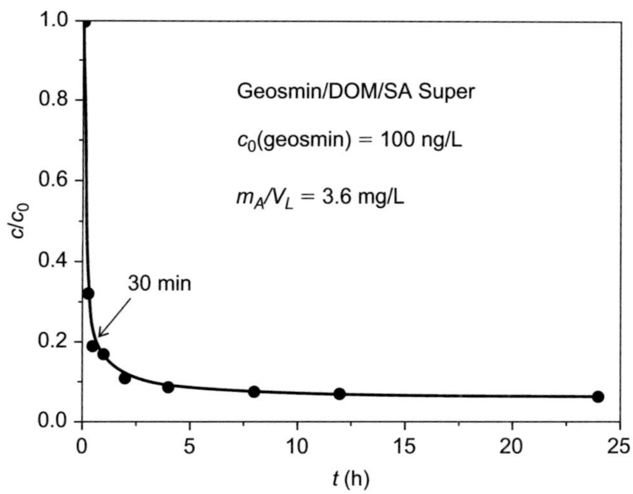

Figure 4.26: Kinetics of geosmin adsorption from DOM-containing water (activated carbon SA Super).

For the system geosmin/DOM, short-term isotherms (contact time: 30 minutes) were determined for an initial geosmin concentration of 100 ng/L. The experimental short-term data can be plotted according to eq. (4.68) in the same way as equilibrium data (Figure 4.27). Obviously, eq. (4.68) also holds for the nonequilibrium data. Furthermore, taking the parameters ln A and n_1 estimated from the plot, the removal curve for other initial concentrations can be predicted as exemplarily shown for an initial concentration of 20 ng/L (Figure 4.28). The diagram demonstrates that the removal curve is independent from the initial concentration not only for equilibrium but also for nonequilibrium conditions.

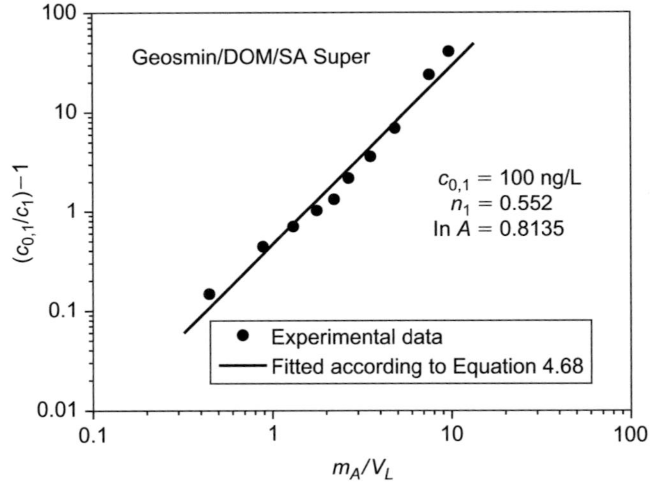

Figure 4.27: Application of the simplified EBCM to short-term data (30 min contact time) of geosmin adsorption from DOM-containing water.

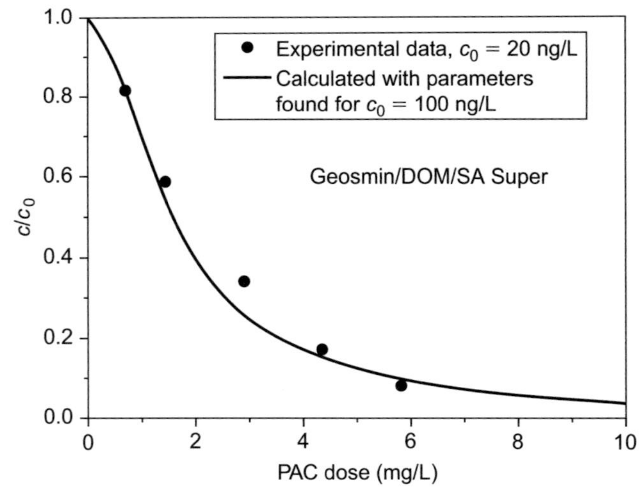

Figure 4.28: Prediction of the geosmin adsorption from DOM-containing water in a slurry reactor with short contact time (30 min).

In general, using short-term isotherms within the equilibrium models can be a feasible alternative to the application of more complex kinetic models, at least for slurry adsorber design.

4.9 Special applications of the fictive component approach

During drinking water treatment, different processes of the treatment train contribute to the removal of DOM. The fictive component approach (adsorption analysis) shown in Section 4.7.2 can be used as a helpful tool to characterize the removal efficiency with regard to changes in the DOM composition. In particular, which fractions will be removed by other treatment steps and the consequence for the removal efficiency during the subsequent adsorption can be evaluated. The general principle is schematically shown in Figure 4.29. It has to be noted that this method only works if the same definition of the fictive components is used for all adsorption analyses (i.e. the same number of fictive components, the same Freundlich parameters), so that the characteristic changes in the concentration distribution can be observed.

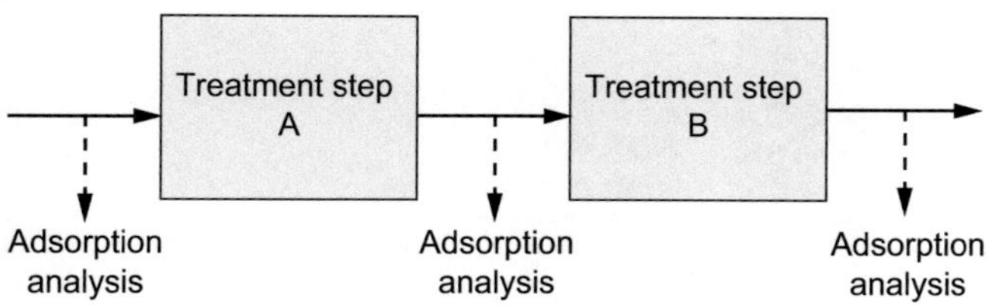

Figure 4.29: Application of the adsorption analysis to evaluate the removal of DOM fractions by different treatment steps.

This method will be demonstrated by using raw water from a drinking water reservoir as an example (Zoschke et al. 2011). In the waterworks, this raw water is treated by coagulation/flocculation prior to the application of powdered activated carbon. Figure 4.30 shows the concentration distribution of the fictive components (absolute and relative) before and after the coagulation/flocculation process as found from adsorption analyses with four fictive components. As can be derived from the diagrams, only mainly weakly and moderately adsorbable DOM fractions are removed during the coagulation/flocculation process. These fractions are expected to be of minor relevance in view of competition with micropollutants. The stronger adsorbable fraction, which is a stronger competitor, is nearly unaffected by coagulation/flocculation. These findings from the adsorption analyses are confirmed by the comparison of the adsorption behavior of the trace compound geosmin before and after coagulation/flocculation (Figure 4.31), which shows that the competitive effect of DOM is independent of the pretreatment of the water by coagulation/flocculation.

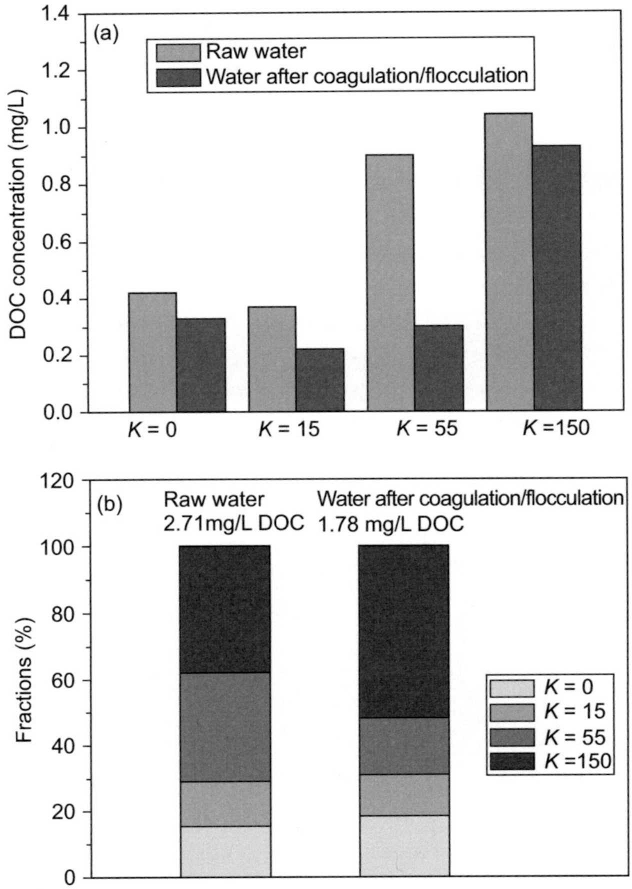

Figure 4.30: Absolute (a) and relative (b) change of the concentration distribution of the DOM fractions as a result of coagulation/flocculation (K in [mg C/g]/[mg C/L]^n).

Adsorbent screening, in particular comparison of adsorbents in view of their efficiency in DOM removal, is another interesting application of the adsorption analysis. Taking the same definition of the fictive components and using the same DOM-containing water, the concentration distribution found from the adsorption analysis can be taken as a quality parameter for the considered adsorbent. In Figure 4.32, activated carbon B is much better suited for the DOM removal than activated carbon A. This can be derived from the higher fraction of strongly adsorbable compounds with $K = 150$ (mg/g)/(mg/L)^n.

These examples make clear that the fictive component approach is not only a modeling approach for adsorber design but also a helpful assessment tool for different purposes.

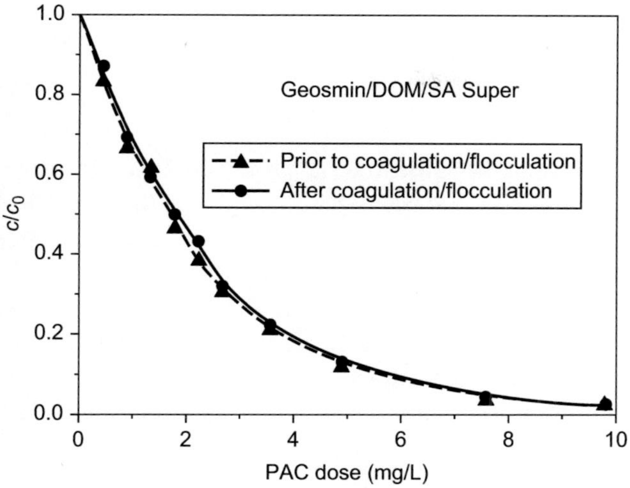

Figure 4.31: Adsorption of geosmin from raw water and from water pretreated by coagulation/flocculation.

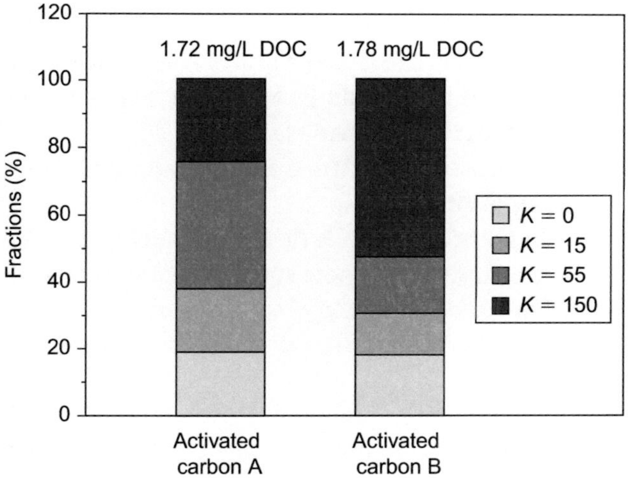

Figure 4.32: Application of the adsorption analysis for comparison of adsorbents (K in [mg C/g]/[mg C/L]^n).

5 Adsorption kinetics

5.1 Introduction

Typically, adsorption equilibria are not established instantaneously. This is in particular true for porous adsorbents. The mass transfer from the solution to the adsorption sites within the adsorbent particles is constrained by mass transfer resistances that determine the time required to reach the state of equilibrium. The time progress of the adsorption process is referred to as adsorption kinetics. The rate of adsorption is usually limited by diffusion processes toward the external adsorbent surface and within the porous adsorbent particles. Investigations into the adsorption kinetics are necessary to clarify the rate-limiting mass transfer mechanisms and to evaluate the characteristic mass transfer parameters. The mass transfer parameters, together with the equilibrium data, are essential input data for determination of the required contact times in slurry reactors as well as for fixed-bed adsorber design.

5.2 Mass transfer mechanisms

The progress of the adsorption process can be characterized by four consecutive steps:

1. Transport of the adsorbate from the bulk liquid phase to the hydrodynamic boundary layer localized around the adsorbent particle
2. Transport through the boundary layer to the external surface of the adsorbent, termed "film diffusion" or "external diffusion"
3. Transport into the interior of the adsorbent particle (termed "intraparticle diffusion" or "internal diffusion") by diffusion in the pore liquid (pore diffusion) and/or by diffusion in the adsorbed state along the internal surface (surface diffusion)
4. Energetic interaction between the adsorbate molecules and the final adsorption sites

It is a generally accepted assumption that the first and the fourth step are very fast and the total rate of the adsorption process is determined by film and/or intraparticle diffusion. Since film diffusion and intraparticle diffusion act in series, the slower process determines the total adsorption rate. It is therefore interesting to look at the main influence factors and their impact on the diffusion rates.

A basic difference between film and intraparticle diffusion consists in the dependence on the hydrodynamic conditions, in particular stirrer velocity in slurry reactors or flow velocity in fixed-bed adsorbers. This difference allows differentiating between the transport mechanisms and provides the opportunity to influence their relative impact on the total adsorption rate. An increase in the stirrer or flow velocity increases the rate of film diffusion due to the reduction of the boundary layer

https://doi.org/10.1515/9783110715507-005

thickness. In contrast, the intraparticle diffusion is independent of the stirrer or flow velocity. The particle radius influences the film diffusion as well as the intraparticle diffusion due to the change of surface area and diffusion paths.

The mass transfer within the adsorbent particle takes place normally by pore diffusion and surface diffusion in parallel, but their portions are difficult to separate. Therefore, often only one intraparticle diffusion mechanism is assumed as predominant and considered in the kinetic model. In most adsorption processes from aqueous solutions onto porous adsorbents, the intraparticle diffusion can be described successfully by a surface diffusion approach.

5.3 Experimental determination of kinetic curves

In order to study the adsorption kinetics, a solution volume, V_L, is brought in contact with the adsorbent mass, m_A, and the resulting change of concentration with time is measured. In most cases, the concentration cannot be measured in situ. Therefore, samples have to be taken after defined time intervals. That causes a disturbance of the kinetic measurement because a portion of liquid is removed from the system with each sampling. To overcome this problem, a sufficiently large solution volume has to be chosen for the experiment, so that the loss of volume and adsorbate can be neglected. In the case of a direct analytical method (without enrichment or transformation step), the sample volume can be returned to reduce the disturbance of the kinetic measurement. An arithmetical consideration of the changes in the solution volume and in the amount of adsorbate caused by sampling is a theoretical alternative but often too complicated in practice.

As a result of a kinetic experiment, the kinetic curve is found in the form

$$c = \mathrm{f}(t) \tag{5.1}$$

where c is the concentration and t is the time.

During the adsorption process, the concentration decreases from the initial value, c_0, to the equilibrium concentration, c_{eq}. Since for each time during the experiment the material balance equation

$$\bar{q}(t) = \frac{V_L}{m_A}[c_0 - c(t)] \tag{5.2}$$

holds (see Chapter 3, Section 3.2), the kinetic curve can also be expressed as

$$\bar{q} = \mathrm{f}(t) \tag{5.3}$$

where $\bar{q}$ is the mean solid-phase concentration (adsorbed amount). Typical kinetic curves according to eqs. (5.1) and (5.3) are shown in Figure 5.1.

The experimentally determined kinetic curve provides the data for a fitting procedure with a kinetic model (Section 5.4), where the respective mass transfer

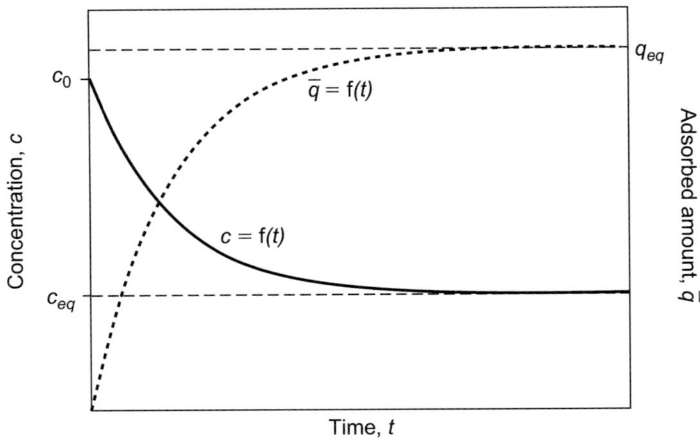

Figure 5.1: Kinetic curves. Progress of concentration and adsorbent loading with time.

coefficients are the fitting parameters. Prior to the selection of a specific kinetic model that should be used for fitting, assumptions concerning the dominating transport mechanisms have to be made (transport hypothesis). The verification of the transport hypothesis is then carried out by comparing the calculated and experimental kinetic curves. In this way, the kinetic measurements give information on the relevant mechanisms and provide the values of the transport parameters.

Due to the dependence of the film diffusion on the hydrodynamic conditions, already mentioned in Section 5.2, the experimentally determined film diffusion mass transfer coefficients are only valid for the given experimental setup and cannot be applied to other experimental conditions or technical adsorbers. For adsorption processes in slurry reactors, film diffusion typically determines only the very first time of the process and therefore can be neglected in many cases. In fixed-bed adsorbers, however, the film diffusion determines the shape of the breakthrough curve to some extent and should be considered in the breakthrough curve model. For this application, the required mass transfer coefficients can be estimated from hydrodynamic parameters by using empirical correlations (see Chapter 7). Therefore, the aim of most kinetic experiments is to determine only the mass transfer parameters for the internal diffusion processes, which do not depend on the hydrodynamic conditions and therefore are transferable to other process conditions. Accordingly, kinetic experiments are often carried out in a manner that the film diffusion becomes very fast and therefore does not need to be considered in the mathematical model. Such an experimental approach simplifies considerably the determination of the desired mass transfer coefficients of the internal diffusion processes. The requirements of this approach have to be considered in the choice of the reactor for kinetic experiments.

There are different types of adsorbers that can be used for kinetic experiments: slurry batch reactors, basket reactors, and differential column batch reactors (Figures 5.2 and 5.3).

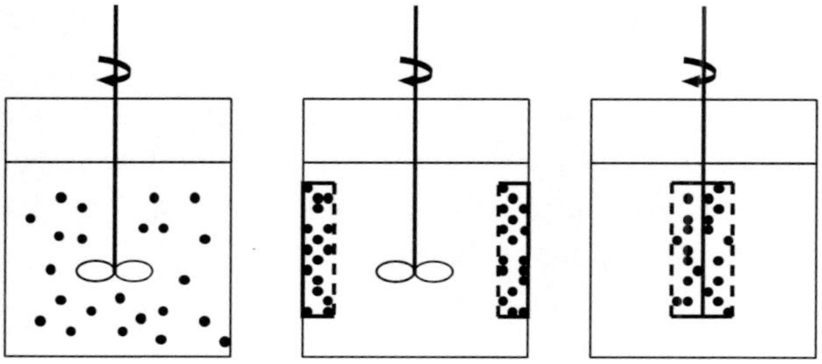

Figure 5.2: Experimental setup for kinetic experiments. Slurry batch reactor (left) and different types of basket reactors (middle and right).

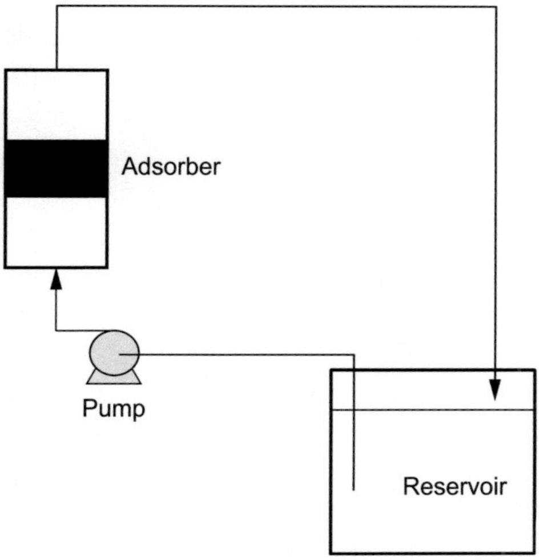

Figure 5.3: Experimental setup for kinetic experiments. Differential column batch reactor.

Slurry batch reactors are typically used to determine equilibrium data but can also be applied for kinetic experiments. A slurry batch reactor consists of a tank with an electrical stirrer. If the influence of film diffusion should be minimized, high stirrer velocities are necessary. With increasing stirrer velocity, however, the risk of destruction of the adsorbent particle increases. This can lead to errors due to the particle size dependence of the mass transfer coefficients.

A possible way to avoid the destruction of the adsorbent particles caused by high stirrer velocities is to fix the adsorbent in baskets. These baskets can be located directly on the stirrer shaft (this reactor is referred to as a spinning basket reactor or Carberry reactor) or on the vessel walls.

In a differential column batch reactor, the solution flows through a fixed-bed of small height. High flow velocities prevent the influence of film diffusion. The placement of the adsorbent particles in the fixed bed shields them from destruction. Due to the recirculation of the solution, the reactor acts like a batch reactor, and therefore batch adsorption models can be used for fitting the kinetic curves and evaluating the mass transfer coefficients. The samples are taken from the reservoir.

To find the minimum stirrer or flow velocity that is necessary to prevent the influence of film diffusion, several kinetic curves with increasing stirrer or flow velocities have to be measured. If the adsorption rate is no more influenced by film diffusion, the shape of the kinetic curve remains constant.

5.4 Mass transfer models

5.4.1 General considerations

In the following sections, the basics of the most important kinetic models will be presented. These models are important not only for estimation of transport coefficients from kinetic curves but also as basic constituents of fixed-bed adsorber models (Chapter 7). Prior to the presentation of the different kinetic approaches, some general aspects will be discussed in this section.

In general, a kinetic model includes mass transfer equations, equilibrium relationships, and the material balance for the reactor applied (Figure 5.4).

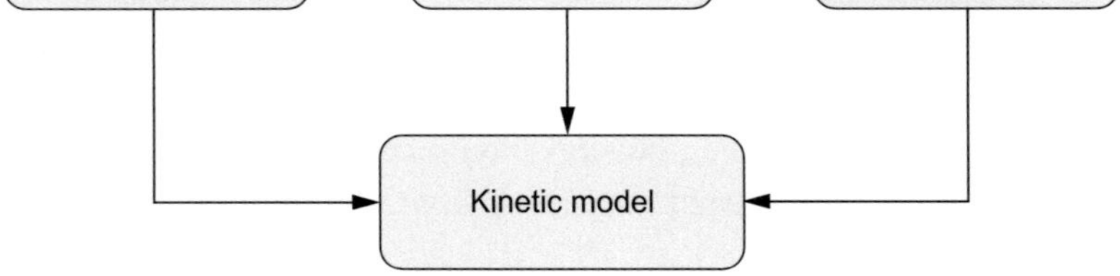

Figure 5.4: Modeling of adsorption kinetics. Model constituents and input data.

The common assumptions in kinetic models are as follows: (a) the temperature is assumed to be constant, (b) the bulk solution is assumed to be completely mixed, (c) the mass transfer into and within the adsorbent can be described as diffusion processes, (d) the attachment of the adsorbate onto the adsorbent surface is much faster than the diffusion processes, and (e) the adsorbent is assumed to be spherical and isotropic.

An important aspect to be considered is that adsorption kinetics is not independent of the adsorption equilibrium. Therefore, kinetic models can be applied only if the required equilibrium parameters are known. If necessary, an isotherm has to be measured prior to the kinetic experiment.

The question of how to deal with multicomponent adsorbate systems is another important aspect. A widely used simplification in view of multicomponent adsorption consists of the assumption that the competition influences only the equilibrium but not the mass transfer. As a consequence, the mass transfer equations for single-solute and multisolute adsorption are the same, and the kinetic models for single-solute and multisolute adsorption differ only in the description of the equilibrium. A further consequence of this assumption is that the mass transfer coefficients can be determined in single-solute experiments because the coefficients for single-solute and multisolute adsorption are assumed to be identical. Some pros and cons of this simplifying assumption are discussed in Section 5.4.8.

In this chapter, only batch systems will be considered. The integration of the kinetic approaches into breakthrough curve models for fixed-bed adsorbers is the subject matter of Chapter 7.

The differential material balance for a batch system reads

$$m_A \frac{d\bar{q}}{dt} = -V_L \frac{dc}{dt} \tag{5.4}$$

where m_A is the adsorbent mass, and V_L is the liquid volume in the reactor. This equation links the change of the mean adsorbent loading with time to the change of the liquid-phase concentration with time. Integration of eq. (5.4) with the initial conditions $c(t=0) = c_0$ and $\bar{q}(t=0) = 0$ leads to the material balance equation in the form

$$\bar{q}(t) = \frac{V_L}{m_A}[c_0 - c(t)] \tag{5.5}$$

In the development of kinetic models, it is often reasonable to use dimensionless quantities. After introducing the distribution parameter for the batch reactor, D_B,

$$D_B = \frac{m_A\, q_0}{V_L\, c_0} \tag{5.6}$$

and the dimensionless concentration and adsorbent loading, according to the definitions given in Chapter 3, Section 3.7,

$$X = \frac{c}{c_0}, \quad \bar{Y} = \frac{\bar{q}}{q_0} \tag{5.7}$$

the dimensionless material balance equation can be expressed as

$$X + D_B \bar{Y} = 1 \tag{5.8}$$

The loading q_0 in eqs. (5.6) and (5.7) is the equilibrium adsorbent loading related to the initial concentration, c_0.

5.4.2 Film diffusion

Basics

The film diffusion (also referred to as external diffusion) comprises the transport of the adsorbate from the bulk liquid to the external surface of the adsorbent particle. As long as the state of equilibrium is not reached, the concentration at the external adsorbent surface is always lower than in the bulk liquid due to the continuing adsorption process. As a consequence, a concentration gradient results that extends over a boundary layer of thickness δ. The difference between the concentration in the bulk solution, c, and the concentration at the external surface, c_s, acts as a driving force for the mass transfer through the boundary layer. Figure 5.5 shows the typical concentration profile for the limiting case where the adsorption rate is determined only by film diffusion and the diffusion within the particle is very fast ($\bar{q} = q_s$, no

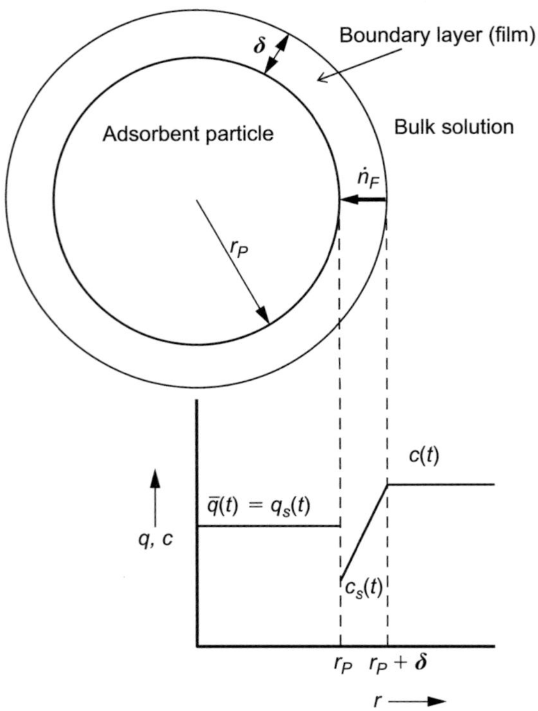

Figure 5.5: Concentration profiles in the case of rate-limiting film diffusion (no internal mass transfer resistance).

gradient within the particle). Note that the adsorbent loading at the external surface of the adsorbent particle, q_s, is the equilibrium loading related to the concentration at the surface, c_s.

The mass transfer equation for the film diffusion can be derived from Fick's law

$$\dot{n}_F = D_L \frac{\mathrm{d}c}{\mathrm{d}\delta} \tag{5.9}$$

where $\dot{n}_F$ is the flux, for instance, given in mol/(m$^2\cdot$s) or g/(m$^2\cdot$s), and D_L is the diffusion coefficient in the aqueous phase (m^2/s). Integration under the assumption of a linear gradient within the boundary layer leads to

$$\dot{n}_F = k_F(c - c_s) \tag{5.10}$$

with

$$k_F = \frac{D_L}{\delta} \tag{5.11}$$

where k_F is the film mass transfer coefficient (m/s). The amount of adsorbate that is removed from the liquid phase and adsorbed onto the solid per unit of time, $\dot{N}_F$, can be expressed by means of the differential material balance (eq. (5.4))

$$\dot{N}_F = m_A \frac{\mathrm{d}\bar{q}}{\mathrm{d}t} = -V_L \frac{\mathrm{d}c}{\mathrm{d}t} \tag{5.12}$$

The relationship between $\dot{N}_F$ and the flux, $\dot{n}_F$, is given by

$$\dot{n}_F = \frac{\dot{N}_F}{A_s} \tag{5.13}$$

where A_s is the total external surface area of all adsorbent particles within the reactor. With eq. (5.13), $\dot{n}_F$ can be related to the differential mass balance

$$\dot{n}_F = \frac{m_A}{A_s}\frac{\mathrm{d}\bar{q}}{\mathrm{d}t} = -\frac{V_L}{A_s}\frac{\mathrm{d}c}{\mathrm{d}t} \tag{5.14}$$

and the following mass transfer equation can be derived

$$\frac{\mathrm{d}\bar{q}}{\mathrm{d}t} = \frac{k_F A_s}{m_A}(c - c_s) = k_F\, a_m (c - c_s) \tag{5.15}$$

where a_m is the total surface area related to the adsorbent mass available in the reactor ($a_m = A_s/m_A$). The expression for the concentration decay can be derived from the equation for the adsorbate uptake by using the material balance (eq. (5.4)) in the form

$$-\frac{\mathrm{d}c}{\mathrm{d}t} = \frac{m_A}{V_L}\frac{\mathrm{d}\bar{q}}{\mathrm{d}t} \tag{5.16}$$

The resulting mass transfer equation reads

$$-\frac{dc}{dt} = k_F\, a_m \frac{m_A}{V_L}(c - c_s) \tag{5.17}$$

According to the relationship given in Table 2.1, the ratio m_A/V_L can be expressed by

$$\frac{m_A}{V_L} = \rho_P \frac{1-\varepsilon_B}{\varepsilon_B} \tag{5.18}$$

Thus, eq. (5.17) can be written in the alternative form

$$-\frac{dc}{dt} = k_F\, a_m \frac{\rho_P(1-\varepsilon_B)}{\varepsilon_B}(c - c_s) \tag{5.19}$$

As shown in Chapter 2 (Section 2.5.3), a_m is given for spherical particles by

$$a_m = \frac{3}{r_P\, \rho_P} \tag{5.20}$$

where r_P is the particle radius and ρ_P is the particle density. Accordingly, the respective mass transfer equations read

$$\frac{d\bar{q}}{dt} = \frac{3\, k_F}{r_P\, \rho_P}(c - c_s) \tag{5.21}$$

$$-\frac{dc}{dt} = \frac{3\, k_F}{r_P} \frac{(1-\varepsilon_B)}{\varepsilon_B}(c - c_s) \tag{5.22}$$

Frequently, the external adsorbent surface area is also related to the reactor volume, V_R, or to the volume of the adsorbent, V_A,

$$a_{VR} = \frac{A_s}{V_R} \tag{5.23}$$

$$a_{VA} = \frac{A_s}{V_A} \tag{5.24}$$

The respective mass transfer equations for the different forms of the normalized surface area are summarized in Table 5.1.

To calculate the kinetic curve for an adsorption process controlled by film diffusion, one of the equations given in Table 5.1 has to be solved together with the equilibrium relationship and the material balance and under additional consideration of the initial condition. Since the equations in Table 5.1 are equivalent, it is irrelevant which of them is used for the model development. Here, the film mass transfer equation will be used in the form

$$\frac{d\bar{q}}{dt} = \frac{k_F\, a_{VR}}{\rho_B}(c - c_s) = k_F\, a_{VR} \frac{V_R}{m_A}(c - c_s) \tag{5.25}$$

Table 5.1: Different forms of the film mass transfer equation. For the meaning of the terms ε_B, $(1-\varepsilon_B)$, $(1-\varepsilon_B)/\varepsilon_B$, and $\rho_P(1-\varepsilon_B)/\varepsilon_B$, see also Table 2.1.

External surface area related to	General film mass transfer equations	Specific particle surface in the case of spherical particles	Mass transfer equations for spherical particles
Adsorbent volume	$\frac{d\bar{q}}{dt}=\frac{k_F\,a_{VA}}{\rho_P}(c-c_S)$ $-\frac{dc}{dt}=k_F\,a_{VA}\frac{1-\varepsilon_B}{\varepsilon_B}(c-c_S)$	$a_{VA}=\frac{A_S}{V_A}=\frac{3}{r_P}$	$\frac{d\bar{q}}{dt}=\frac{3\,k_F}{r_P\,\rho_P}(c-c_S)$ $-\frac{dc}{dt}=\frac{3\,k_F}{r_P}\frac{1-\varepsilon_B}{\varepsilon_B}(c-c_S)$
Reactor volume	$\frac{d\bar{q}}{dt}=\frac{k_F\,a_{VR}}{\rho_B}(c-c_S)$ $-\frac{dc}{dt}=\frac{k_F\,a_{VR}}{\varepsilon_B}(c-c_S)$	$a_{VR}=\frac{A_S}{V_R}=\frac{3}{r_P}(1-\varepsilon_B)$	$\frac{d\bar{q}}{dt}=\frac{3\,k_F}{r_P\,\rho_P}(c-c_S)$ $-\frac{dc}{dt}=\frac{3\,k_F}{r_p}\frac{1-\varepsilon_B}{\varepsilon_B}(c-c_S)$
Adsorbent mass	$\frac{d\bar{q}}{dt}=k_F\,a_m(c-c_S)$ $-\frac{dc}{dt}=k_F\,a_m\frac{\rho_P(1-\varepsilon_B)}{\varepsilon_B}(c-c_S)$	$a_m=\frac{A_S}{m_A}=\frac{3}{r_P\,\rho_P}$	$\frac{d\bar{q}}{dt}=\frac{3\,k_F}{r_P\,\rho_P}(c-c_S)$ $-\frac{dc}{dt}=\frac{3\,k_F}{r_P}\frac{1-\varepsilon_B}{\varepsilon_B}(c-c_S)$

Since the intraparticle diffusion is assumed to be fast and therefore not rate limiting, the equilibrium relationship in general form reads

$$q_S=\bar{q}=\mathrm{f}(c_S) \tag{5.26}$$

The initial condition for the mass transfer equation is

$$c=c_0,\ \bar{q}=0 \quad \text{at} \quad t=0 \tag{5.27}$$

For the sake of simplification, the kinetic model is formulated by using dimensionless quantities. The dimensionless concentrations and adsorbent loadings are given as

$$X=\frac{c}{c_0},\quad X_S=\frac{c_S}{c_0} \tag{5.28}$$

$$\bar{Y}=\frac{\bar{q}}{q_0},\quad Y_S=\frac{q_S}{q_0} \tag{5.29}$$

The adsorbent loading q_0 is the equilibrium loading related to c_0. Furthermore, a dimensionless time, T_B, is defined by using the distribution parameter, D_B, introduced in Section 5.4.1

$$T_B=\frac{k_F\,a_{VR}}{\varepsilon_B\,D_B}t \tag{5.30}$$

With these definitions, the following set of equations can be derived.
Material balance:

$$X + D_B \bar{Y} = 1 \tag{5.31}$$

Mass transfer equations:

$$-\frac{dX}{dT_B} = D_B(X - X_s) \tag{5.32}$$

$$\frac{d\bar{Y}}{dT_B} = X - X_s \tag{5.33}$$

Initial condition:

$$X = 1, \ \bar{Y} = 0 \quad \text{at} \quad T_B = 0 \tag{5.34}$$

Equilibrium relationship:

$$\bar{Y} = Y_s = f(X_s) \tag{5.35}$$

To solve the set of equations, in general, the application of numerical methods is necessary. An analytical solution can be found only for the special case of a linear isotherm.

Special case: Linear isotherm

In the case of a linear isotherm, the equilibrium relationship can be derived from eq. (3.80) or (3.82) (Chapter 3) with $n = 1$ or $R^* = 1$ and under consideration of eq. (5.35)

$$\bar{Y} = Y_s = X_s \tag{5.36}$$

Combining eq. (5.36) with eq. (5.31) gives

$$X_s = \frac{1 - X}{D_B} \tag{5.37}$$

and substituting X_s in eq. (5.32) by eq. (5.37) leads to

$$\frac{dX}{dT_B} = 1 - (D_B + 1)X \tag{5.38}$$

Finally, integrating eq. (5.38) gives the equation of the kinetic curve

$$X = \frac{1}{D_B + 1} + \frac{D_B}{D_B + 1} e^{-(D_B+1)T_B} \tag{5.39}$$

Determination of the film mass transfer coefficient, k_F

In principle, the kinetic model described previously could be used to fit the experimental data in order to determine the film diffusion mass transfer coefficient. That would require that the film diffusion alone determines the adsorption rate and the influence of intraparticle diffusion is negligible over the entire contact time. In practice, however, this condition is rarely fulfilled. In most practical cases, the film diffusion influences only the beginning of the adsorption process. Later, the intraparticle diffusion becomes more important. Therefore, it is necessary to determine the mass transfer coefficient for the film diffusion from the initial part of the kinetic curve.

With the condition

$$c = c_0, \ c_s = 0 \quad \text{at} \quad t = 0 \tag{5.40}$$

eq. (5.17) can be written as

$$k_F = -\frac{V_L}{a_m \, m_A \, c_0} \left(\frac{\mathrm{d}c}{\mathrm{d}t} \right)_{t=0} \tag{5.41}$$

According to eq. (5.41), k_F can be found by drawing a tangent to the kinetic curve at $t = 0$ and estimating the slope of this tangent (Figure 5.6). The mass-related surface area can be calculated for spherical particles by using the equation given in Table 5.1.

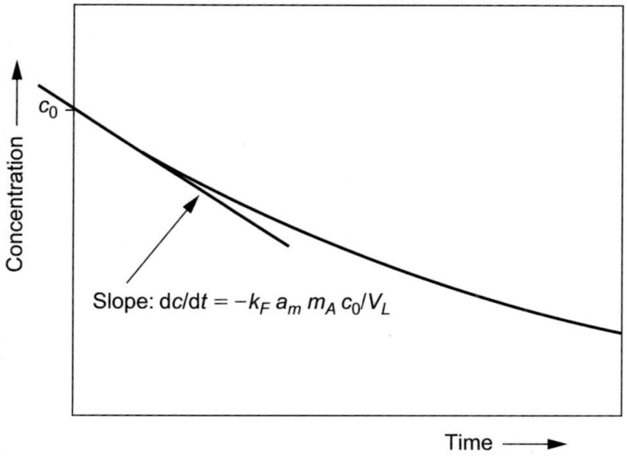

Figure 5.6: Determination of k_F according to eq. (5.41).

Another, more exact, method results from integration of eq. (5.17), neglecting c_s for the first short time period and using the initial condition $c = c_0$ at $t = 0$

$$\ln \frac{c}{c_0} = -\frac{m_A}{V_L} k_F \, a_m \, t \tag{5.42}$$

Plotting the kinetic curve according to eq. (5.42) allows estimating k_F from the initial linear part of the curve (Figure 5.7).

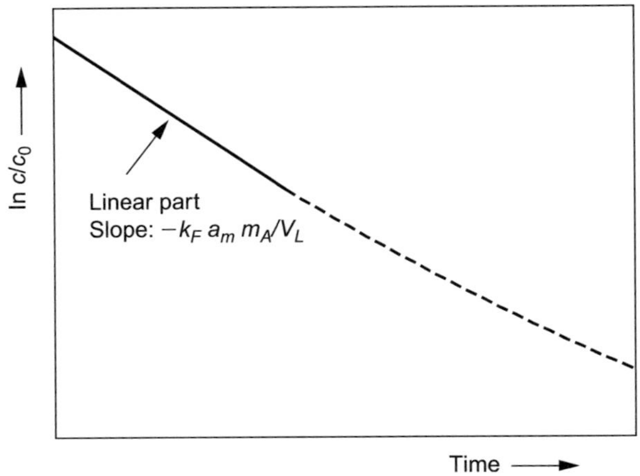

Figure 5.7: Determination of k_F according to eq. (5.42).

Instead of eq. (5.17), the other forms of the film mass transfer equation listed in Table 5.1 can also be used to determine k_F in an analogous manner.

It has to be noted that film mass transfer coefficients can also be determined by using lab-scale fixed-bed adsorbers and applying a respective breakthrough curve model or from empirical correlations (Chapter 7).

Factors influencing the film mass transfer coefficient, k_F

As follows from the definition (eq. (5.11)), k_F depends on the same influence factors as the diffusion coefficient in the free liquid, D_L, which are in particular temperature and molecule size. The free liquid diffusion coefficient, D_L, and therefore also k_F, increases with increasing temperature and decreasing molecule size. Furthermore, the value of k_F depends on the film thickness, δ. The film thickness decreases with increasing stirrer velocity in slurry reactors or increasing flow velocity in fixed-bed adsorbers, and therefore k_F increases under these conditions. Based on these well-known dependences of k_F, empirical correlations between k_F and the influence factors were established, which can be used to predict film mass transfer coefficients. These correlations were proposed in particular for fixed-bed conditions and will therefore be discussed in Chapter 7.

5.4.3 Surface diffusion

Basics

In the surface diffusion approach, it is assumed that the mass transfer occurs in the adsorbed state along the internal surface of the adsorbent particle. Here, the gradient of the solid-phase concentration within the particle acts as driving force for the transport. In the surface diffusion model, the adsorbent is considered a homogeneous medium. This model is therefore also referred to as the homogeneous surface diffusion model (HSDM). Regarding the model derivation, a distinction has to be made between two different cases:

1. The film diffusion is relatively slow and has to be considered in the model as a previous transport step (film and homogeneous surface diffusion model).
2. The film diffusion is much faster than the surface diffusion, and the mass transfer resistance in the boundary layer can be neglected. In this case, there is no concentration difference between the external surface and the bulk solution, and the adsorption rate can be described by the HSDM alone.

The concentration profiles for both cases are shown in Figures 5.8 and 5.9

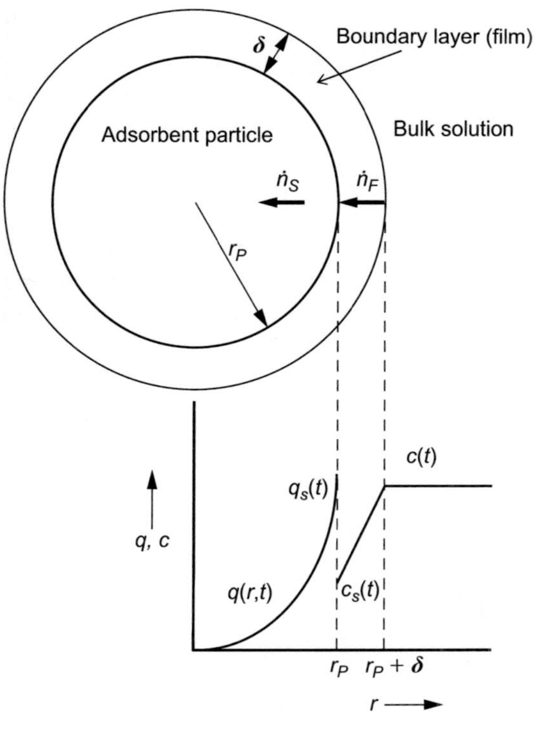

Figure 5.8: Concentration profiles in the case of film and surface diffusion.

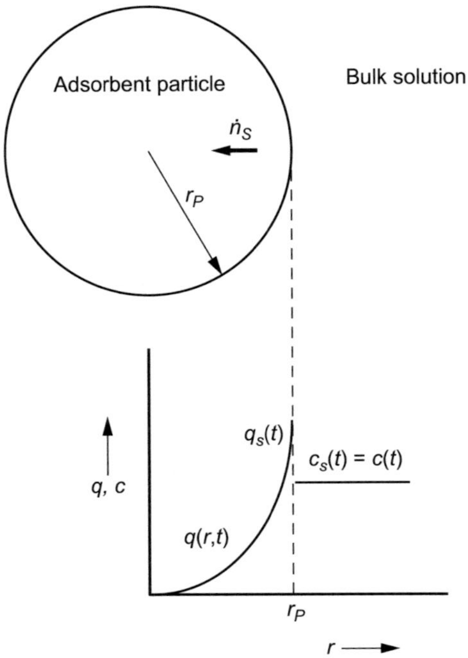

Figure 5.9: Concentration profiles in the case of surface diffusion (no external mass transfer resistance).

For surface diffusion, the mass transfer rate per unit of surface area, $\dot{n}_S$, is given by Fick's law as

$$\dot{n}_S = \rho_P\, D_S \frac{\partial q}{\partial r} \tag{5.43}$$

where D_S is the surface diffusion coefficient and r is the radial coordinate. The material balance for a thin spherical shell of thickness Δr can be derived from the condition that the amount of substance fed to the shell equals the amount of substance adsorbed

$$\Delta \dot{n}_S\, 4\pi r^2 \Delta t = \Delta q\, 4\pi r^2 \Delta r\, \rho_P \tag{5.44}$$

In differential form, eq. (5.44) reads

$$\frac{\partial (r^2 \dot{n}_S)}{\partial r} = \rho_P\, r^2 \frac{\partial q}{\partial t} \tag{5.45}$$

Combining eqs. (5.43) and (5.45) gives

$$\frac{\partial q}{\partial t} = \frac{1}{r^2}\frac{\partial}{\partial r}\left(r^2 D_S \frac{\partial q}{\partial r}\right) \tag{5.46}$$

If the surface diffusion coefficient, D_S, is assumed to be constant, eq. (5.46) simplifies to

$$\frac{\partial q}{\partial t} = \frac{D_S}{r^2}\frac{\partial}{\partial r}\left(r^2 \frac{\partial q}{\partial r}\right) \tag{5.47}$$

or

$$\frac{\partial q}{\partial t} = D_S \left(\frac{\partial^2 q}{\partial r^2} + \frac{2}{r} \frac{\partial q}{\partial r} \right) \tag{5.48}$$

The respective initial and boundary conditions are

$$q = 0 \quad \text{at} \quad t = 0 \quad \text{and} \quad 0 \leq r \leq r_P \tag{5.49}$$

$$c = c_0 \quad \text{at} \quad t = 0 \tag{5.50}$$

$$\frac{\partial q}{\partial r} = 0 \quad \text{at} \quad t > 0 \quad \text{and} \quad r = 0 \tag{5.51}$$

$$\rho_P \, D_S \frac{\partial q}{\partial r} = k_F (c - c_s) \quad \text{at} \quad t > 0 \quad \text{and} \quad r = r_P \tag{5.52}$$

The boundary condition given by eq. (5.52) follows from the continuity of the mass transfer

$$\dot{n}_S = \dot{n}_F \tag{5.53}$$

and is valid for the case where both film and surface diffusion are relevant. If the film diffusion is very fast, only surface diffusion determines the adsorption rate. In this case, eq. (5.52) has to be substituted by

$$\rho_P \, D_S \frac{\partial q}{\partial r} = -\frac{\varepsilon_B}{a_{VR}} \frac{\partial c}{\partial t} \quad \text{at} \quad t > 0 \quad \text{and} \quad r = r_P \tag{5.54}$$

where a_{VR} is the external surface area related to the reactor volume (see also Table 5.1).

Furthermore, the adsorbent loading at the outer surface, q_s, is related to the concentration at the outer surface, c_s, by the equilibrium relationship

$$q(r = r_P) = q_s = \mathrm{f}[c(r = r_P)] = \mathrm{f}(c_s) \tag{5.55}$$

The relationship between the concentration in the bulk liquid and the mean adsorbent loading, $\bar{q}$, is given by the material balance (eq. (5.5)), and the mean adsorbent loading results from the integration over all spherical shells of the adsorbent particle

$$\bar{q} = \frac{3}{r_P^3} \int_0^{r_P} q \, r^2 \mathrm{d}r \tag{5.56}$$

By using the dimensionless parameters X, Y, and D_B, defined in Section 5.4.1, and after introducing a specific dimensionless time, T_B, and a dimensionless radial coordinate, R,

$$T_B = \frac{D_S \, t}{r_P^2} \tag{5.57}$$

$$R = \frac{r}{r_P} \tag{5.58}$$

the model equations can be written in dimensionless form

$$\frac{\partial Y}{\partial T_B} = \frac{\partial^2 Y}{\partial R^2} + \frac{2}{R}\frac{\partial Y}{\partial R} \tag{5.59}$$

$$Y = 0 \quad \text{at} \quad T_B = 0 \quad \text{and} \quad 0 \leq R \leq 1 \tag{5.60}$$

$$X = 1 \quad \text{at} \quad T_B = 0 \tag{5.61}$$

$$\frac{\partial Y}{\partial R} = 0 \quad \text{at} \quad T_B > 0 \quad \text{and} \quad R = 0 \tag{5.62}$$

$$\frac{\partial Y}{\partial R} = Bi\,(X - X_s) \quad \text{at} \quad T_B > 0 \quad \text{and} \quad R = 1 \tag{5.63}$$

(for film and surface diffusion)

or

$$\frac{\partial Y}{\partial R} = -\frac{1}{3\,D_B}\frac{\partial X}{\partial T_B} \quad \text{at} \quad T_B > 0 \quad \text{and} \quad R = 1 \tag{5.64}$$

(for surface diffusion)

For the isotherm, the material balance, and the mean adsorbent loading, the following dimensionless equations can be derived:

$$Y(R=1) = Y_s = \mathrm{f}[X(R=1)] = \mathrm{f}(X_s) \tag{5.65}$$

$$X + D_B\,\bar{Y} = 1 \tag{5.66}$$

$$\bar{Y} = 3\int_{R=0}^{1} Y\,R^2\,\mathrm{d}R \tag{5.67}$$

The dimensionless Biot number, Bi, used in eq. (5.63) characterizes the ratio of internal and external mass transfer resistances. In the case of surface diffusion as internal mass transfer, Bi is defined as

$$Bi = \frac{k_F\,r_P\,c_0}{D_S\,\rho_P\,q_0} \tag{5.68}$$

The higher the Biot number, the higher is the rate of film diffusion in comparison to surface diffusion. If $Bi > 50$, the influence of the film diffusion on the overall adsorption rate is negligible, and the surface diffusion alone is rate limiting.

Usually, the set of equations given previously has to be solved by numerical methods in order to calculate kinetic curves and to find the diffusion coefficient, D_S, by curve fitting. Analytical solutions are only available for special cases.

Analytical solutions for special cases

In the case of a linear isotherm and negligible film diffusion, the following analytical solution to the HSDM (Crank 1975) can be used to describe the adsorption kinetics in a batch reactor:

$$X = 1 - \frac{1}{1+D_B}\left[1 - \sum_{n=1}^{\infty} \frac{6\,D_B(D_B+1)\exp(-u_n^2\,T_B)}{9+9\,D_B+u_n^2\,D_B^2}\right] \tag{5.69}$$

where u_n is the nth nonzero positive solution of eq. (5.70)

$$\tan u_n = \frac{3\,u_n}{3+D_B\,u_n^2} \tag{5.70}$$

For the special case of an infinite volume where the concentration outside the particle is constant, the surface diffusion can be described by Boyd's equation (Boyd et al. 1947)

$$F = \frac{q}{q_{eq}} = \frac{c_0 - c}{c_0 - c_{eq}} = 1 - \frac{6}{\pi^2}\sum_{n=1}^{\infty}\frac{1}{n^2}\exp\left(-n^2\,\pi^2\,T_B\right) \tag{5.71}$$

where F is the fractional uptake. It is evident that the condition of constant concentration is not fulfilled in a batch reactor. On the other hand, this equation can be used to develop a design equation for a completely mixed flow-through reactor (CMFR) where the concentration in the reactor can be assumed to be constant over time (steady-state condition). This case is considered separately in Section 5.5.

Determination of the surface diffusion coefficient, D_S

The estimation of surface diffusion coefficients can be carried out on the basis of experimental kinetic curves and under the condition that the equilibrium isotherm is known. For the determination of D_S, it is recommended that the influence of the film diffusion be eliminated by increasing the stirrer velocity in the batch reactor or increasing the flow velocity in the differential column batch reactor as described in Section 5.3. In this case, the kinetic model simplifies to a pure HSDM consisting of eqs. (5.59)–(5.62) and (5.64)–(5.67). The general way to find D_S is to fit calculated kinetic curves to the experimental data. That requires a repeated calculation under variation of D_S in order to minimize the deviations between the experimental and calculated concentrations. Despite the elimination of film diffusion, the kinetic model remains complex and can only be solved with numerical methods with exception of the special case of the linear isotherm described in the previous section.

In order to further simplify the calculation, several authors have published user-friendly standard solutions in the form of diagrams or empirical polynomials. In particular, solutions based on the frequently used Freundlich isotherm were provided. If the Freundlich isotherm is taken as the equilibrium relationship, the dimensionless kinetic curves depend only on two parameters, the relative equilibrium concentration, c_{eq}/c_0, and the Freundlich exponent, n. The first parameter includes the combined influence of the adsorbent dose and the equilibrium position. The equilibrium concentration that will be reached during the kinetic experiment can be calculated by combining the material balance and the equilibrium isotherm as shown in Chapter 3. Suzuki and Kawazoe (1974a) have published a set of diagrams for different Freundlich exponents, n. Each diagram contains an array of curves

$$X = \mathrm{f}(T_B) \tag{5.72}$$

with c_{eq}/c_0 as curve parameter. By comparing the standard curve $X = \mathrm{f}(T_B)$ for the given c_{eq}/c_0 with the experimental data $X = \mathrm{f}(t)$ for different X values, pairs of values (t, T_B) can be found that allow calculating the surface diffusion coefficient by using eq. (5.57).

Zhang et al. (2009) have approximated the exact solutions to the HSDM by empirical polynomials of the general form

$$\bar{C} = A_0 + A_1 \ln T_B + A_2 (\ln T_B)^2 + A_3 (\ln T_B)^3 \tag{5.73}$$

Herein, the dimensionless concentration $\bar{C}$ is defined as

$$\bar{C} = \frac{c - c_{eq}}{c_0 - c_{eq}} \qquad 0 \le \bar{C} \le 1 \tag{5.74}$$

The empirical coefficients A_i are available for different n and c_{eq}/c_0; respective tables are given in the Appendix (Table 10.4).

Since D_S is included in the dimensionless time, T_B, eq. (5.73) can be used to determine D_S from experimental kinetic data by a fitting procedure. At first, the equilibrium concentration that results from the applied adsorbent dose has to be calculated by an iteration method based on the material balance and the isotherm

$$\frac{V_L}{m_A}(c_0 - c_{eq}) = K\, c_{eq}^n \tag{5.75}$$

If c_{eq}/c_0 and n are known, the respective parameters A_i can be identified. If the A_i for the exact process parameters c_{eq}/c_0 and n are not available, the values for conditions closest to the experiment have to be used. After that, kinetic curves for different values of D_S (different T_B) have to be calculated to find the value of D_S that best fits the experimental data.

For illustration, Figure 5.10 shows the application of eq. (5.73) to the experimentally investigated system 4-chlorophenol/activated carbon F300. The n value found from isotherm measurement was $n = 0.4$. Since the relative equilibrium concentration

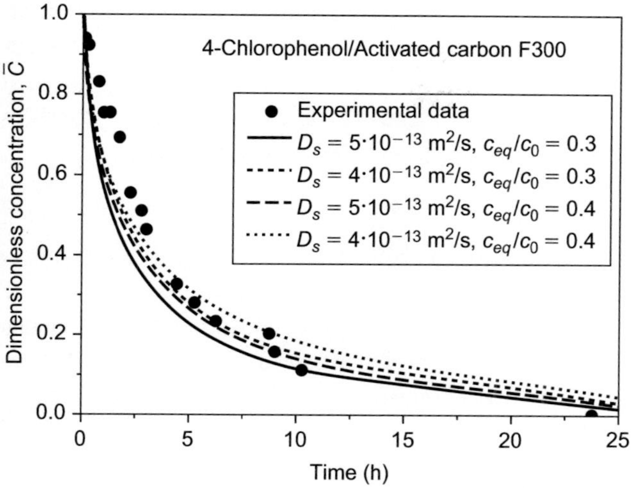

Figure 5.10: Application of eq. (5.73) to describe the kinetic curve of 4-chlorophenol adsorption onto activated carbon F300. Variation of D_S and c_{eq}/c_0. Experimental data from Heese (1996).

was 0.342, the calculations were carried out with the parameters given for $c_{eq}/c_0 = 0.3$ and 0.4, respectively. The kinetic curves were exemplarily calculated for only two diffusion coefficients. This example illustrates the sensitivity of the calculated curves in view of changes of D_S and c_{eq}/c_0 and also the significance of experimental errors. It is recommended that kinetic experiments be repeated to minimize the errors in determination of D_S.

To find approximate D_S values without time-consuming kinetic experiments, an indirect prediction method can be applied. This method is based on a simplified surface diffusion model (linear driving force [LDF] model) and will be presented in Section 5.4.6.

Surface diffusion coefficient, D_S: Influence factors and typical values

The main factors that influence the value of D_S are the temperature of the aqueous solution and the molecular weight of the adsorbate. D_S increases with increasing temperature and decreases with increasing molecular weight. In accordance with the transport mechanism, it can also be expected that the surface diffusion coefficient decreases with increasing adsorption strength (decreasing mobility in the adsorbed state). The typical range of surface diffusion coefficients found for activated carbons is between 10^{-11} m^2/s for small molecules and 10^{-15} m^2/s for larger molecules such as humic substances.

In a number of studies, it was found that D_S also depends on the adsorbate concentration. This is a secondary effect and a consequence of the surface-loading dependence. The dependence of D_S on the surface loading may occur in the case of energetically heterogeneous adsorbents and can be explained by the decrease of the adsorption energy with increasing surface loading that leads to an increase of adsorbate mobility. In contrast, pore diffusion coefficients are independent of the adsorbate concentration (Section 5.4.4).

The dependence of D_S on the adsorbent loading can be described by exponential equations, (Neretnieks 1976; Sudo et al. 1978). A suitable approach is

$$D_S = D_S(0)\exp(\omega q) \tag{5.76}$$

where $D_S(0)$ is the intrinsic surface diffusion coefficient at $q = 0$, and ω is an empirical parameter that has to be determined from kinetic experiments.

Although eq. (5.76) allows considering the loading (or concentration) dependence of D_S in the mathematical description of adsorption kinetics, it makes the diffusion model even more complex. For practical purposes, the concentration dependence is therefore often neglected. This is acceptable if only small concentration ranges are considered. In this case, the estimated diffusion coefficients have to be considered average values that are valid for the given concentration range. If D_S values are determined for later application in fixed-bed adsorber modeling, it is recommended that the kinetic experiments be carried out with the same concentration as occurs in the fixed-bed adsorption process.

It has to be noted that concentration dependence can also be observed if the adsorption rate is determined by a combined surface and pore diffusion mechanism. This effect is independent of a possible loading dependence of D_S and can be observed even if D_S is constant (Section 5.4.5).

5.4.4 Pore diffusion

Basics

Instead of or in addition to surface diffusion, the adsorbate transport within the adsorbent particles can also take place in the pore liquid. The concentration profiles for film and pore diffusion as well as for pore diffusion alone (fast film diffusion) are given in Figures 5.11 and 5.12.

In comparison to surface diffusion, the model development for pore diffusion is more complicated. In the surface diffusion model, the adsorbent is considered to be homogeneous, and the adsorption equilibrium is assumed to exist only at the outer surface of the adsorbent particle. In the case of pore diffusion, however, the adsorption equilibrium has to be considered at each point of the pore system. In general, it is assumed that there is a local equilibrium between the pore fluid concentration and the solid-phase concentration. Therefore, the material balance for a thin shell of the adsorbent particle has to account for simultaneous change of concentration and adsorbent loading. The material balance equation reads

$$\Delta \dot{n}_P \, 4\pi r^2 \, \Delta t = \Delta q \, 4\pi r^2 \Delta r \, \rho_P + \varepsilon_P \, \Delta c_p \, 4\pi r^2 \, \Delta r \tag{5.77}$$

where $\dot{n}_P$ is the mass transfer rate per unit of surface area, r is the radial coordinate, c_p is the adsorbate concentration in the pore fluid, q is the adsorbent loading, ε_P is the particle (internal) porosity, and ρ_P is the particle density.

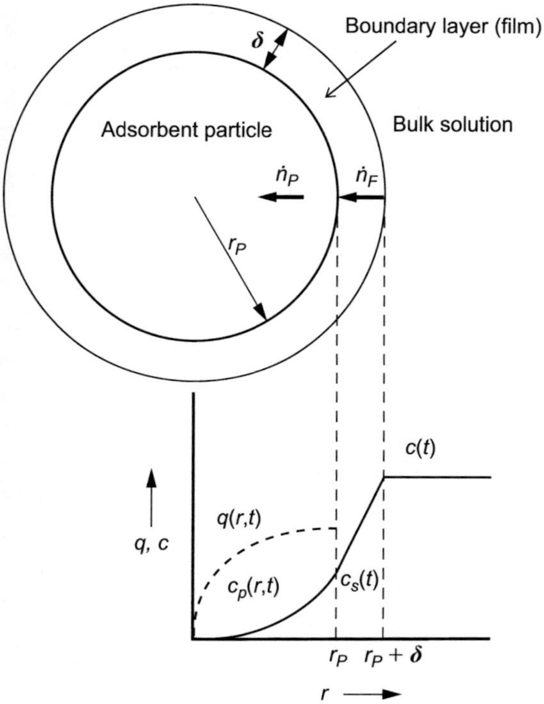

Figure 5.11: Concentration profiles in the case of film and pore diffusion.

For pore diffusion, the mass transfer rate per unit of surface area is given by

$$\dot{n}_P = D_P \frac{\partial c_p}{\partial r} \tag{5.78}$$

where D_P is the pore diffusion coefficient. Combining eqs. (5.77) and (5.78) and following the procedure described for surface diffusion leads to

$$\rho_P \frac{\partial q}{\partial t} + \varepsilon_P \frac{\partial c_p}{\partial t} = D_P \left(\frac{\partial^2 c_p}{\partial r^2} + \frac{2}{r} \frac{\partial c_p}{\partial r} \right) \tag{5.79}$$

where D_P is assumed to be constant. Applying the chain rule gives

$$\rho_P \frac{\partial q}{\partial c_p} \frac{\partial c_p}{\partial t} + \varepsilon_P \frac{\partial c_p}{\partial t} = D_P \left(\frac{\partial^2 c_p}{\partial r^2} + \frac{2}{r} \frac{\partial c_p}{\partial r} \right) \tag{5.80}$$

$$\left(\rho_P \frac{\partial q}{\partial c_p} + \varepsilon_P \right) \frac{\partial c_p}{\partial t} = D_P \left(\frac{\partial^2 c_p}{\partial r^2} + \frac{2}{r} \frac{\partial c_p}{\partial r} \right) \tag{5.81}$$

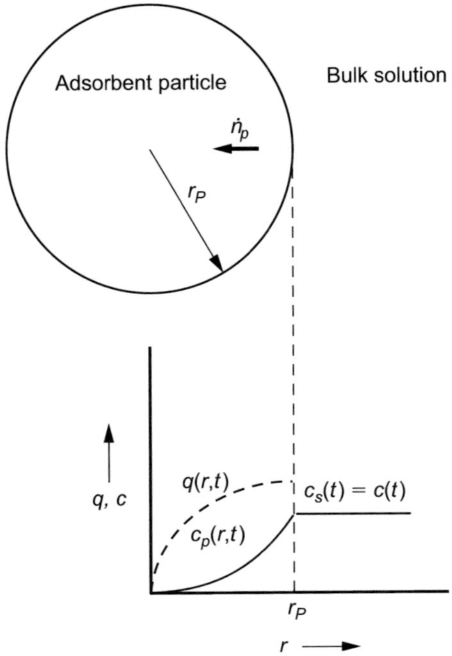

Figure 5.12: Concentration profiles in the case of pore diffusion (no external mass transfer resistance).

Introducing an apparent pore diffusion coefficient, D_a,

$$D_a = \frac{D_P}{\rho_P \dfrac{\partial q}{\partial c_p} + \varepsilon_P} \tag{5.82}$$

simplifies eq. (5.79) to

$$\frac{\partial c_p}{\partial t} = D_a \left(\frac{\partial^2 c_p}{\partial r^2} + \frac{2}{r} \frac{\partial c_p}{\partial r} \right) \tag{5.83}$$

It has to be noted that D_a contains the slope of the isotherm $\partial q / \partial c_p$, which has a constant value only in the case of the linear isotherm ($\partial q / \partial c_p = K_H$). For nonlinear isotherms, D_a is concentration dependent. In the case of the Freundlich isotherm, for instance, the slope becomes

$$\frac{\partial q}{\partial c_p} = n K c_p^{n-1} \tag{5.84}$$

From eq. (5.79), an expression for $\partial q / \partial t$ can be derived by applying the chain rule

$$\left(\rho_P + \varepsilon_P \frac{\partial c_p}{\partial q} \right) \frac{\partial q}{\partial t} = D_P \left(\frac{\partial^2 c_p}{\partial r^2} + \frac{2}{r} \frac{\partial c_p}{\partial r} \right) \tag{5.85}$$

Again, an apparent pore diffusion coefficient could be defined by dividing D_P by the term within the brackets on the left-hand side of eq. (5.85). However, the term $\varepsilon_P\, \partial c_p/\partial q$ can often be neglected, in particular if the concentration is low and the adsorbate is strongly adsorbed. Therefore, pore diffusion is frequently described by the simplified diffusion equation

$$\rho_P \frac{\partial q}{\partial t} = D_p\left(\frac{\partial^2 c_p}{\partial r^2} + \frac{2}{r}\frac{\partial c_p}{\partial r}\right) \tag{5.86}$$

The respective initial and boundary conditions for the batch reactor are

$$q = 0,\ c_p = 0 \quad \text{at} \quad t = 0 \quad \text{and} \quad 0 \le r \le r_P \tag{5.87}$$

$$c = c_0 \quad \text{at} \quad t = 0 \tag{5.88}$$

$$\frac{\partial c_p}{\partial r} = 0 \quad \text{at} \quad t > 0 \quad \text{and} \quad r = 0 \tag{5.89}$$

$$D_P \frac{\partial c_p}{\partial r} = k_F(c - c_s) \quad \text{at} \quad t > 0 \quad \text{and} \quad r = r_P \tag{5.90}$$

(for film and pore diffusion)

or

$$D_P \frac{\partial c_p}{\partial r} = -\frac{\varepsilon_B}{a_{VR}}\frac{\partial c}{\partial t} \quad \text{at} \quad t > 0 \quad \text{and} \quad r = r_P \tag{5.91}$$

(for pore diffusion)

where a_{VR} is the external surface area related to the reactor volume (see Table 5.1).

As in the case of surface diffusion, the basic model equations can be formulated with dimensionless parameters. For pore diffusion, the dimensionless time, T_B, is defined as

$$T_B = \frac{D_P\, t\, c_0}{r_P^2\, \rho_P\, q_0} \tag{5.92}$$

and the respective Biot number is

$$Bi = \frac{k_F\, r_P}{D_P} \tag{5.93}$$

The definitions of dimensionless concentration (X), adsorbent loading (Y), radial coordinate (R), and distribution parameter (D_B) are the same as used in the surface diffusion model.

By using these definitions, the pore diffusion equation reads

$$\frac{\partial Y}{\partial T_B} = \frac{\partial^2 X_p}{\partial R^2} + \frac{2}{R}\frac{\partial X_p}{\partial R} \tag{5.94}$$

with the initial and boundary conditions

$$Y=0,\ X_p=0 \quad \text{at} \quad T_B=0 \quad \text{and} \quad 0 \le R \le 1 \tag{5.95}$$

$$X=1 \quad \text{at} \quad T_B=0 \tag{5.96}$$

$$\frac{\partial X_p}{\partial R}=0 \quad \text{at} \quad T_B>0 \quad \text{and} \quad R=0 \tag{5.97}$$

$$\frac{\partial X_p}{\partial R}=Bi\,(X-X_s) \quad \text{at} \quad T_B>0 \quad \text{and} \quad R=1 \tag{5.98}$$

(for film and pore diffusion)

$$\frac{\partial X_p}{\partial R}=-\frac{1}{3\,D_B}\frac{\partial X}{\partial T_B} \quad \text{at} \quad T_B>0 \quad \text{and} \quad R=1 \tag{5.99}$$

(for pore diffusion)

This set of equations has to be solved by numerical methods. Analytical solutions of the pore diffusion model exist only for the linear and the irreversible isotherm.

Special cases: Linear and irreversible isotherms

In the case of a linear isotherm, the dimensionless isotherm reads

$$Y=X_p \tag{5.100}$$

and eqs. (5.59) and (5.94) become identical. The solution given for surface diffusion and linear isotherm (eq. (5.69)) is therefore also valid for pore diffusion and linear isotherm. The only difference consists in the different definitions of T_B.

For the irreversible isotherm, Suzuki and Kawazoe (1974b) have given the following solution:

$$T_B=\frac{1+\alpha^3}{3\alpha}\left[\alpha\ln\frac{\beta^3+\alpha^3}{1+\alpha^3}+\ln\frac{\beta+\alpha}{1+\alpha}-\frac{1}{2}\ln\frac{\beta^2-\alpha\beta+\alpha^2}{1-\alpha+\alpha^2}+\sqrt{3}\left(\tan^{-1}\frac{2-\alpha}{\sqrt{3}}-\tan^{-1}\frac{2\beta-\alpha}{\sqrt{3}}\right)\right] \tag{5.101}$$

with

$$\alpha=\left(\frac{c_{eq}}{c_0-c_{eq}}\right)^{1/3} \tag{5.102}$$

and

$$\beta=\left(\frac{c-c_{eq}}{c_0-c_{eq}}\right)^{1/3} \tag{5.103}$$

Determination of the pore diffusion coefficient, D_P

Although the diffusion in the pore liquid follows, in principle, the same mechanism as the diffusion in the bulk liquid, free-liquid molecular diffusion coefficients, which are known for many solutes, cannot be used as pore diffusion coefficients. The reason for that is an additional hindrance of the adsorbate transport caused by pore restrictions and pore intersections. Therefore, the pore diffusion coefficient is generally smaller than the diffusion coefficient in the free liquid ($D_P < D_L$).

This deviation can be considered by introducing a labyrinth factor (tortuosity), τ_P, in addition to the internal porosity (particle porosity), ε_P

$$D_P = \frac{D_L\,\varepsilon_P}{\tau_P} \tag{5.104}$$

While the porosity can be easily determined, there is no method for an independent estimation of the tortuosity. Therefore, eq. (5.104) can only be used to calculate a theoretical maximum value of D_P under the assumption $\tau_P = 1$. On the other hand, eq. (5.104) can be used to find the tortuosity by comparing experimentally determined D_P with D_L. In the literature, tortuosity values between 2 and 6 are frequently reported for microporous adsorbents such as activated carbons. Empirical equations for estimating D_L are given in Chapter 7 (Table 7.8).

Since D_P cannot be evaluated independently, it has to be determined by kinetic experiments. The general procedure is the same as in the case of surface diffusion; the only difference consists in the kinetic model to be applied. As for surface diffusion, standard solutions also exist for pore diffusion. For instance, Suzuki and Kawazoe (1974b) have published diagrams, $X = f(T_B)$, for different Freundlich exponents and equilibrium concentrations, c_{eq}/c_0.

Pore diffusion coefficient, D_P: Influence factors and typical values

It can be derived from eq. (5.104) that D_P depends on the same influence factors as the aqueous-phase diffusion coefficient, in particular molecular weight and temperature. On the other hand, it can also be derived from eq. (5.104) that D_P should be independent of the adsorbate concentration. However, in contrast to D_P, the apparent diffusion coefficient, D_a, defined in eq. (5.82), depends on the concentration because the isotherm slope is included in the definition of D_a.

Given that for small- and medium-size molecules the aqueous-phase diffusion coefficients are in the range of 10^{-10} m^2/s to 10^{-9} m^2/s, and considering typical values of ε_P and τ_P, the pore diffusion coefficients, D_P, are expected to fall into the range of 10^{-10} m^2/s to 10^{-8} m^2/s.

5.4.5 Combined surface and pore diffusion

Basics

In the previous sections, the intraparticle transport was considered to be based on only a single mechanism, either surface or pore diffusion. This simplified description of the adsorption kinetics is an appropriate approach for most practical cases; in particular, the HSDM was proved to be an adequate model to describe the adsorption of strongly adsorbable substances on porous adsorbents. A more general approach takes into account that surface and pore diffusion may act in parallel. The total flux is then given as the sum of the fluxes caused by surface diffusion (eq. (5.43)) and by pore diffusion (eq. (5.78))

$$\dot{n}_T = \dot{n}_S + \dot{n}_P = \rho_P D_S \frac{\partial q}{\partial r} + D_P \frac{\partial c_p}{\partial r} \tag{5.105}$$

According to eqs. (5.48) and (5.86), the differential adsorbate uptake can be expressed as

$$\frac{\partial q}{\partial t} = D_S \left(\frac{\partial^2 q}{\partial r^2} + \frac{2}{r} \frac{\partial q}{\partial r} \right) + \frac{D_P}{\rho_P} \left(\frac{\partial^2 c_p}{\partial r^2} + \frac{2}{r} \frac{\partial c_p}{\partial r} \right) \tag{5.106}$$

Introducing the dimensionless time, T_B,

$$T_B = \frac{D_S\, t}{r_P^2} + \frac{D_P\, t\, c_0}{r_P^2\, \rho_P\, q_0} \tag{5.107}$$

and a parameter λ that gives the ratio of pore and surface diffusion

$$\lambda = \frac{D_P\, c_0}{D_S\, \rho_P\, q_0} \tag{5.108}$$

leads to the dimensionless form of eq. (5.106)

$$\frac{\partial Y}{\partial T_B} = \frac{1}{1+\lambda} \left(\frac{\partial^2 Y}{\partial R^2} + \frac{2}{R} \frac{\partial Y}{\partial R} \right) + \frac{\lambda}{1+\lambda} \left(\frac{\partial^2 X_p}{\partial R^2} + \frac{2}{R} \frac{\partial X_p}{\partial R} \right) \tag{5.109}$$

The parameter λ describes the contributions of the different diffusion mechanisms to the total intraparticle flux. Since surface diffusion and pore diffusion act in parallel, the faster process is dominating and determines the total adsorption rate. At $\lambda = 1$, both diffusion mechanisms contribute to the total flux to the same extent; $\lambda = 0$ ($D_P = 0$) represents the limiting case of pure surface diffusion, whereas $\lambda = \infty$ ($D_S = 0$) represents the limiting case of pure pore diffusion. For the limiting cases, eq. (5.109) reduces to eqs. (5.59) and (5.94), respectively.

A solution of the combined diffusion equation is possible by defining a fictive adsorbent loading, Y^*,

$$Y^* = \frac{1}{1+\lambda}\left(Y + \lambda X_p\right) \tag{5.110}$$

that allows reducing eq. (5.109) to an equation analogous to eq. (5.59) for surface diffusion (Fritz et al. 1981)

$$\frac{\partial Y}{\partial T_B} = \frac{\partial^2 Y^*}{\partial R^2} + \frac{2}{R}\frac{\partial Y^*}{\partial R} \tag{5.111}$$

If the film diffusion should be considered an additional transport step in the model, a Biot number has to be applied to characterize the ratio of external and internal diffusion processes. The definition of the Biot number in the case of combined surface and pore diffusion is given by

$$Bi = \frac{k_F\, r_P\, c_0}{D_P\, c_0 + D_S\, \rho_P\, q_0} \tag{5.112}$$

A simpler approach to describe the combined surface and pore diffusion is based on the introduction of effective diffusion coefficients for surface or pore diffusion and applying the mathematical models for pure surface diffusion or pure pore diffusion presented in the previous sections. By applying the chain rule, eq. (5.105) can be written in the form

$$\dot{n}_T = \rho_P\, D_S \frac{\partial q}{\partial r} + D_P \frac{\partial c_p}{\partial q}\frac{\partial q}{\partial r} = \rho_P\, D_{S,eff} \frac{\partial q}{\partial r} \tag{5.113}$$

with

$$D_{S,eff} = D_S + D_P \frac{1}{\rho_P}\frac{\partial c_p}{\partial q} \tag{5.114}$$

where $D_{S,eff}$ is an effective surface diffusion coefficient that includes possible contributions of pore diffusion.

Alternatively, the combined pore and surface diffusion equation can be transformed into the mathematical expression for pore diffusion by introducing an effective pore diffusion coefficient, $D_{P,eff}$,

$$\dot{n}_T = \rho_P\, D_S \frac{\partial q}{\partial c_p}\frac{\partial c_p}{\partial r} + D_P \frac{\partial c_p}{\partial r} = D_{P,eff} \frac{\partial c_p}{\partial r} \tag{5.115}$$

$$D_{P,eff} = \rho_P\, D_S \frac{\partial q}{\partial c_p} + D_P \tag{5.116}$$

As can be seen from eqs. (5.114) and (5.116), the effective diffusion coefficients depend on the slope of the isotherm, which has a constant value only in the case of a linear isotherm. In all other cases, the effective coefficients are not constant over the considered concentration range due to the change of the isotherm slope with concentration. In the case of a favorable isotherm, for instance, the slope of the isotherm decreases with increasing concentration. For this case, it can be derived from the given equations that the influence of the surface diffusion on the total adsorption kinetics will decrease with increasing concentration, whereas the influence of pore diffusion will increase.

In order to simplify the concentration dependence of the effective diffusion coefficients, the differential quotient $\partial q/\partial c_p$ can be replaced by the ratio q_0/c_0, which was shown to be a reasonable approximation (Neretnieks 1976). Here, c_0 is the initial concentration, and q_0 is the equilibrium adsorbent loading related to c_0.

Under the assumption of a constant D_S, eqs. (5.114) and (5.116) can be used to separate the contributions of pore and surface diffusion to the total transport and to determine the respective diffusion coefficients. For this purpose, kinetic curves have to be determined experimentally for different initial concentrations. From these kinetic curves, effective diffusion coefficients can be found by applying either the pore diffusion model or the surface diffusion model. The estimated effective diffusion coefficients can then be plotted against the isotherm slope (or approximately q_0/c_0) to find D_S and D_P as shown in Figures 5.13 and 5.14.

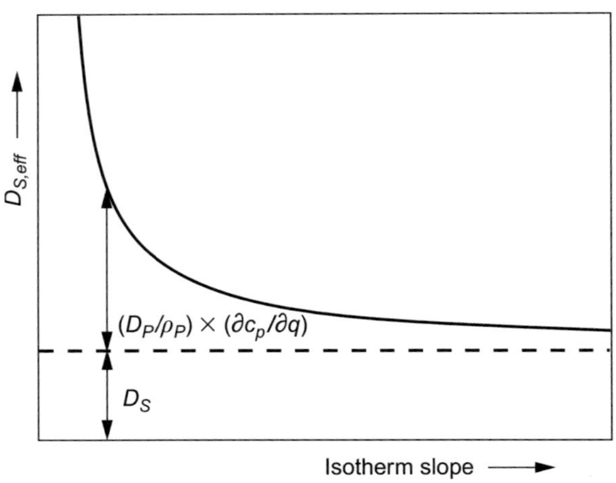

Figure 5.13: Dependence of the effective surface diffusion coefficient, $D_{S,eff}$, on the isotherm slope.

It is recommended that the plausibility of the determined pore diffusion coefficient be checked. Since the tortuosity, τ_P, must be greater than 1, pore diffusion coefficients greater than the product $D_L \times \varepsilon_P$ are not plausible (eq. (5.104)). If this case occurs, an additional concentration dependence of D_S has to be taken into account (Section 5.4.3).

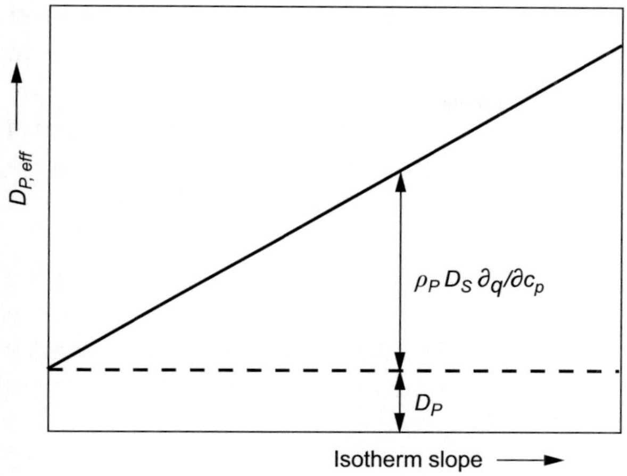

Figure 5.14: Dependence of the effective pore diffusion coefficient, $D_{P,eff}$, on the isotherm slope.

Special case: Linear isotherm

In the case of a linear isotherm, the isotherm slope becomes constant

$$\frac{\partial q}{\partial c_p} = K_H = \text{constant} \tag{5.117}$$

and the effective diffusion coefficients, defined by eqs. (5.114) and (5.116), read

$$D_{S,eff} = D_S + \frac{D_P}{\rho_P K_H} \tag{5.118}$$

$$D_{P,eff} = \rho_P K_H D_S + D_P \tag{5.119}$$

Combining eqs. (5.118) and (5.119) leads to the relationship

$$D_{P,eff} = \rho_P K_H D_{S,eff} \tag{5.120}$$

which demonstrates that in the case of a linear isotherm each effective surface coefficient can be converted to an equivalent effective pore diffusion coefficient and vice versa by a constant conversion factor. Thus, both surface and pore diffusion models can be used to describe the kinetic curve, and the solutions are identical as described before. Consequently, it is not possible to distinguish between the mechanisms and to separate the contributions of pore and surface diffusion to the overall transport.

Determination of effective diffusion coefficients

The general way to determine effective diffusion coefficients that consider the contributions of both intraparticle diffusion mechanisms is to measure kinetic curves and to fit the curves with either the pore or the surface diffusion model. If the contributions of both mechanisms to the overall kinetics should be evaluated, kinetic curves at different concentrations have to be measured, and the estimation method demonstrated in Figures 5.13 and 5.14 has to be applied.

In order to reduce the experimental effort, Crittenden et al. (1987b) have proposed a method that allows estimating the effective surface diffusion coefficient. Here, the effective surface diffusion coefficient is related to the pore diffusion coefficient by

$$D_{S,eff} = SPDFR \frac{D_P\, c_0}{\rho_P\, q_0} \tag{5.121}$$

where *SPDFR* is the surface-to-pore diffusion flux ratio. If D_P is expressed according to eq. (5.104) and the unknown tortuosity is set to $\tau_P = 1$ (maximum D_P) or integrated into the parameter *SPDFR*, the equation can be written as

$$D_{S,eff} = SPDFR \frac{\varepsilon_P\, D_L\, c_0}{\rho_P\, q_0} \tag{5.122}$$

D_L can be found from databases or, alternatively, from empirical correlations (Chapter 7, Table 7.8). From a series of experiments with different activated carbons and different adsorbates, a mean value of 6.6 was found for *SPDFR*. However, it has to be noted that the *SPDFR* value may vary over a relatively broad range, depending on the nature of the adsorbate and the type of adsorbent. Consequently, the application of eq. (5.122) allows only a rough estimate of $D_{S,eff}$.

5.4.6 Simplified intraparticle diffusion model (LDF model)

Basics

The intraparticle diffusion models discussed previously include partial derivatives with respect to time and radial coordinate, which causes an increased effort for the numerical solution. In order to reduce the mathematical effort, the linear driving force (LDF) approach (Glueckauf and Coates 1947; Glueckauf 1955) can be used. Due to its simpler mathematical structure, this approach has found widespread use for modeling adsorption processes in slurry and fixed-bed adsorbers. The LDF model can be considered a simplification of the surface diffusion model. Its basic concept is comparable to the model approach used for describing the film diffusion where the concentration gradient in Fick's law is replaced by a linear concentration difference.

In the LDF model approach, it is formally assumed that the decrease of adsorbent loading takes place within a fictive solid film comparable to the solution-side film in the film diffusion model. Accordingly, the solid-phase concentration gradient in eq. (5.43) is replaced by a linear difference between the equilibrium loading at the outer particle surface and the mean loading of the particle. Thus, the equation for the flux is approximated by

$$\dot{n}_S = \rho_P k_S (q_s - \bar{q}) \tag{5.123}$$

where ρ_P is the particle density ($\rho_P = m_A/V_A$), k_S is the intraparticle mass transfer coefficient, q_s is the adsorbent loading at the external surface of the adsorbent particle, and $\bar{q}$ is the mean loading in the particle.

The concentration profiles for the simplified surface diffusion model with and without consideration of film diffusion are shown schematically in Figures 5.15 and 5.16.

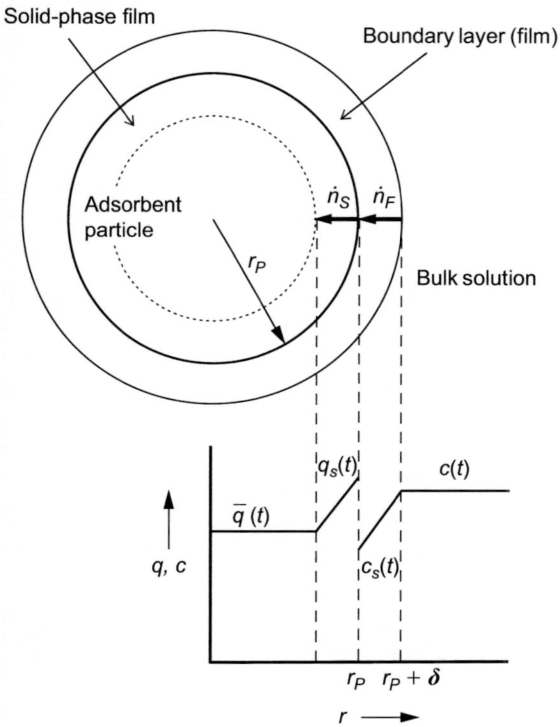

Figure 5.15: Concentration profiles according to the LDF model with external mass transfer resistance.

With the material balance equation

$$\dot{n}_S = \frac{m_A}{A_s}\frac{d\bar{q}}{dt} = -\frac{V_L}{A_s}\frac{dc}{dt} \tag{5.124}$$

the following mass transfer equation can be derived from eq. (5.123):

$$\frac{d\bar{q}}{dt} = k_S \frac{A_s}{V_A}(q_s - \bar{q}) = k_S a_{VA}(q_s - \bar{q}) \tag{5.125}$$

where a_{VA} is the external surface area available for mass transfer related to the adsorbent volume (A_s/V_A).

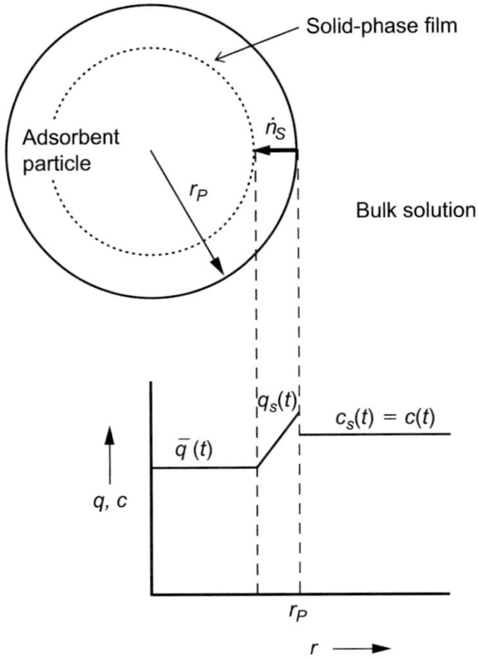

Figure 5.16: Concentration profiles according to the LDF model without external mass transfer resistance.

As in the case of film diffusion, the mass transfer equation can be formulated in different forms depending on the definition of the specific surface area (see Section 5.4.2). In principle, the surface area can be related to the adsorbent mass ($a_m = A_s/m_A$), the adsorbent volume (as in eq. (5.125)), or the total volume of the reactor ($a_{VR} = A_s/V_R$). The different forms of the equation for the adsorbate uptake are summarized in Table 5.2, together with the respective equations for the concentration decay, which can be derived under consideration of the material balance (eq. (5.124)). The specific mass transfer equations for uniform spherical adsorbent particles are also given in the table.

The different forms of the mass transfer equation can be generalized by introducing a modified intraparticle mass transfer coefficient, k_S^*,

$$\frac{\mathrm{d}q}{\mathrm{d}t} = k_S^*(q_s - \bar{q}) \tag{5.126}$$

with

$$k_S^* = k_S\, a_{VA} = \frac{k_S\, a_{VR}}{1-\varepsilon_B} = k_S\, a_m\, \rho_P \tag{5.127}$$

Glueckauf (1955) has found the following equivalence relationship between the mass transfer coefficient, k_S^*, and the surface diffusion coefficient, D_S:

Table 5.2: Different forms of the intraparticle mass transfer equation. For the meaning of the terms ε_B, $(1-\varepsilon_B)$, $(1-\varepsilon_B)/\varepsilon_B$, and $\rho_P(1-\varepsilon_B)/\varepsilon_B$, see also Table 2.1.

External surface area related to	General mass transfer equations	Specific particle surface in the case of spherical particles	Mass transfer equations for spherical particles
Adsorbent volume	$\frac{d\bar{q}}{dt} = k_S\, a_{VA}(q_s - \bar{q})$	$a_{VA} = \frac{A_S}{V_A} = \frac{3}{r_P}$	$\frac{d\bar{q}}{dt} = \frac{3\,k_S}{r_P}(q_s - \bar{q})$
	$-\frac{dc}{dt} = k_S\, a_{VA}\frac{\rho_P(1-\varepsilon_B)}{\varepsilon_B}(q_s - \bar{q})$		$-\frac{dc}{dt} = \frac{3\,k_S}{r_P}\frac{\rho_P(1-\varepsilon_B)}{\varepsilon_B}(q_s - \bar{q})$
Reactor volume	$\frac{d\bar{q}}{dt} = k_S\, a_{VR}\frac{1}{1-\varepsilon_B}(q_s - \bar{q})$	$a_{VR} = \frac{A_S}{V_R} = \frac{3}{r_P}(1-\varepsilon_B)$	$\frac{d\bar{q}}{dt} = \frac{3\,k_S}{r_P}(q_s - \bar{q})$
	$-\frac{dc}{dt} = k_S\, a_{VR}\frac{\rho_P}{\varepsilon_B}(q_s - \bar{q})$		$-\frac{dc}{dt} = \frac{3\,k_S}{r_P}\frac{\rho_P(1-\varepsilon_B)}{\varepsilon_B}(q_s - \bar{q})$
Adsorbent mass	$\frac{d\bar{q}}{dt} = k_S\, a_m\, \rho_P(q_s - \bar{q})$	$a_m = \frac{A_S}{m_A} = \frac{3}{r_P\,\rho_P}$	$\frac{d\bar{q}}{dt} = \frac{3\,k_S}{r_P}(q_s - \bar{q})$
	$-\frac{dc}{dt} = k_S\, a_m\frac{\rho_P\,\rho_B}{\varepsilon_B}(q_s - \bar{q})$		$-\frac{dc}{dt} = \frac{3\,k_S}{r_P}\frac{\rho_P(1-\varepsilon_B)}{\varepsilon_B}(q_s - \bar{q})$

$$k_S^* = \frac{15\,D_S}{r_P^2} \tag{5.128}$$

As shown in Table 5.2, for spherical adsorbent particles, k_S^* is given by

$$k_S^* = \frac{3\,k_S}{r_P} \tag{5.129}$$

Accordingly, the equivalence relationship can also be written as

$$k_S = \frac{5\,D_S}{r_P} \tag{5.130}$$

In order to set up a simplified kinetic model for intraparticle diffusion, eq. (5.126) has to be combined with the material balance and the isotherm, $q_s = \mathrm{f}(c_s)$. The initial and boundary conditions are

$$\bar{q} = 0, \quad c = c_0 \quad \text{at} \quad t = 0 \tag{5.131}$$

$$\rho_P\, k_S\,(q_s - \bar{q}) = k_F(c - c_s) \quad \text{at} \quad t > 0 \tag{5.132}$$

Equation (5.132) follows from the continuity of the material flux ($\dot{n}_S = \dot{n}_F$) and has to be considered only if both film and intraparticle diffusion are relevant for the adsorption kinetics.

The mass transfer equation as well as the initial and boundary conditions can be written in dimensionless form after introducing a dimensionless time, T_B,

$$T_B = k_S^* t \tag{5.133}$$

and a Biot number that gives the ratio of external and internal mass transfer

$$Bi = \frac{k_F c_0}{k_S \rho_P q_0} \tag{5.134}$$

Furthermore, the dimensionless concentration (X) and adsorbent loading (Y) as well as the distribution factor (D_B), as defined in Section 5.4.1, have to be considered.

The resulting mass transfer equation is

$$\frac{d\bar{Y}}{dT_B} = Y_s - \bar{Y} \tag{5.135}$$

with

$$\bar{Y} = 0, \quad X = 1 \quad \text{at} \quad T_B = 0 \tag{5.136}$$

and

$$Y_s - \bar{Y} = Bi\,(X - X_s) \quad \text{at} \quad T_B > 0 \tag{5.137}$$

The mass transfer equation has to be solved together with the dimensionless isotherm

$$Y_s = f(X_s) \tag{5.138}$$

and the material balance equation

$$X + D_B\,\bar{Y} = 1 \tag{5.139}$$

to obtain the kinetic curve $X = f(T_B)$.

If the film diffusion is fast enough, the conditions given in eqs. (5.132) and (5.137) can be omitted. In this case, the isotherm reads

$$Y_s = f(X) \tag{5.140}$$

because c equals c_s (see Figure 5.16).

In most cases, the LDF model is a good approximation to the exact but more complicated surface diffusion model. For comparison, Figure 5.17 shows kinetic curves calculated with the LDF model and the HSDM, in both cases without considering film diffusion. For the application of the LDF model, the mass transfer coefficient equivalent to D_S was estimated by using eq. (5.128). Although the curvatures of the curves are slightly different, the LDF model reflects the general trend of the kinetic curve in sufficient quality.

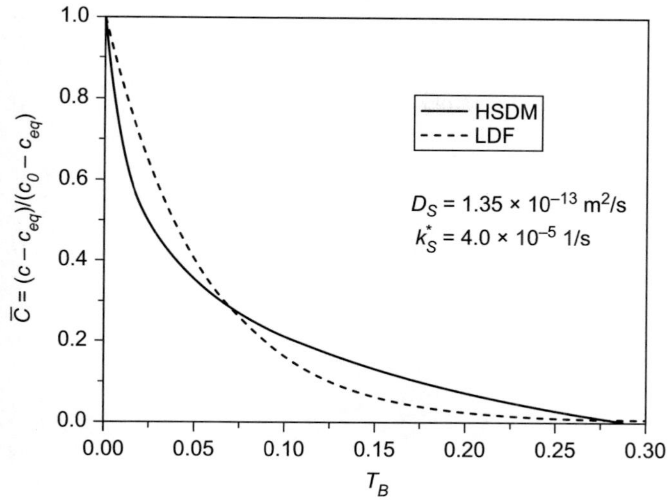

Figure 5.17: Comparison of kinetic curves calculated by the LDF model and the HSDM.

A better approximation can be achieved if the intraparticle mass transfer coefficient is not considered a constant but a parameter that depends on the adsorbed amount. This dependence can be described analogously to the loading dependence of D_S (Section 5.4.3) by

$$k_S^* = k_S^*(0)\ \exp(\omega\,\bar{q}) \tag{5.141}$$

where $k_S^*(0)$ is the intrinsic mass transfer coefficient and ω is an empirical parameter that describes the strength of the influence of the adsorbed amount, $\bar{q}$. The comparison made in Figure 5.18 shows the improved approximation to the HSDM. On the other side, the application of the LDF model becomes more difficult because two parameters, $k_S^*(0)$ and ω, are needed to describe the intraparticle mass transfer. It has to be decided on a case-by-case basis whether the effort is worth it or not.

Although the LDF model was originally developed as a simplified version of the surface diffusion model, it can also be related to the pore diffusion model. As shown for the combined surface and pore diffusion mechanism (Section 5.4.5), an effective surface diffusion coefficient can be defined as

$$D_{S,eff} = D_S + \frac{D_P}{\rho_P}\frac{\partial c_p}{\partial q} \approx D_S + \frac{D_P\, c_0}{\rho_P\, q_0} \tag{5.142}$$

Accordingly, it is possible to link the mass transfer coefficient used in the LDF model with the effective surface diffusion coefficient by

$$k_{S,eff}^* = \frac{15\, D_{S,eff}}{r_P^2} = \frac{15\, D_S}{r_P^2} + \frac{15\, D_P}{r_P^2}\frac{c_0}{\rho_P\, q_0} \tag{5.143}$$

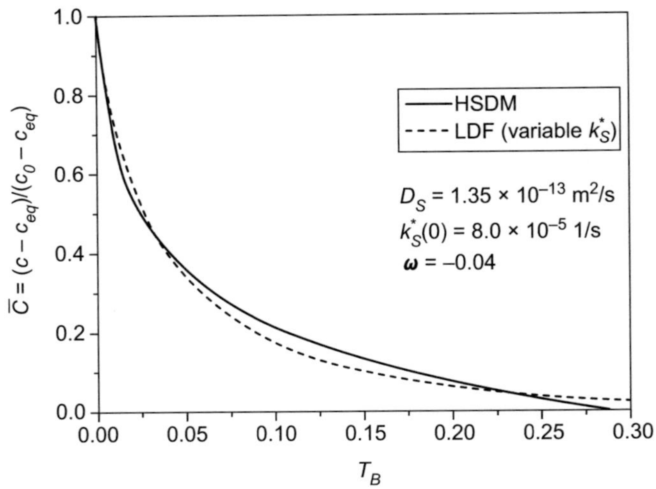

Figure 5.18: Kinetic curve calculated by the LDF model with a variable mass transfer coefficient in comparison with the kinetic curve calculated by the HSDM.

It can be seen from eq. (5.142) that in the case of $D_S = 0$ an effective surface diffusion coefficient can be calculated only on the basis of D_P. Accordingly, for the limiting case of pore diffusion, eq. (5.143) reduces to

$$k_{S,eff}^* = \frac{15\,D_P}{r_P^2}\frac{c_0}{\rho_P\,q_0} \tag{5.144}$$

Given that the particle density can be expressed by the bulk density ($\rho_B = m_A/V_R$) and the void fraction (bulk porosity), ε_B,

$$\rho_P = \frac{\rho_B}{1-\varepsilon_B} \tag{5.145}$$

the relationship between $k_{S,eff}^*$ and D_P can also be written in the form

$$k_{S,eff}^* = \frac{15\,D_P\,(1-\varepsilon_B)}{r_P^2}\frac{c_0}{\rho_B\,q_0} \tag{5.146}$$

In summary, it has to be stated that surface diffusion as well as pore diffusion can be approximated by the LDF model.

Special case: Linear isotherm

In the case of a linear isotherm, the following analytical solution to the LDF model (without consideration of film diffusion) is found:

$$X = \frac{1}{D_B + 1} + \frac{D_B}{D_B + 1} e^{-(D_B+1)T_B} \tag{5.147}$$

Equation (5.147) is identical in form to the solution to the film diffusion model with linear isotherm (eq. (5.39)). However, it has to be considered that the definitions of T_B are different in both cases.

Determination of the intraparticle mass transfer coefficient, k_S (k_S^*)

As for the other models, a general way to estimate the mass transfer coefficients from experimental kinetic curves consists of a fitting procedure based on model calculations with varying mass transfer coefficients. For this purpose, it is recommended that the impact of the film diffusion be eliminated through a high stirrer velocity (batch reactor) or a high flow rate (differential column batch reactor). To illustrate the principle of the fitting procedure, Figure 5.19 shows kinetic curves calculated under variation of the mass transfer coefficient.

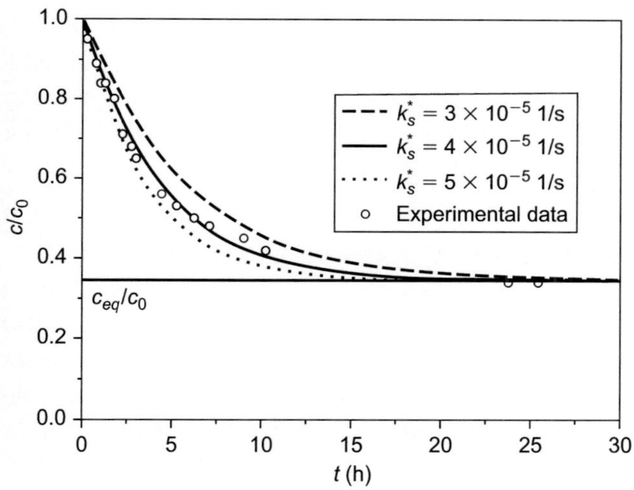

Figure 5.19: Estimation of the internal mass transfer coefficient, exemplarily shown for the adsorption of 4-chlorophenol onto activated carbon F300. Experimental data from Heese (1996).

A combined graphical/analytical method for estimating k_S^* can be derived from eqs. (5.124) and (5.126). Introducing eq. (5.126) into eq. (5.124) and rearranging gives

$$k_S^* = -\frac{V_L}{m_A} \cdot \frac{1}{q_s(t) - \bar{q}(t)} \left(\frac{dc}{dt}\right)_t \tag{5.148}$$

The procedure based on eq. (5.148) is as follows: read $c(t)$ and $(dc/dt)_t$ for a selected time t from the kinetic curve, set $c(t) = c_s(t)$ (fast film diffusion), calculate $q_s(t)$ related to $c_s(t)$ by using the isotherm equation, calculate $\bar{q}(t)$ by means of the material balance (eq. (5.5)), and finally calculate k_S^* by eq. (5.148). To find an average value for k_S^*, the procedure has to be repeated for different pairs of values (c, t).

The graphical/analytical method can also be applied if both film diffusion and intraparticle diffusion are relevant in the given system. In this case, an equation for k_S can be derived by considering the continuity of the material fluxes ($\dot{n}_F = \dot{n}_S$). Combining eqs. (5.10) and (5.123) gives

$$k_S = \frac{k_F}{\rho_P} \frac{[c(t) - c_S(t)]}{[q_S(t) - \bar{q}(t)]} \tag{5.149}$$

The film mass transfer coefficient k_F can be determined from the initial part of the kinetic curve as described in Section 5.4.2. For a given concentration $c(t)$, the related $c_S(t)$ can be found from the mass transfer equation for film diffusion and the slope of the kinetic curve at time t

$$c_S(t) = c(t) + \frac{V_L}{m_A\, k_F\, a_m} \left(\frac{\partial c}{\partial t}\right)_t \tag{5.150}$$

The adsorbent loadings $q_S(t)$ and $\bar{q}(t)$ are calculated from the isotherm and the material balance equation as shown before; a_m is available from the equation given in Table 5.1 or Table 5.2. The intraparticle mass transfer coefficient k_S^* can be found from eq. (5.127).

The intraparticle mass transfer coefficient, k_S^*: Influence factors and prediction methods

The relationships between k_S^* and the intraparticle diffusion coefficients (eqs. (5.128) and (5.144)) discussed previously imply that k_S^* (and therefore also k_S) depends on the same influence factors as these diffusion coefficients, in particular on temperature, molecule size, and particle radius. Furthermore, a dependence on concentration or loading may occur as a result of a combined mechanism (see eq. (5.143)) or as a result of a loading dependence of D_S. From a systematic study on the influence factors carried out with a number of different adsorbates and activated carbons, the following empirical equation was found (Worch 2008):

$$k_S^* = 0.00129 \sqrt{\frac{D_L\, c_0}{q_0\, r_P^2}} \tag{5.151}$$

The empirical factor in eq. (5.151) is valid under the condition that the following units are used: m^2/s for D_L, mg/L for c_0, mg/g for q_0, and m for r_P. The unit of the resulting mass transfer coefficient is 1/s. The correlation is depicted in Figure 5.20. Equation (5.151) together with the equivalence relationships (eqs. (5.128) and (5.144))

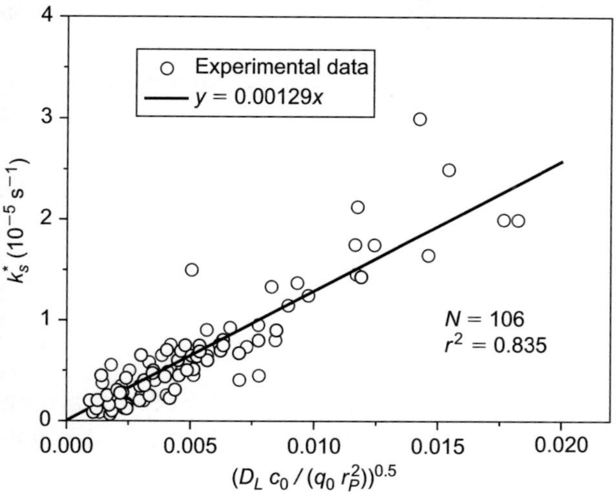

Figure 5.20: Correlation of internal mass transfer coefficients with adsorbate and adsorbent parameters according to eq. (5.151).

can also be used to predict the respective diffusion coefficients. For estimation of D_L, required in eq. (5.151), see Table 7.8 in Chapter 7.

While this correlation gives reasonable results for the adsorption of defined micropollutants onto activated carbon, it fails for natural organic matter (NOM) adsorption. For NOM, the mass transfer coefficients calculated by eq. (5.151) are typically too high. Instead of using eq. (5.151), a rough estimate of an average mass transfer coefficient for NOM fractions can be found from the simple correlation

$$k_S^* = a + b\frac{c_0}{r_P^2} \tag{5.152}$$

which was established on the basis of batch and column experiments with NOM-containing water samples from different sources. Here, c_0 is the total concentration of all adsorbable NOM fractions. The empirical parameters were found to be $a = 3 \cdot 10^{-6}$ 1/s and $b = 3.215 \cdot 10^{-14}$ (m^2 L)/(mg s).

5.4.7 Reaction kinetic models

Although diffusion models are widely accepted as appropriate models to describe adsorption kinetics for porous adsorbents, a number of papers have been published in recent years in which the adsorption kinetics is described by simple models based on chemical reaction kinetics. These papers deal with the adsorption of different adsorbates (not only organic substances but also heavy metals) mainly onto alternative adsorbents (biosorbents, low-cost adsorbents) but also onto activated carbon. Although these models must be viewed very critically for several reasons, for the sake of completeness they will be briefly presented here.

Under the assumption that the adsorbate uptake follows a first-order rate law, the adsorption kinetics can be described by

$$\frac{d\bar{q}}{dt} = k_1 (q_{eq} - \bar{q}) \tag{5.153}$$

where k_1 is the first-order rate constant. Integration with the condition $\bar{q} = 0$ at $t = 0$ gives

$$\ln \frac{(q_{eq} - \bar{q})}{q_{eq}} = -k_1 t \tag{5.154}$$

The first-order rate constant can be easily estimated by plotting the data according to eq. (5.154). The equilibrium loading is available from the isotherm and the material balance. The frequently mentioned alternative equation

$$\ln(q_{eq} - \bar{q}) = \ln q_{eq} - k_1 t \tag{5.155}$$

is inappropriate because here q_{eq} is both a fitting parameter and part of the dependent variable.

At first view, eq. (5.153) seems to be identical with (eq. (5.126)) of the LDF model, but there is a decisive difference in the driving forces. In eq. (5.153), the driving force is expressed as the difference between the (final) equilibrium loading (which is constant for a given initial concentration and adsorbent dose) and the loading at time t, whereas the driving force in the LDF model is given by the difference between the equilibrium adsorbent loading at the outer surface at time t (related to the concentration at the outer surface at time t) and the mean adsorbent loading at the same time, t. Thus, in the reaction kinetic model, only the mean adsorbent loading is time dependent; whereas in the LDF model, both loadings, q_s and $\bar{q}$, change with time (Figure 5.21). Consequently, kinetic curves calculated with these models are slightly different.

Another reaction kinetic approach is based on a pseudo second-order rate law. The basic equation is

$$\frac{d\bar{q}}{dt} = k_2 (q_{eq} - \bar{q})^2 \tag{5.156}$$

where k_2 is the second-order rate constant. Integration with the initial condition $\bar{q} = 0$ at $t = 0$ gives

$$\frac{t}{\bar{q}} = \frac{1}{k_2 q_{eq}^2} + \frac{t}{q_{eq}} \tag{5.157}$$

To find the rate constant, a linear regression according to eq. (5.157) has to be carried out. If q_{eq} is given from the isotherm, it is appropriate to run the linear regression with t/q_{eq} as the independent variable and with the fixed slope of 1.

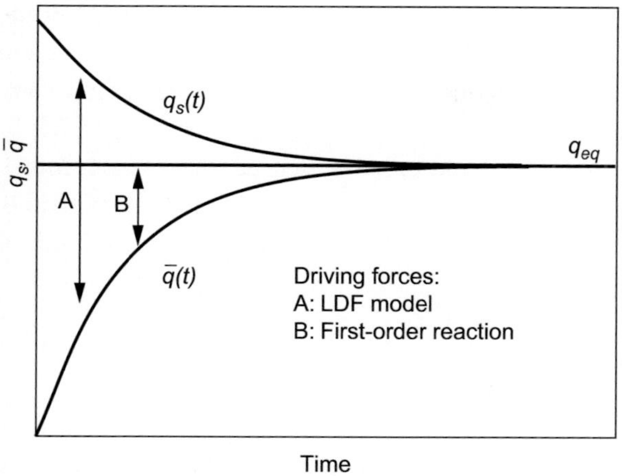

Figure 5.21: Linear driving forces in the LDF model and in the first-order reaction kinetic model.

As already mentioned, the reaction kinetic models have to be viewed critically. From long-term experience, it is well known that intraparticle diffusion (surface and/or pore diffusion) plays an important role in adsorption kinetics and is typically rate limiting for adsorption onto porous adsorbents. It is also a widely accepted assumption that the final adsorption step is much faster than the previous adsorbate transport by diffusion. A strong argument for considering adsorption kinetics as diffusion controlled is the general validity of the diffusion approaches, in particular the possibility of estimating diffusion coefficients in separate batch experiments and applying these coefficients and the related diffusion models to describe breakthrough behavior in fixed-bed adsorbers. On the contrary, the reaction kinetic models described previously were not shown to be applicable to conditions other than those present in the studied batch adsorption process. Furthermore, no theoretically founded relations of the rate coefficients to process conditions are known. Moreover, the intraparticle diffusion models can be extended to account for an additional impact of film diffusion, whereas no such extension has been reported for reaction kinetics.

It follows from general theoretical considerations that reaction kinetic models are reasonable, if any, only for weakly porous adsorbents where slow surface reactions (chemisorption) play a major role and film diffusion resistance does not exist.

Even if the reaction kinetic equations are able to describe the batch experiments with porous adsorbents satisfactorily, the weak theoretical background of the models in view of the adsorption process and the missing transferability of the rate constants to other conditions make them empirical equations. This empirical character has to be taken into account if these approaches are to be applied. The increasing application of these reaction kinetic models in practice-oriented adsorption studies is possibly attributed to their much simpler structure in comparison to the exact diffusion models.

It has to be noted that the reaction kinetic models are often compared with a simplified pore diffusion equation (linear relationship between q and $t^{0.5}$) in order to demonstrate that diffusion is not relevant in the considered system. However, this linear relationship is an approximate solution to the diffusion model, which is valid only for the initial stage of the adsorption process. Consequently, a nonlinear run of the q–$t^{0.5}$ plot over a broader time interval can also be expected for diffusion-controlled processes and is therefore not evidence for missing intraparticle diffusion impact on adsorption kinetics.

5.4.8 Adsorption kinetics in multicomponent systems

In principle, the adsorption of a considered component in a multicomponent system can be influenced by the other components as a result of

– competition for the existing adsorption sites (equilibrium effect).
– interactions during transport to the adsorption sites (kinetic effect).

While the influence of competing components on the adsorption equilibrium can be easily proved and quantified by appropriate models (Chapter 4), the evaluation of the impact of cosolutes on adsorption kinetics is more complicated. Typically, kinetic models include simplifying assumptions, and the possible effect of adsorbate interactions during the adsorption process may be masked by model uncertainties.

For film diffusion, it is generally assumed that the mass transfer of the individual components proceeds independently of each other. Therefore the single-solute model equations as well as the single-solute film mass transfer coefficients can be used unmodified to describe the film diffusion in multisolute systems. The competition effect is considered only by using multisolute equilibrium relationships.

In the case of intraparticle diffusion, interactions between the adsorbates during the mass transfer cannot be excluded. To account for these interactions, the single-solute diffusion equations for surface or pore diffusion have to be extended

$$\dot{n}_{S,i} = \rho_P \sum_{j=1}^{N} D_{S,ij} \frac{\partial q_j}{\partial r} \tag{5.158}$$

$$\dot{n}_{P,i} = \sum_{j=1}^{N} D_{P,ij} \frac{\partial c_{p,j}}{\partial r} \tag{5.159}$$

Taking a bisolute system with surface diffusion as an example, the respective diffusion equations are

$$\dot{n}_{S,1} = \rho_P \left(D_{S,11} \frac{\partial q_1}{\partial r} + D_{S,12} \frac{\partial q_2}{\partial r} \right) \tag{5.160}$$

$$\dot{n}_{S,2} = \rho_P \left(D_{S,22} \frac{\partial q_2}{\partial r} + D_{S,21} \frac{\partial q_1}{\partial r} \right) \tag{5.161}$$

Hence, the flux of a component in a bisolute system depends not only on its own solid-phase concentration gradient but also on the solid-phase concentration gradient of the second component. A complete bisolute kinetic model would consist of the diffusion equations (eqs. (5.160) and (5.161)), the material balance equations for both components, and the equilibrium model (ideal adsorbed solution theory [IAST] or mixture isotherm). While the diffusion coefficients of the single components, $D_{S,11}$ and $D_{S,22}$, can be found from single-solute kinetic experiments, the cross coefficients, $D_{S,12}$ and $D_{S,21}$, must be determined from experiments with the bisolute adsorbate system.

If the number of components exceeds $N = 2$, the experimental effort to determine the cross coefficients becomes unacceptably high. Moreover, the mathematical effort to solve the equations of such a mixture adsorption kinetic model is much higher than for a single-solute adsorption model.

There are different approaches to dealing with this problem. The most rigorous simplification is to assume that the cross coefficients are much lower than the diffusion coefficients for single-solute adsorption and therefore can be neglected

$$D_{P,ij}(i \neq j) = 0 \tag{5.162}$$

$$D_{S,ij}(i \neq j) = 0 \tag{5.163}$$

Under this assumption, eqs. (5.158) and (5.159) simplify to the diffusion equations for single-solute adsorption, and only kinetic experiments in single-solute systems have to be carried out to find the diffusion coefficients. In this case, only the equilibrium relationship accounts for competitive adsorption effects.

An alternative approach is to apply the single-solute diffusion equations to kinetic curves measured in multisolute systems and to determine apparent diffusion coefficients, which include possible deviations caused by interaction effects. The apparent coefficients can be compared to coefficients found from single-solute experiments to verify whether there are interactions during the adsorbate transport into the particle or not. If there are interactions, the validity of the apparent diffusion coefficients is limited to the conditions of the kinetic experiment. In the opposite case, the kinetics in the multisolute system can be described on the basis of single-solute coefficients and single-solute diffusion equations without any limitations.

A simplified approach must also be used for unknown multicomponent systems such as DOM. In this case, an adsorption analysis (Chapter 4, Section 4.7.2) has to be carried out prior to the kinetic experiments. The kinetic curves can then be modeled on the basis of the fictive component approach and by using single-solute diffusion equations for all fictive components.

5.5 Practical aspects: Slurry adsorber design

In the previous sections, it was shown how characteristic mass transfer or diffusion coefficients can be estimated by applying kinetic models to experimental data determined in batch experiments. In most cases, these experiments are carried out with the aim of determining characteristic kinetic parameters that can be used to predict the adsorption behavior of solutes in slurry or fixed-bed adsorbers. Since fixed-bed adsorption modeling is the subject of Chapter 7, only slurry reactors will be considered here.

In view of slurry reactor modeling, a distinction has to be made between completely mixed batch reactors (CMBRs) and completely mixed flow-through reactors (CMFRs). In practice, the design of batch reactors is mostly based on equilibrium relationships and material balances as already shown in Chapters 3 and 4. If the contact time is too short to reach the equilibrium state, the kinetic models described in the previous sections together with the experimentally determined mass transfer or diffusion coefficients can be used to find the residual concentration for a given contact time. In this case, the calculation methods are the same as used for estimating the mass transfer parameters in kinetic experiments.

Alternatively, short-term isotherms for the given contact time can be applied as the basis for a simplified process modeling (Chapter 4, Section 4.8.4).

In the case of flow-through reactors, the specific process conditions, which are quite different from batch reactors, have to be taken into account. In contrast to the batch reactor where the concentration decreases during the adsorption process, the concentration in the flow-through reactor is constant over the time if steady-state conditions can be assumed. The steady-state concentration is equal to the outlet concentration.

For an ideal CMFR, the mean constituent residence time is equal to the mean hydraulic residence time, $\bar{t}_r$, which is given by the ratio of reactor volume, V_R, and volumetric flow rate, $\dot{V}$,

$$\bar{t}_r = \frac{V_R}{\dot{V}} \tag{5.164}$$

The material balance equation for the CMFR can be derived from eq. (3.66) by setting $c_0 = c_{in}$ and $q_0 = 0$ and by replacing c_{eq} by the steady-state concentration, c_{out},

$$\bar{q} = \frac{\dot{V}}{\dot{m}_A}(c_{in} - c_{out}) \tag{5.165}$$

where $\dot{m}_A$ is the mass of adsorbent added to the reactor per time unit. The mean fractional uptake, $\bar{F}$, is then given by

$$\bar{F} = \frac{\bar{q}}{q_{eq}} = \frac{\dot{V}}{\dot{m}_A}\frac{(c_{in} - c_{out})}{q_{eq}} \tag{5.166}$$

where q_{eq} is the equilibrium adsorbent loading related to the steady-state concentration, c_{out}. From eq. (5.166), the following general reactor design equation can be derived:

$$c_{out} = c_{in} - q_{eq} \frac{\dot{m}_A}{\dot{V}} \bar{F} \tag{5.167}$$

To solve eq. (5.167), an expression for the mean fractional uptake is required. A simple solution can be obtained for surface diffusion if the fractional uptake at constant concentration at the external adsorbent surface is expressed by eq. (5.71) (Boyd's equation). Considering the age distribution of the particles, the average fractional uptake can be found from

$$\bar{F} = \int_0^\infty F \frac{e^{-t/\bar{t}_r}}{\bar{t}_r} \mathrm{d}t \tag{5.168}$$

Introducing eq. (5.71) into eq. (5.168) and solving the integral gives (Traegner et al. 1996)

$$\bar{F} = 1 - \frac{6}{\pi^2} \sum_{n=1}^{\infty} \frac{1}{n^2 \, (n^2 \, \pi^2 \, T_B + 1)} \tag{5.169}$$

with

$$T_B = \frac{D_S \, \bar{t}_r}{r_P^2} \tag{5.170}$$

Finally, the equilibrium loading, q_{eq}, in eq. (5.167) has to be replaced by the respective isotherm equation – for instance, by the Freundlich isotherm. In this case, the resulting design equation reads

$$c_{out} = c_{in} - K \, c_{out}^n \frac{\dot{m}_A}{\dot{V}} \left[1 - \frac{6}{\pi^2} \sum_{n=1}^{\infty} \frac{1}{n^2 \, (n^2 \, \pi^2 \, T_B + 1)} \right] \tag{5.171}$$

Figure 5.22 illustrates the influence of adsorbent particle size and adsorbent dosage on the achievable outlet concentration by means of model calculations for a fictive adsorbate ($K = 50$ [mg/g]/[mg/L]^n, $n = 0.4$, $c_{in} = 1$ µg/L, $D_S = 1 \cdot 10^{-14}$ m^2/s). As expected, the efficiency of the adsorption process for a given mean residence time in the CMFR is higher the smaller the particles are and the higher the adsorbent dosage is. It has to be noted that a constant surface diffusion coefficient was used in the calculations. If D_S increases with decreasing particle diameter as can be expected as a result of shorter diffusion paths, the positive effect of smaller particles would be even stronger.

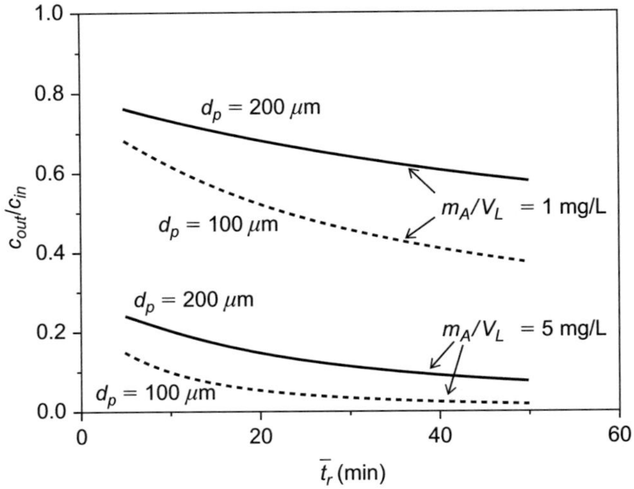

Figure 5.22: Influence of adsorbent particle size and adsorbent dosage on the outlet concentration of a CMFR. Model calculations based on the surface diffusion model with Freundlich isotherm.

6 Adsorption dynamics in fixed-bed adsorbers

6.1 Introduction

For engineered adsorption processes, besides slurry reactors, fixed-bed adsorbers are frequently used. In contrast to slurry reactors, which are appropriate to the application of powdered adsorbents in particular, fixed-bed adsorbers (or adsorption filters) are suitable for granular adsorbents. In comparison to adsorption in slurry reactors, the fixed-bed adsorption process is more complex. In this chapter, some general aspects and basic principles of fixed-bed adsorption are discussed, whereas fixed-bed adsorber modeling and design is the subject matter of Chapter 7.

Adsorption in a fixed-bed adsorber is a time- and distance-dependent process. During the adsorption process, each adsorbent particle in the bed accumulates adsorbate from the percolating solution as long as the state of equilibrium is reached. This equilibration process proceeds successively, layer by layer, from the column inlet to the column outlet. However, due to the slow adsorption kinetics, there is no sharp boundary between loaded and unloaded adsorbent layers. Instead of that, the equilibration takes place in a more or less broad zone of the adsorbent bed, referred to as the mass transfer zone (MTZ) or adsorption zone. This MTZ is characterized by typical concentration and adsorbent loading profiles.

In the case of single-solute adsorption, at a given time, a distinction can be made between three different zones within the adsorbent bed (Figure 6.1).

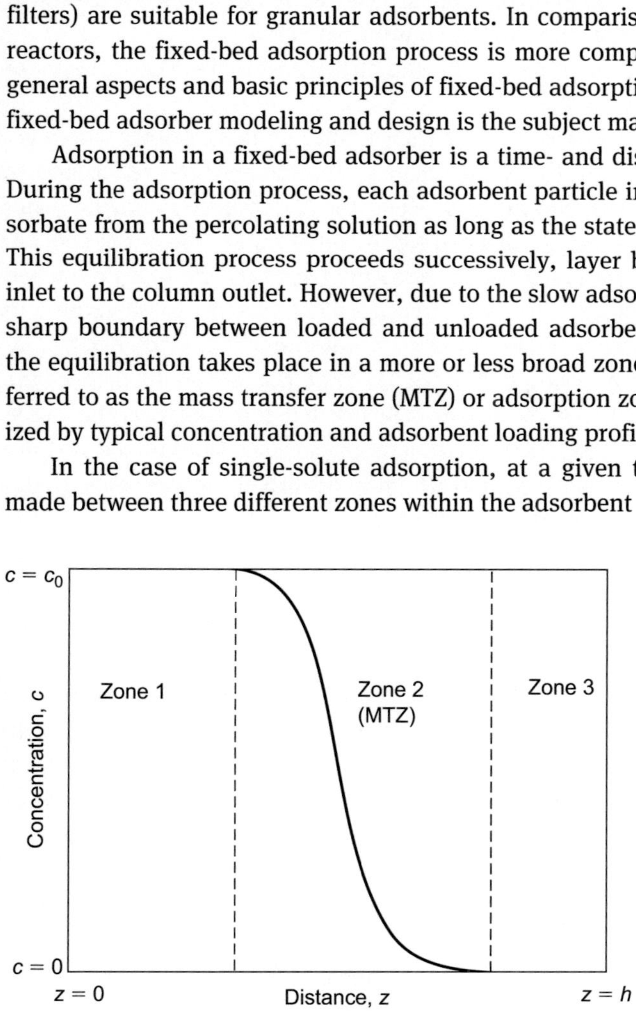

Figure 6.1: Concentration profile during single-solute adsorption in a fixed-bed adsorber of bed height h.

In the first zone between the adsorber inlet and the MTZ, the adsorbent is already loaded with the adsorbate to the adsorbed amount, q_0, which is in equilibrium

https://doi.org/10.1515/9783110715507-006

with the inlet concentration, c_0. The available adsorption capacity in this zone is exhausted, and no more mass transfer from the liquid phase to the adsorbent particles takes place. Therefore, the concentration in the liquid phase is constant and equals c_0.

In the second zone (MTZ), the mass transfer from the liquid phase to the solid phase just takes place. Due to the mass transfer from the liquid to the solid phase, the concentration in this zone decreases from $c = c_0$ to $c = 0$, and the adsorbed amount increases from $q = 0$ to $q = q_0(c_0)$. The shape and length of the MTZ depend on the adsorption rate and the shape of the equilibrium curve.

The adsorbent in the third zone is still free of the adsorbate. The fluid-phase concentration in this zone is $c = 0$.

During the adsorption process, the MTZ travels through the adsorber with a velocity that is much slower than the water velocity. The stronger the adsorption of the adsorbate, the greater the difference between the MTZ velocity and the water velocity. As long as the MTZ has not reached the adsorber outlet, the outlet concentration is $c = 0$. The adsorbate occurs in the adsorber outlet for the first time when the MTZ reaches the end of the adsorber. This time is referred to as breakthrough time, t_b. After the breakthrough time, the concentration in the adsorber outlet increases due to the progress of adsorption in the MTZ and the related decrease of the remaining adsorbent capacity. If the entire MTZ has left the adsorber, the outlet concentration equals c_0. At this point, all adsorbent particles in the fixed bed are saturated to the equilibrium loading, and no more adsorbate uptake takes place. The related time is referred to as saturation time, t_s.

The concentration versus time curve, which is measurable at the adsorber outlet, is referred to as the breakthrough curve (BTC). The BTC is a mirror of the MTZ and is therefore affected by the same factors, in particular adsorption rate and shape of the equilibrium curve. The position of the BTC on the time axis depends on the traveling velocity of the MTZ, which in turn depends on the flow velocity and, as mentioned previously, on the strength of adsorption. For a given flow velocity, the better adsorbable the solute is, the later the breakthrough occurs. The relation between the traveling of the MTZ and the development of the BTC is schematically shown in Figure 6.2.

The spreading of the MTZ is mainly determined by the mass transfer resistances. In principle, dispersion also leads to a spreading of the MTZ, but this effect is usually negligible under the typical conditions of engineered fixed-bed adsorber processes.

In the limiting case of infinitely fast mass transfer processes and missing dispersion, the length of the MTZ reduces to zero and the sigmoid BTC becomes a concentration step. The concentration step is referred to as ideal BTC, and the time after that the concentration step occurs is termed “ideal breakthrough time”, t_b^{id}. Each real BTC can be approximated by a related ideal BTC. As required by the material balance (Section 6.4.2), the ideal BTC must intersect the related real BTC at its center of mass.

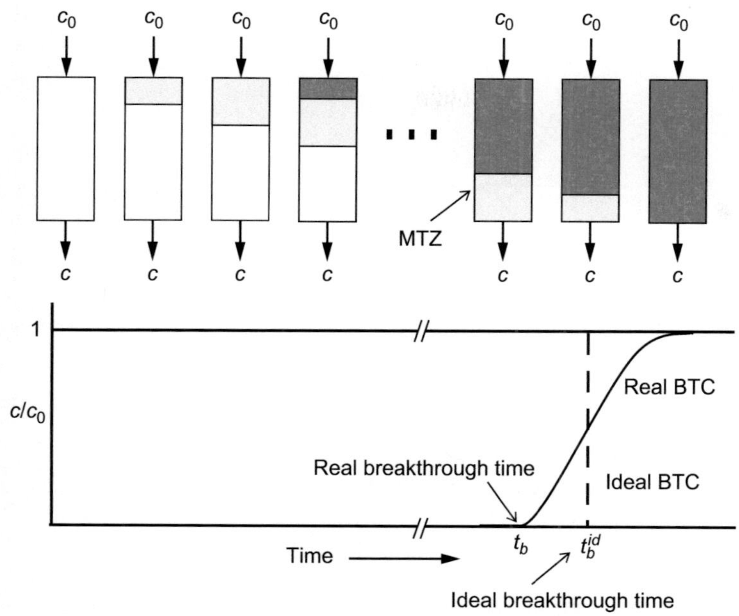

Figure 6.2: Traveling of the mass transfer zone (MTZ) through the adsorber bed and development of the breakthrough curve (BTC).

From the previous general considerations, two important advantages of fixed-bed adsorption in comparison to adsorption in batch reactors can be derived.

- While in batch reactors the mass transfer driving force, and therefore also the adsorption rate, decreases during the process due to the decreasing concentration in the reactor, the adsorbent in the fixed-bed adsorber is always in contact with the inlet concentration, c_0, which results in a high driving force over the whole process.
- In a batch reactor, very low residual concentrations can only be achieved if very high adsorbent doses are applied. In contrast, in a fixed-bed adsorber, the adsorbate will be completely removed until the breakthrough occurs.

So far, only single-solute adsorption was considered. In the case of a multisolute adsorbate system, individual MTZs for all components occur, which travel – according to their different adsorption strengths – with different velocities through the adsorbent bed. As a result, displacement processes take place leading to quite different breakthrough behavior in comparison to single-solute adsorption.

Figure 6.3 shows the breakthrough behavior of a two-component system. As a typical result of competition and displacement, a concentration overshoot can be observed for the weaker adsorbable component 1. Since the traveling velocity of the MTZ depends on the adsorption strength, the MTZ of the weaker adsorbable component 1

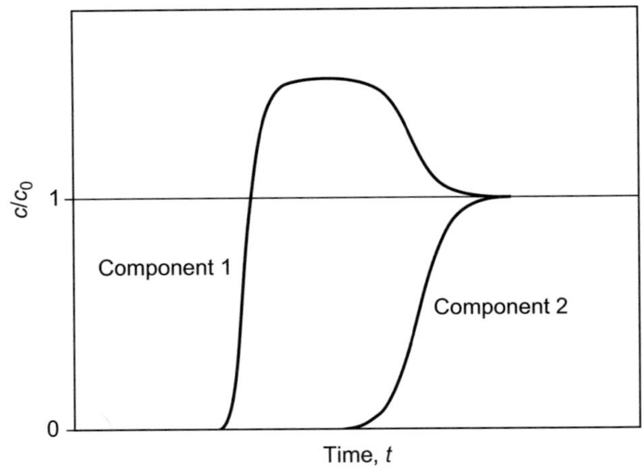

Figure 6.3: Breakthrough curves of a bisolute adsorbate system. Component 1: weaker adsorbable; component 2: stronger adsorbable.

travels faster through the adsorber. It always reaches the layers of fresh adsorbent as the first component and is therefore adsorbed in these layers as a single solute. Later, when the stronger adsorbable component 2 reaches the same layers, a new (bisolute) equilibrium state is established. This is connected with partial displacement of the previously adsorbed component 1. The displaced amount of component 1 equals the difference between the equilibrium adsorbent loadings in single-solute and bisolute adsorption. As a result of this displacement process, the concentration of component 1 in the region between both MTZs is higher than its initial concentration. If the difference between the MTZ velocities of the components is large enough, a plateau zone with a constant concentration of $c_{pl,1} > c_{0,1}$ will occur.

An analogous behavior can be observed for adsorbate mixtures with more than two components. In multicomponent systems, except for the strongest adsorbable component, all other components are subject to displacement processes. As an example, Figure 6.4 shows the BTC of a three-component system. Here, the first component shows two concentration plateaus located above the inlet concentration; the first results from the displacement by component 2, and the second from the displacement by component 3. Component 2 shows one plateau concentration as a result of displacement by component 3.

Generally, in a system consisting of N components, the number of plateau zones, P, that can be expected for each component is given by

$$P = N - C \tag{6.1}$$

where C is the ordinal number of the component in order of increasing adsorbability. In the example given previously, the number of plateaus, P, for component 1 is given by $P = 3 - 1 = 2$, for component 2 by $P = 3 - 2 = 1$, and for component 3 by $P = 3 - 3 = 0$.

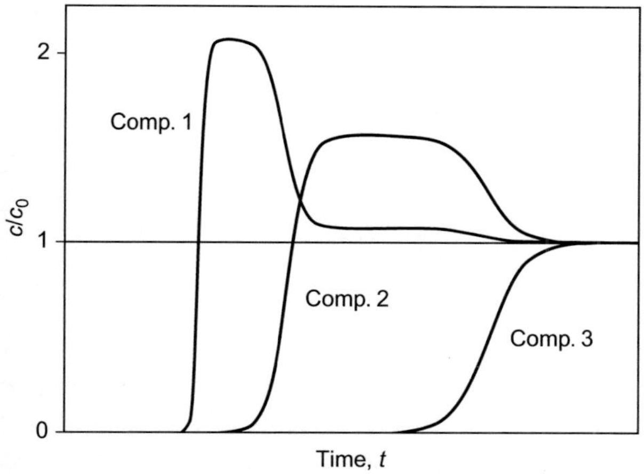

Figure 6.4: Breakthrough curves of a three-component adsorbate mixture.

The different MTZ velocities and the resulting displacement processes are also reflected in the total BTCs that can be obtained by the addition of the BTCs of the mixture constituents. Such total BTCs are characterized by concentration steps as shown exemplarily in Figure 6.5 for a three-component system.

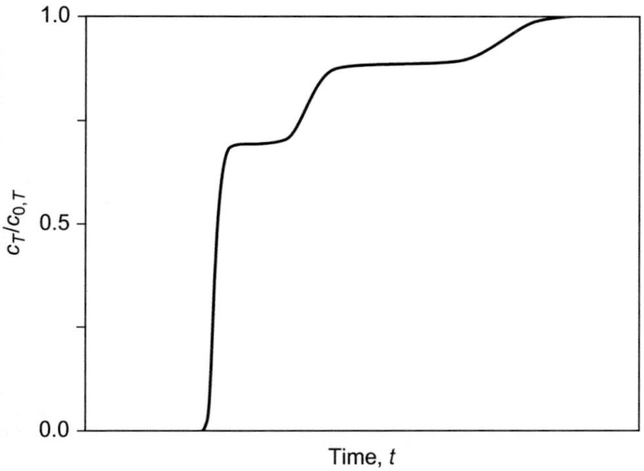

Figure 6.5: Total (summary) breakthrough curve of a three-component adsorbate mixture.

It has to be noted that the plateau zones in the component BTCs and the steps in the total BTC are only fully developed if the MTZs are completely separated. This is a special case that does not necessarily occur in practice. Frequently, the MTZs of

the components overlap. In particular, overlapping can be expected if one or several of the following factors apply:
- Short adsorber
- High flow velocity
- Small differences in the adsorption strengths and therefore also in the MTZ velocities
- Large number of components
- Slow adsorption processes (broad MTZs)

Figure 6.6 shows the BTCs of a three-component system with overlapping MTZs. A typical effect of MTZ overlapping is that the breakthrough of the better adsorbable component occurs before the concentration plateau of the displaced component is fully established. Consequently, the concentration overshoot is not as high as in the case of completely separated MTZs. Furthermore, the concentration steps in the corresponding total BTC are not so clearly visible (Figure 6.7).

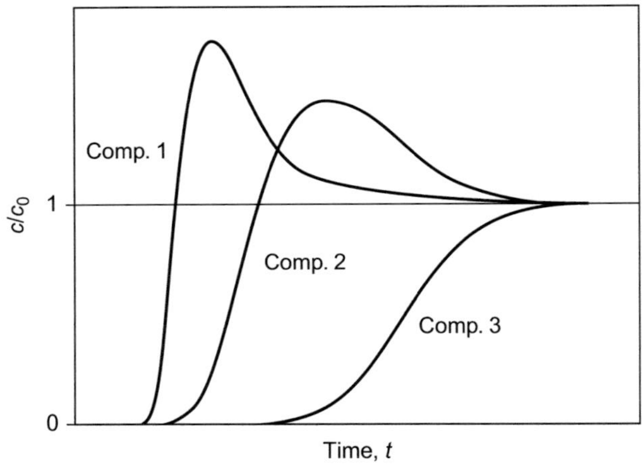

Figure 6.6: Breakthrough curves of a three-component adsorbate mixture with overlapping mass transfer zones.

For multicomponent systems of unknown composition, such as natural organic matter or effluent organic matter, only total BTCs can be measured, typically by using the collective parameter dissolved organic carbon (DOC). Since the MTZs of the components in such a multicomponent system normally overlap, a DOC BTC looks similar to the curve shown in Figure 6.7 with the exception that the curve typically starts at a concentration higher than zero as a consequence of the existence of a nonadsorbable fraction (see Section 4.7 in Chapter 4) that breaks through instantaneously.

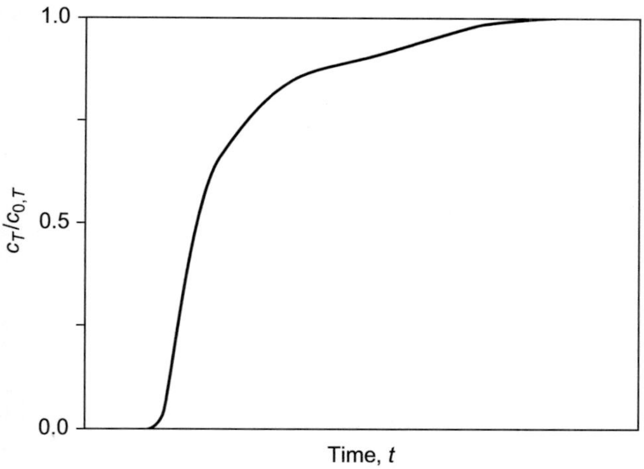

Figure 6.7: Total (summary) breakthrough curve of a three-component adsorbate mixture with overlapping mass transfer zones.

6.2 Experimental determination of breakthrough curves

There are different reasons to determine fixed-bed adsorber BTCs experimentally in the laboratory. Experimental BTCs are necessary to verify the applicability of a chosen adsorption model for a given adsorbent/adsorbate system and to estimate the related mass transfer coefficients, in case they are not known from separate kinetic measurements. The model as well as the related kinetic parameters can then be used to predict the adsorption behavior in a full-size adsorber. Another objective of BTC determination could be the application of a scale-up method. Simple scale-up methods are used if full-size adsorbers should be designed without the application of a complicated BTC model (Chapter 7, Section 7.2).

Laboratory-scale fixed-bed adsorbers are typically made of glass or stainless steel. To avoid wall effects, which could influence the shape of the BTC, the ratio of column diameter (d_R) and particle diameter (d_P) should not be too low (d_R:d_P > 10 is recommended). In laboratory-scale fixed-bed adsorbers, the flow direction is typically from the bottom to the top of the column. That ensures a uniform streaming and avoids channeling. A solution reservoir, an adjustable pump, a flow meter as well as sampling points before and after the column complete the experimental setup (Figure 6.8).

The BTC measurement is carried out by taking samples at the column outlet after defined time intervals and by subsequently determining the adsorbate concentration. Besides the determination of the outlet concentration, a periodic control of the initial concentration is essential.

In order to analyze the experimental BTC or to fit the experimental data by means of a BTC model, a number of process parameters have to be known. The most

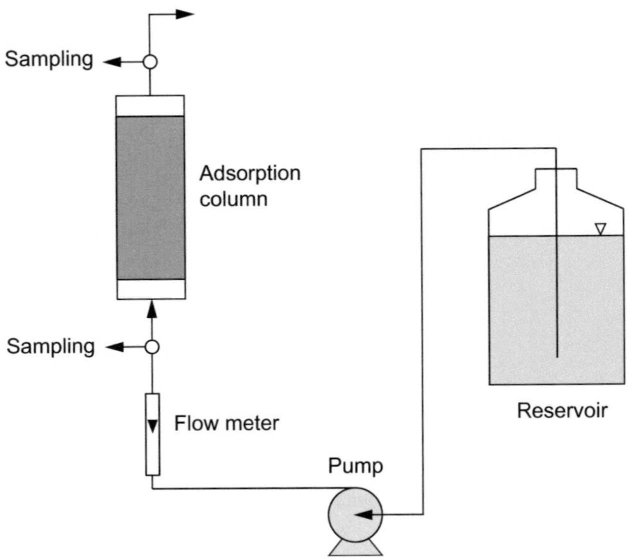

Figure 6.8: Experimental setup for breakthrough curve determination.

important parameters, besides the initial concentration and the flow rate, are the adsorbent mass, the mean adsorbent particle diameter, the adsorbent density, the bed density, and the bed porosity.

6.3 Fixed-bed process parameters

In fixed-bed adsorption models, a number of different process parameters are used to characterize the adsorbent bed and the flow conditions within the bed. Below, the most important process parameters will be defined and relations between these parameters will be presented (see also Chapter 2, Section 2.5).

Usually, the adsorbent bed is characterized by the parameters bed density, ρ_B; bed porosity, ε_B; adsorbent mass, m_A; adsorbent volume, V_A; and adsorber volume, V_R.

The bed density, ρ_B, is defined as the ratio of the adsorbent mass in the reactor and the volume of the reactor. The adsorber volume can be expressed as the sum of the volume of the adsorbent particles, V_A, and the liquid-filled void volume, V_L. It has to be noted that the term "reactor volume" is here used in the sense of volume of the adsorbent bed

$$\rho_B = \frac{m_A}{V_R} = \frac{m_A}{V_A + V_L} \tag{6.2}$$

In contrast, the particle density is given by

$$\rho_P = \frac{m_A}{V_A} \tag{6.3}$$

The bed porosity is the void fraction of the reactor volume

$$\varepsilon_B = \frac{V_L}{V_R} = \frac{V_R - V_A}{V_R} = 1 - \frac{V_A}{V_R} \tag{6.4}$$

Substituting the volumes V_A and V_R in eq. (6.4) by the densities given in eqs. (6.2) and (6.3), the bed porosity can also be expressed as

$$\varepsilon_B = 1 - \frac{\rho_B}{\rho_P} \tag{6.5}$$

The adsorber volume, V_R, can be written as the product of the cross-sectional area, A_R, and the height of the adsorber (more precisely, the height of the adsorbent bed), h,

$$V_R = A_R\, h \tag{6.6}$$

Together with eq. (6.2), a relationship between adsorbent mass and adsorber height can be derived

$$m_A = V_R\, \rho_B = A_R\, h\, \rho_B \tag{6.7}$$

The relationship between the adsorbent volume, V_A, and the adsorbent particle radius, r_P, is given by

$$V_A = Z_T \frac{4\, \pi\, r_P^3}{3} \tag{6.8}$$

where Z_T is the total number of the (spherical) adsorbent particles in the adsorbent bed. Equation (6.8) can be used to find an equivalent radius for cylindrical or irregularly shaped adsorbent particles. Taking into consideration eq. (6.3) and after setting $Z_T = m_A\, Z_{sp}$, the following equation results

$$r_P = \sqrt[3]{\frac{3}{4\, \pi\, \rho_P\, Z_{sp}}} \tag{6.9}$$

where Z_{sp} is the average number of particles per gram adsorbent (specific number of particles).

As already discussed in Chapter 5, the external surface area of the adsorbent particles is an important parameter in the mass transfer equations for both film and intraparticle diffusion. The total surface area in the adsorbent bed is given by

$$A_s = Z_T\, 4\, \pi\, r_P^2 \tag{6.10}$$

Equating eqs. (6.8) and (6.10) gives

$$A_S = \frac{3\,V_A}{r_P} \tag{6.11}$$

In the mass transfer equations used in fixed-bed adsorber models, the adsorbent surface area is often related to the reactor volume (see also Chapter 2, Section 2.5.3)

$$a_{VR} = \frac{A_S}{V_R} = \frac{3\,V_A}{V_R\,r_P} = \frac{3}{r_P}(1-\varepsilon_B) \tag{6.12}$$

Flow velocity and residence time are other important process parameters. For these parameters, different definitions are in use. The linear filter velocity (superficial velocity), v_F, is given as the quotient of the volumetric flow rate, $\dot{V}$, and the cross-sectional area of the adsorber, A_R,

$$v_F = \frac{\dot{V}}{A_R} \tag{6.13}$$

This formal definition is based on the assumption of an empty adsorber where the total cross-sectional area is available for the water flow. In reality, the adsorber is filled with adsorbent particles, and only the void fraction of the bed is available for the water flow. Therefore, at the same volumetric flow rate, the effective velocity in the bed must be higher than the filter velocity, v_F. The effective flow velocity (interstitial velocity), u_F, can be found by dividing the volumetric flow rate by the cross-sectional area available for water flow, which is given by the product of the cross-sectional area, A_R, and the bed porosity, ε_B,

$$u_F = \frac{\dot{V}}{A_R\,\varepsilon_B} = \frac{v_F}{\varepsilon_B} \tag{6.14}$$

Generally, the residence time can be calculated from the flow velocity and the adsorber height. According to the different definitions for the flow velocity, two different residence times can be defined. The residence time for an empty reactor is referred to as empty bed contact time, *EBCT*, and given by

$$EBCT = \frac{h}{v_F} = \frac{h\,A_R}{\dot{V}} = \frac{V_R}{\dot{V}} \tag{6.15}$$

The effective residence time, t_r, related to the effective flow velocity, u_F, is lower than the *EBCT* and given by

$$t_r = \frac{h}{u_F} = \frac{h\,A_R\,\varepsilon_B}{\dot{V}} = \frac{V_R\,\varepsilon_B}{\dot{V}} = EBCT\,\varepsilon_B \tag{6.16}$$

For presentation of BTCs, it may sometimes be advantageous to use a time-proportional relative parameter instead of the absolute time elapsed – for instance, in order to normalize the BTCs with respect to the bed size. The throughput given as the number of bed volumes fed to the adsorber, BV, is such a parameter that is frequently used in practice. The dimensionless parameter BV is defined as the volume of water fed to the adsorber (V_{Feed} = volumetric flow rate × time) divided by the bed volume of the adsorber. According to eq. (6.15), BV can also be expressed as time divided by the $EBCT$,

$$BV = \frac{V_{Feed}}{V_R} = \frac{\dot{V}\,t}{V_R} = \frac{t}{EBCT} \tag{6.17}$$

Instead of BV, sometimes the specific throughput, V_{sp}, is used as a normalized parameter to describe the adsorber runtime. In this case, the volume fed to the adsorber is related to the adsorbent mass. The unit is therefore L water treated per kg adsorbent,

$$V_{sp} = \frac{V_{Feed}}{m_A} = \frac{\dot{V}\,t}{m_A} = \frac{\dot{V}\,t}{V_R\,\rho_B} = \frac{t}{EBCT\,\rho_B} = \frac{BV}{\rho_B} \tag{6.18}$$

The specific throughput until the breakthrough, $V_{sp,b}$, can be used to quantify the performance of an adsorber. $V_{sp,b}$ specifies, which volume of water can be treated until the time of breakthrough at the treatment objective is reached. It can be found from eq. (6.18) by setting the time equal to the breakthrough time ($t = t_b$). Often, also the reciprocal of $V_{sp,b}$ – the carbon usage rate (CUR) – is used as a performance parameter ($CUR = 1/V_{sp,b}$).

6.4 Material balances

6.4.1 Types of material balances

Material balances are essential relationships for describing adsorption processes in fixed-bed adsorbers. They can be written in integral or in differential form.

Integral material balances are established for the entire adsorber and allow deriving simple BTC models in which adsorption kinetics is neglected. Furthermore, integral material balance equations provide the mathematical basis for determination of both breakthrough and equilibrium adsorbent loadings from experimental BTCs.

On the contrary, differential material balance equations are established for a differential adsorbent layer. They are necessary to formulate more sophisticated BTC models that include adsorption kinetics.

6.4.2 Integral material balance

Prior to the derivation of mass balance equations for the fixed-bed adsorber, it is necessary to recall the relationship between ideal and real BTCs. As shown in Section 6.1, a real BTC is a mirror of the MTZ and typically exhibits a sigmoid shape. For given process conditions, the location of the center of the BTC with respect to the time axis is determined by the adsorption equilibrium, whereas the steepness of the BTC is determined by both adsorption equilibrium and adsorption kinetics. The ideal BTC represents the limiting case of the real BTC for an infinitely fast adsorption rate. Under this limiting condition, the MTZ reduces to a sharp boundary between loaded and unloaded adsorbent layers, and the BTC reduces to a concentration step from $c/c_0 = 0$ to $c/c_0 = 1$.

The area between the concentration axis, the time axes, the line $c/c_0 = 1$, and the BTC is proportional to the equilibrium adsorbent loading (Figure 6.9a). Therefore, for each real BTC, an equivalent ideal BTC can be constructed by locating the ideal BTC at the center of mass of the real curve (Figure 6.9b). The ideal BTC at the corresponding ideal breakthrough time, t_b^{id}, divides the real BTC into equal areas (A1 and A2 in Figure 6.9b). In the special case of a symmetrical BTC, t_b^{id} is located at $c/c_0 = 0.5$.

The integral material balance for a fixed-bed adsorber can be derived from the condition that the amount of adsorbate that is fed to the adsorber until a defined time must equal the sum of the amount adsorbed and the amount accumulated in the fluid phase within the void fraction of the bed. Accordingly, the balance equation for the ideal BTC reads

$$c_0\,\dot{V}\,t_b^{id} = q_0\,m_A + c_0\,\varepsilon_B\,V_R \tag{6.19}$$

where t_b^{id} is the ideal breakthrough time, the time corresponding to the center of mass of the real BTC. The term on the left-hand side of eq. (6.19) represents the amount of adsorbate fed to the adsorber until the ideal breakthrough time. On the right-hand side, the first term represents the amount of adsorbate adsorbed and the second term represents the amount of adsorbate that is accumulated in the void fraction of the adsorbent bed. Rearranging this equation gives an expression for the ideal breakthrough time

$$t_b^{id} = \frac{q_0\,m_A}{c_0\,\dot{V}} + \frac{V_R\,\varepsilon_B}{\dot{V}} = t_{st} + t_r \tag{6.20}$$

As can be seen from eq. (6.20), the ideal breakthrough time is the sum of two terms. The first term includes the process conditions adsorbent mass, m_A, and volumetric flow rate, $\dot{V}$, as well as the inlet concentration, c_0, and the corresponding equilibrium loading, $q_0(c_0)$. Since this term is related to the equilibrium data, it is referred to as the stoichiometric time, t_{st}. The second term is the effective residence time, t_r, as defined in eq. (6.16).

For strongly adsorbable substances, the stoichiometric time is some orders of magnitude greater than the residence time of the aqueous solution in the adsorber.

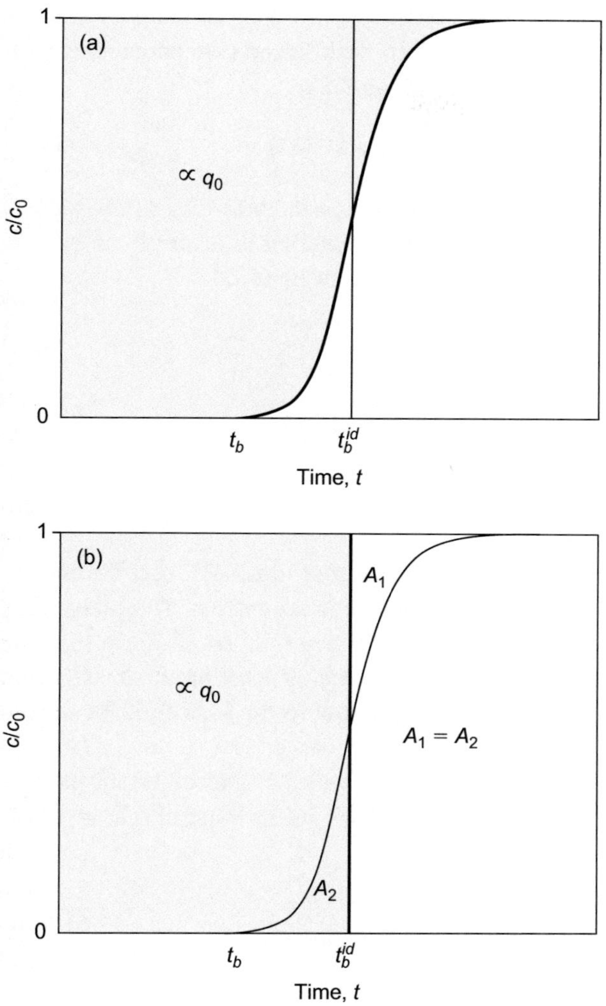

Figure 6.9: Graphical representation of the equilibrium adsorbent loading for a real breakthrough curve (a) and for the corresponding ideal breakthrough curve (b).

Accordingly, the residence time can be neglected, and the ideal breakthrough time is approximately equal to the stoichiometric time

$$t_b^{id} \approx t_{st} = \frac{q_0\, m_A}{c_0\, \dot{V}} \tag{6.21}$$

On the other hand, if the solute is not adsorbable ($q_0 = 0$), no retardation takes place, and the ideal breakthrough time is the same as the effective residence time of the aqueous solution.

By using the relationships given in Section 6.3, the integral material balance equation can be formulated in an alternative form with the process parameters filter velocity, v_F, and adsorber height, h, instead of $\dot{V}$ and m_A,

$$c_0\, v_F\, t_b^{id} = q_0\, \rho_B\, h + c_0\, \varepsilon_B\, h = (q_0\, \rho_B + c_0\, \varepsilon_B)\, h \tag{6.22}$$

Given that the ideal breakthrough time is the time that the MTZ (more precisely, its center of mass) needs to travel through the entire adsorber of height h, an expression for the velocity of the MTZ, v_z, can be derived from eq. (6.22)

$$v_z = \frac{h}{t_b^{id}} = \frac{v_F\, c_0}{q_0\, \rho_B + c_0\, \varepsilon_B} \tag{6.23}$$

For strongly adsorbable substances ($t_{st} \gg t_r$), eq. (6.23) reduces to

$$v_z \approx \frac{h}{t_{st}} = \frac{v_F\, c_0}{q_0\, \rho_B} \tag{6.24}$$

The simple integral material balance equation for the ideal BTC can be used to check the plausibility of the results of adsorption measurements. The plausibility condition is that the ideal breakthrough time calculated from eq. (6.20) or (6.21) by using independently determined isotherm data meets, at least approximately, the center of mass of the experimental BTC. Larger deviations are indicators for serious errors in isotherm or BTC determination.

An integral material balance equation for the real BTC can be established by substituting the ideal breakthrough time in eq. (6.19) by an integral that accounts for the concentration profile

$$c_0\, \dot{V} \int_{t=0}^{t=\infty} \left(1 - \frac{c}{c_0}\right) dt = q_0\, m_A + c_0\, \varepsilon_B\, V_R \tag{6.25}$$

If neglecting the storage capacity in the void volume, eq. (6.25) reduces to

$$c_0\, \dot{V} \int_{t=0}^{t=\infty} \left(1 - \frac{c}{c_0}\right) dt \approx q_0\, m_A \tag{6.26}$$

Until the breakthrough time, t_b, the concentration is zero, and after the saturation time, t_s, the concentration equals the initial concentration. Therefore, eqs. (6.25) and (6.26) can also be written as

$$c_0\, \dot{V}\, t_b + c_0\, \dot{V} \int_{t=t_b}^{t=t_s} \left(1 - \frac{c}{c_0}\right) dt = q_0\, m_A + c_0\, \varepsilon_B\, V_R \tag{6.27}$$

and

$$c_0 \dot{V} t_b + c_0 \dot{V} \int_{t=t_b}^{t=t_s} \left(1 - \frac{c}{c_0}\right) \mathrm{d}t \approx q_0 m_A \tag{6.28}$$

The first terms in eqs. (6.27) and (6.28) are related to the adsorbent loading until breakthrough (breakthrough loading), q_b,

$$c_0 \dot{V} t_b = q_b m_A + c_0 V_R \varepsilon_B \approx q_b m_A \tag{6.29}$$

Since in a single fixed-bed adsorber typically not the equilibrium capacity but only the breakthrough capacity can be utilized, the ratio of breakthrough and equilibrium loading describes the degree of efficiency of the adsorber, η_A,

$$\eta_A = \frac{q_b}{q_0} \tag{6.30}$$

Accordingly, all other conditions being equal, the efficiency of an adsorber decreases as the BTC becomes flatter (i.e. as the MTZ becomes broader).

The ideal BTC and the related material balance equation can also be used as a reference system for the real BTC. Taking the ideal breakthrough time as a reference parameter, a dimensionless time, referred to as throughput ratio, T, can be introduced. In the general form, T is defined as

$$T = \frac{t}{t_b^{id}} = \frac{t}{t_{st} + t_r} = \frac{t}{\dfrac{m_A q_0}{\dot{V} c_0} + \dfrac{V_R \varepsilon_B}{\dot{V}}} \tag{6.31}$$

Neglecting the short hydraulic residence time, the definition of T reads

$$T = \frac{t}{t_{st}} = \frac{\dot{V} c_0 t}{m_A q_0} = \frac{v_F c_0 t}{\rho_B h q_0} \tag{6.32}$$

The throughput ratio defines the position of a concentration point of the real BTC in relation to the ideal breakthrough time. For $t = t_b^{id}$, T becomes 1. Together with other dimensionless parameters, the throughput ratio can be used to simplify mathematical BTC models as will be shown in Chapter 7.

Another application field of the integral balance equations consists in the determination of equilibrium loadings from experimental BTCs. In principle, two different (but equivalent) methods are applicable:

- Construction of the ideal BTC according to Figure 6.9b and application of eq. (6.19) or (6.22).
- Graphical integration according to eqs. (6.25–6.28) as shown schematically in Figure 6.10.

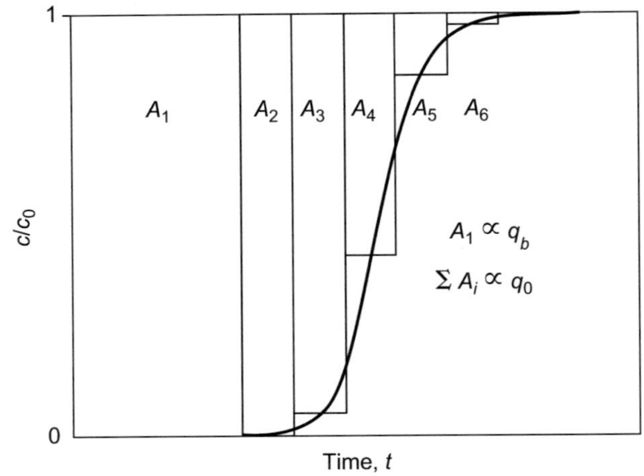

Figure 6.10: Graphical integration of the breakthrough curve.

Equations (6.25) and (6.26) can also be used to describe adsorption processes in multicomponent adsorption systems. As discussed in Section 6.1, competitive adsorption leads to concentration overshoots for all components, except for the strongest adsorbable adsorbate. For all components that show the concentration overshoot, it holds that the area below the BTC and above the line $c = c_0$ represents the mass desorbed due to the displacement process. This is in accordance with eqs. (6.25) and (6.26) in which the value of the integral becomes negative if the concentration, c, is higher than the inlet concentration, c_0.

As an example, eq. (6.26) will be applied to a bisolute system. For the weaker adsorbable component 1, eq. (6.26) has to be written as

$$q_{0,1}\, m_A = c_{0,1}\, \dot{V} \int_{t=0}^{t=\infty} \left(1 - \frac{c_1}{c_{0,1}}\right) dt = c_{0,1}\, \dot{V} \int_{t=0}^{t(c_1=c_{0,1})} \left(1 - \frac{c_1}{c_{0,1}}\right) dt - c_{0,1}\, \dot{V} \int_{t(c_1=c_{0,1})}^{t=\infty} \left(\frac{c_1}{c_{0,1}} - 1\right) dt \tag{6.33}$$

The first term on the right-hand side of eq. (6.33) represents the mass initially adsorbed without the influence of competition, whereas the second term represents the mass desorbed due to the subsequent displacement by component 2. The equilibrium adsorbent loading of component 1 in the bisolute system is given by the difference of both terms. In the BTC diagram shown in Figure 6.11, the terms on the right-hand side of eq. (6.33) correspond to the areas A_1 and A_2, respectively. The equilibrium loading of component 1, $q_{0,1}$, is proportional to the difference A_1–A_2.

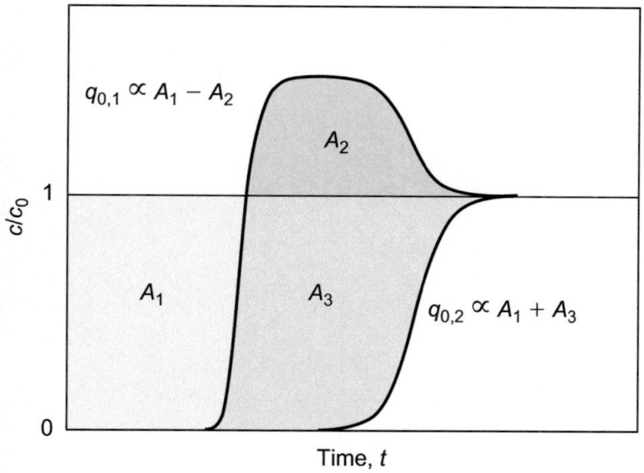

Figure 6.11: Graphical representation of the adsorbed amounts in a bisolute system.

For the stronger adsorbable component 2, which is not subject to a displacement process, the material balance equation reads

$$q_{0,2}\, m_A = c_{0,2}\, \dot{V} \int\limits_{t=0}^{t=\infty} \left(1 - \frac{c_2}{c_{0,2}}\right) \mathrm{d}t \tag{6.34}$$

The corresponding area in Figure 6.11 is $A_1 + A_3$.

The real BTCs in multicomponent systems can be approximated by ideal BTCs in the same way as shown for single-solute systems. Figure 6.12 shows the ideal breakthrough times in a bisolute system. Neglecting the storage capacity in the void volume ($t_b^{id} \approx t_{st}$), the corresponding material balance equations are

$$q_{0,1}\, m_A = c_{0,1}\, \dot{V}\, t_{b,1}^{id} - (c_{pl,1} - c_{0,1})\, \dot{V}\, (t_{b,2}^{id} - t_{b,1}^{id}) \tag{6.35}$$

$$q_{0,2}\, m_A = c_{0,2}\, \dot{V}\, t_{b,2}^{id} \tag{6.36}$$

The concentration $c_{pl,1}$ in eq. (6.35) is the plateau concentration, the maximum concentration overshoot of component 1.

As illustrated in Figure 6.12, a further material balance for a bisolute system can be established by treating the concentration step from $c_1 = 0$ to $c_{pl,1}$ as the ideal BTC of a single solute with the inlet concentration $c_{pl,1}$. Accordingly, the material balance equation reads

$$q_{pl,1}\, m_A = c_{pl,1}\, \dot{V}\, t_{b,1}^{id} \tag{6.37}$$

where $q_{pl,1}$ is the equilibrium adsorbent loading related to $c_{pl,1}$. Substituting $t_{b,1}^{id}$ and $t_{b,2}^{id}$ in eq. (6.35) by use of eqs. (6.36) and (6.37), the following relationship can be found after some rearrangements:

$$\frac{q_{pl,1} - q_{0,1}}{c_{pl,1} - c_{0,1}} = \frac{q_{0,2}}{c_{0,2}} = \frac{\dot{V}\, t_{b,2}^{id}}{m_A} \tag{6.38}$$

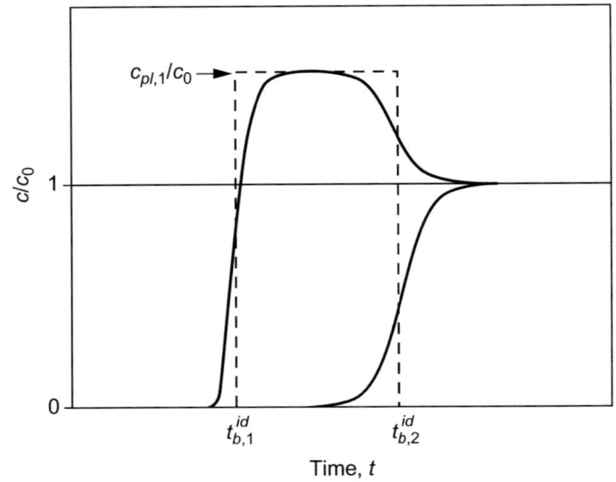

Figure 6.12: Approximation of the real breakthrough curves of a bisolute system by ideal breakthrough curves.

Equation (6.38) relates the changes in the concentrations and in the adsorbent loadings of both components within the adsorption zone of the displacing component 2, which is concurrently the desorption zone of the displaced component 1.

The given set of material balance equations can be used to predict the ideal breakthrough times of both components for a defined volumetric flow rate and adsorbent mass if the equilibrium relationships between the concentrations and adsorbent loadings are known. As equilibrium relationships, either the single-solute isotherm equation or the ideal adsorbed solution theory equations have to be applied, depending on the number of components in the considered MTZ.

This approach, here exemplarily shown for a bisolute system, can be easily extended to multicomponent mixtures. The respective BTC model is known as the equilibrium column model. It will be discussed in more detail in Chapter 7.

6.4.3 Differential material balance

To establish a differential material balance, a differential volume element, $dV = A_R\, dz$, of an adsorber with the cross-sectional area A_R is considered (Figure 6.13). It can be

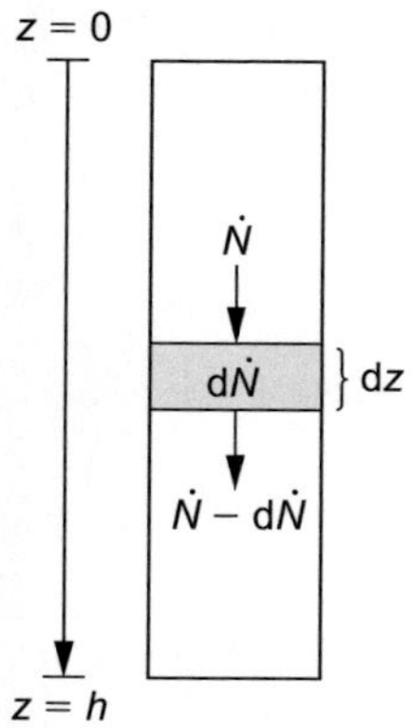

Figure 6.13: Material balance around a differential volume element of the fixed-bed adsorber.

assumed that the amount of adsorbate that is adsorbed onto the adsorbent or accumulated in the void fraction of the volume element must equal the difference between the input and the output of the volume element. Input and output occurs by advection and axial dispersion. Accordingly, the general material balance equation is given by

$$\dot{N}_{accu} + \dot{N}_{ads} = \dot{N}_{disp} + \dot{N}_{adv} \tag{6.39}$$

where $\dot{N}$ represents the change of the amount of adsorbate with time, and the indices indicate the processes accumulation, adsorption, dispersion, and advection.

The accumulation of substance within the void fraction of the volume element, $\mathrm{d}V$, is given by

$$\dot{N}_{accu} = \varepsilon_B\, A_R\, \mathrm{d}z \frac{\partial c(t,z)}{\partial t} = \varepsilon_B\, \mathrm{d}V \frac{\partial c(t,z)}{\partial t} \tag{6.40}$$

and the adsorption onto the adsorbent in the volume element can be written as

$$\dot{N}_{ads} = \rho_B\, A_R\, \mathrm{d}z \frac{\partial \bar{q}(t,z)}{\partial t} = \rho_B\, \mathrm{d}V \frac{\partial \bar{q}(t,z)}{\partial t} \tag{6.41}$$

Here, $\bar{q}$ is the mean adsorbent loading. To describe the advection, the difference between the amount of adsorbate fed to and released by the volume element per unit of time has to be considered

$$\dot{N}_{adv} = v_F\, A_R\, c(t,z) - v_F\, A_R\, c(t,z+\mathrm{d}z) \tag{6.42}$$

In differential form, eq. (6.42) reads

$$\dot{N}_{adv} = -v_F\, A_R \frac{\partial c(t,z)}{\partial z} \mathrm{d}z = -v_F\, \mathrm{d}V \frac{\partial c(t,z)}{\partial z} \tag{6.43}$$

Under the condition that the axial dispersion can be described by Fick's first law, the difference between input and output caused by dispersion is given by

$$\dot{N}_{disp} = D_{ax}\,\varepsilon_B\,A_R\left[\frac{\partial c(t,z)}{\partial z}\right]_{z+\mathrm{d}z} - D_{ax}\,\varepsilon_B\,A_R\left[\frac{\partial c(t,z)}{\partial z}\right]_z \tag{6.44}$$

where D_{ax} is the axial dispersion coefficient. In differential form, this equation reads

$$\dot{N}_{disp} = D_{ax}\,\varepsilon_B\,A_R\,\frac{\partial^2 c(t,z)}{\partial z^2}\,\mathrm{d}z = D_{ax}\,\varepsilon_B\,\mathrm{d}V\,\frac{\partial^2 c(t,z)}{\partial z^2} \tag{6.45}$$

Introducing eqs. (6.40), (6.41), (6.43), and (6.45) into eq. (6.39) and dividing the resulting equation by dV gives the differential material balance equation in its general form

$$v_F\frac{\partial c}{\partial z} + \varepsilon_B\frac{\partial c}{\partial t} + \rho_B\frac{\partial \bar{q}}{\partial t} - D_{ax}\,\varepsilon_B\frac{\partial^2 c}{\partial z^2} = 0 \tag{6.46}$$

In engineered adsorption processes with relatively high flow rates, the dispersion term is usually neglected. Its impact on the spreading of the BTC is negligible in comparison to the influence of slow mass transfer processes

$$v_F\frac{\partial c}{\partial z} + \varepsilon_B\frac{\partial c}{\partial t} + \rho_B\frac{\partial \bar{q}}{\partial t} = 0 \tag{6.47}$$

Sometimes, the accumulation term, $\varepsilon_B\,\partial c/\partial t$, is also neglected. This is based on the assumption that the accumulation in the liquid phase is small in comparison to the adsorption.

The differential material balance equation derived here is valid for both single-solute and multisolute systems. In multisolute systems, material balance equations have to be set up for each component.

From the differential material balance equation (eq. (6.47)), conclusions can be drawn about the influence of the isotherm shape on the velocity of different points of the concentration profile in the adsorber bed. For this purpose, a transformed spatial coordinate, z^*, under consideration of the velocity of a specified concentration point, v_c, is defined

$$z^* = z - v_c\,t \tag{6.48}$$

It has to be noted that for the concentration at the center of mass of the BTC, the velocity of the concentration point, v_c, equals the velocity of the MTZ, v_z. Differentiating eq. (6.48) with respect to z and t and substituting the received differentials into eq. (6.47) gives

$$v_F\frac{\partial c}{\partial z^*} = v_c\,\varepsilon_B\frac{\partial c}{\partial z^*} + v_c\,\rho_B\frac{\partial \bar{q}}{\partial z^*} \tag{6.49}$$

and after rearranging and setting $\bar{q} = q$ (assumption of local equilibrium), we obtain

$$v_c = \frac{v_F}{\varepsilon_B + \rho_B \frac{\partial q}{\partial c}} \tag{6.50}$$

The differential in eq. (6.50) represents the isotherm slope. In the case of a favorable isotherm ($R^* < 1$, $n < 1$, see Chapter 3), the condition

$$\frac{\partial^2 q}{\partial c^2} < 0 \tag{6.51}$$

holds, indicating that the isotherm slope decreases with increasing concentration. According to eq. (6.50), higher concentrations travel with higher velocity than lower concentrations. Therefore, the concentration profile becomes steeper with increasing distance from the adsorber inlet. This effect is referred to as self-sharpening of the concentration profile. Since it is physically impossible that higher concentrations overrun lower concentrations (which would mean that the BTC starts with high concentrations instead of low concentrations), the self-sharpening ends with the limiting case of a vertical BTC (ideal BTC).

In contrast, an unfavorable isotherm ($R^* > 1$, $n > 1$) is characterized by

$$\frac{\partial^2 q}{\partial c^2} > 0 \tag{6.52}$$

According to eq. (6.50), the profile becomes flatter with increasing transport distance.

Finally, in the case of a linear isotherm with

$$\frac{\partial^2 q}{\partial c^2} = 0 \tag{6.53}$$

all points of the concentration profile travel with the same velocity.

So far, the discussion was restricted to the influence of the isotherm shape on the concentration profile. However, under real conditions, the concentration profile is affected not only by the adsorption equilibrium but also by mass transfer processes. The mass transfer resistances lead to a spreading of the profile. In the case of favorable isotherms, this spreading counteracts the self-sharpening effect. As a result, after a certain transport distance, the spreading effect is exactly counterbalanced by the sharpening effect. From this point, all concentration points travel with the same velocity and the height of the MTZ becomes independent of the bed height. This state is referred to as constant pattern (Figure 6.14a). Under constant pattern conditions, BTCs measured at different bed heights run parallel.

The spreading effect caused by mass transfer resistances appears also in the case of linear and unfavorable isotherms. For linear isotherms, therefore, an increasing flattening of the concentration profile instead of a constant concentration profile occurs. In the case of unfavorable isotherms, the spreading effect caused by the isotherm slope will be further enhanced by the mass transfer resistances. However, for strong convex isotherms ($R^* > 2$), this additional effect is negligible. The increase of the MTZ height proportional to the travel distance is referred to as proportionate pattern (Figure 6.14b).

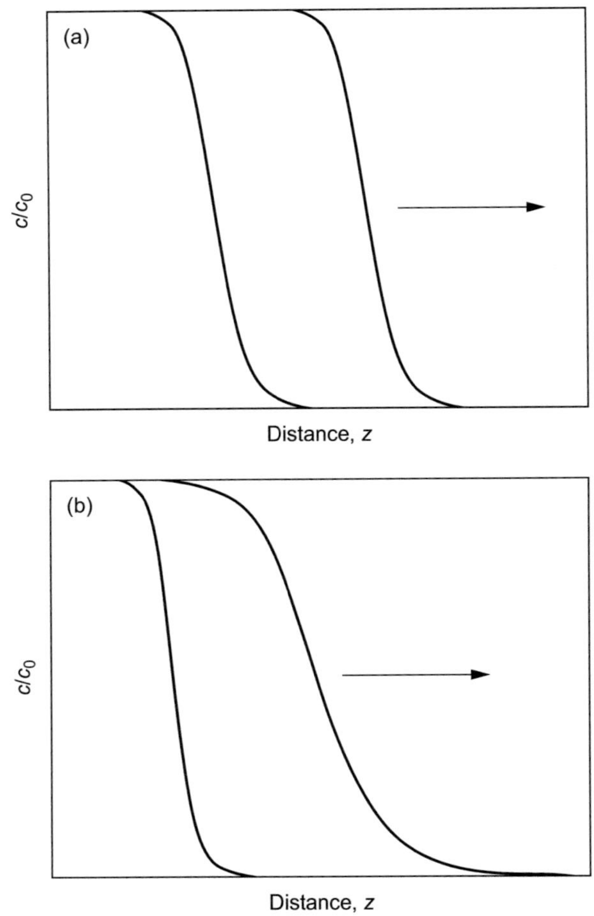

Figure 6.14: Constant (a) and proportionate (b) pattern of the MTZ.

Since adsorption processes are typically used to remove strongly adsorbable substances, the favorable isotherm type is of particular practical relevance. As follows from the previous discussion, the formation of a constant pattern can be expected for substances with favorable isotherms and for sufficiently long adsorbers. Under these conditions, the material balance equation simplifies considerably. In the case of a constant pattern, the velocity v_c is the same for all points of the concentration profile. Since this is also true for the center of the adsorption front, eq. (6.50) can be set equal to eq. (6.23), which is valid for the ideal adsorption front. After integration, the following simple material balance equation for the constant pattern case can be derived:

$$\frac{c}{c_0} = \frac{q}{q_0} \tag{6.54}$$

The assumption of constant pattern conditions allows simplifying the BTC modeling as will be demonstrated in Chapter 7.

6.5 Practical aspects

6.5.1 Introduction

Fixed-bed adsorbers are frequently used in drinking water treatment, but also in wastewater treatment, swimming pool water treatment, groundwater remediation, and, together with membrane processes, in ultrapure water preparation for industrial purposes. In nearly all of these cases, the objective is to remove organic substances from the water. For this, granular activated carbon (GAC) is the adsorbent of choice. Only for few specific purposes, such as removal of arsenic or phosphate from water, other adsorbents like granular ferric hydroxide or aluminum oxide are used. Therefore, the following discussions are focused on GAC application, but most of the conclusions can be transferred to other granular adsorbents.

6.5.2 Typical operating conditions

Fixed-bed adsorbers for drinking water treatment or wastewater treatment are constructed in analogous manner as sand filters used for turbidity removal. The adsorbers can be designed as closed pressure filters or as open gravity filters with circular or rectangular cross section. The filters are typically made of corrosion-resistant steel (stainless steel or steel coated with polymers) or concrete. The adsorbent in a fixed-bed adsorber is located on a perforated bottom, and the water usually streams downward through the adsorbent bed. Often, a small layer (5 to 10 cm) of sand (1 to 2 mm in diameter) is located between the activated carbon and the bottom. This helps to remove carbon fines. Since the pressure loss increases with time due to the accumulation of particles in the sand layer, backwashing is necessary at certain time intervals. This backwashing has to be done very carefully to avoid too strongly mixing the sand and activated carbon. For raw waters with high turbidity, it is recommended that a sand filtration be applied before feeding the water to the GAC adsorber. For the combination of sand filtration and GAC adsorbers, separate filters for each step as well as dual-bed filters are in use. Figure 6.15 shows two types of fixed-bed adsorbers frequently used in water treatment.

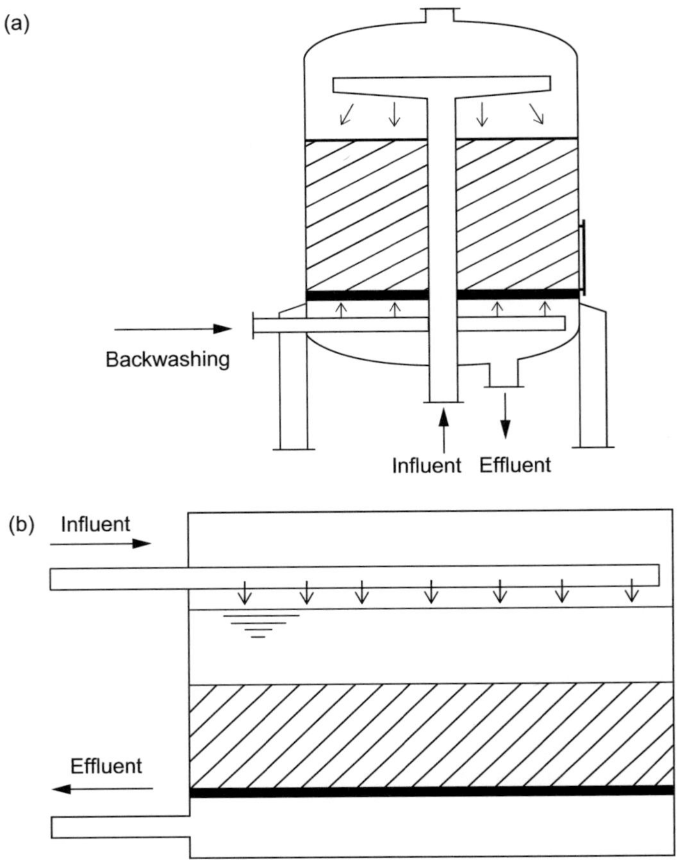

Figure 6.15: Typical fixed-bed adsorbers in water treatment: (a) pressure GAC filter made of corrosion-resistant steel and (b) rectangular gravity filter constructed of concrete.

In Table 6.1, typical values of the main operating parameters are listed. The typical lifetime of such GAC adsorbers is between 100 and 600 days; the bed volumes (BV) treated within this operating time ranges from 2,000 to 20,000.

Table 6.1: Typical operating conditions of GAC adsorbers.

Parameter	Symbol	Unit	Typical values
Bed height	h	m	2–4
Cross-sectional area	A_R	m^2	5–30
Filter velocity	v_F	m/h	5–20
Empty bed contact time	$EBCT$	min	5–30

Table 6.1 (continued)

Parameter	Symbol	Unit	Typical values
Effective contact time	t_r	min	2–10
Particle diameter	d_P	mm	0.5–4
Bed volume	V_R	m^3	10–50
Bed porosity	ε_B	–	0.35–0.45

6.5.3 Fixed-bed versus batch adsorber

Activated carbon, as the most important adsorbent in drinking water and wastewater treatment, is used in two different forms: as powdered activated carbon (PAC) or as granular activated carbon (GAC). PAC is typically applied in batch or flow-through slurry reactors (see Chapters 3 and 5), whereas GAC is applied in fixed-bed adsorbers. Both technical options exhibit advantages as well as disadvantages.

PAC is easy to dose (typically as suspension) and is therefore ideally suited for temporary application. Due to the small particle size, the adsorption rate is very fast. Displacement processes due to competitive adsorption are less pronounced in comparison to fixed-bed adsorption. A concentration increase over the initial concentration is therefore not observed. Disadvantages of the PAC application in slurry reactors consist of the particle discharge from the reactor that requires an additional separation step, the remaining residual (equilibrium) concentration, and the missing regenerability of PAC. PAC has to be burned or deposited after use.

Fixed-bed adsorbers assure low outlet concentrations (zero in ideal case) until the breakthrough. Particle discharge does not need to be suspected. GAC can be regenerated (reactivated) and repeatedly applied. The slower adsorption kinetics, leading to flat BTCs, as well as possible concentration overshoots due to displacement processes are the main disadvantages of GAC application in fixed-bed adsorbers.

An important difference between batch reactors and fixed-bed adsorbers consists in the different exploitation of the adsorption capacity for the same initial adsorbate concentration. As shown in Figure 6.16, the adsorption process in a batch reactor proceeds along the operating line whose slope is given by the reciprocal of the adsorbent dose (Section 3.6). Consequently, the residual concentration is lower than the initial concentration, c_0, and the adsorption capacity that can be utilized in the batch process is the adsorbent loading in equilibrium with the residual concentration. In the case of fixed-bed adsorption, the adsorbate solution is fed continuously to the adsorber, and the adsorbent loading in the zone behind the MTZ is in equilibrium with the inlet concentration, c_0. Accordingly, as can be seen in Figure 6.16, the adsorbent loading for a given c_0 is higher in the case of fixed-bed adsorption than in the case of batch adsorption. However, it has to be noted that this is only true if the total

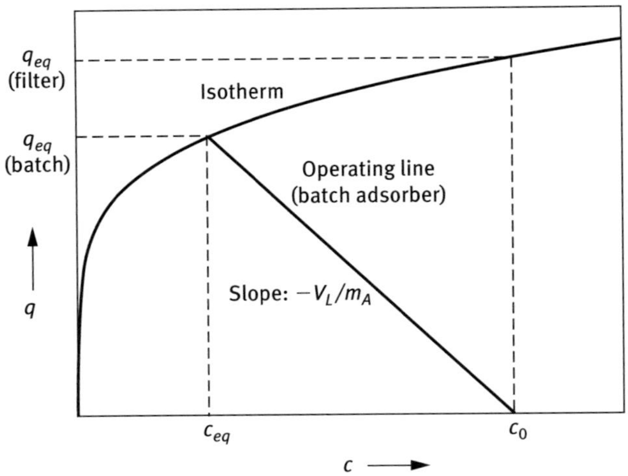

Figure 6.16: Equilibrium adsorbent loadings achievable in fixed-bed and batch adsorption for the same initial concentration.

adsorbent bed can be loaded to the equilibrium. In practice, the fixed-bed adsorption process has to be stopped at a defined breakthrough point, and the achievable breakthrough loading is lower than the equilibrium loading. In particular, when the BTC is very flat, this loss in capacity might be so high that the advantage in adsorbent capacity exploitation compared to the batch process gets lost. To avoid this problem and to attain capacity exploitation near to the ideal case of equilibrium, multiple adsorber systems have to be used as will be described in the next section.

6.5.4 Multiple adsorber systems

The fixed-bed adsorption process in a single adsorber is a semicontinuous process. Until the breakthrough, the water can be continuously treated, but if the breakthrough occurs, the process has to be stopped and the adsorbent has to be regenerated. This is contrary to practical requirements in water treatment plants, where a continuous treatment process is indispensable. This problem can be solved by applying multiple adsorber systems. As an additional effect, multiple adsorber systems can reduce the capacity loss, which results from the need to stop the process at the first adsorbate breakthrough. In principle, there are two different ways to connect single fixed-bed adsorbers to a multiple adsorber system: series connection and parallel connection.

The principle of series connection is demonstrated in Figure 6.17. In this example, the total adsorbent mass is divided between four adsorbers. Only three adsorbers are in operation, whereas one is out of operation in order to regenerate the adsorbent. In the given scheme, the time t_1 shows a point of time where adsorber 1 is out of operation

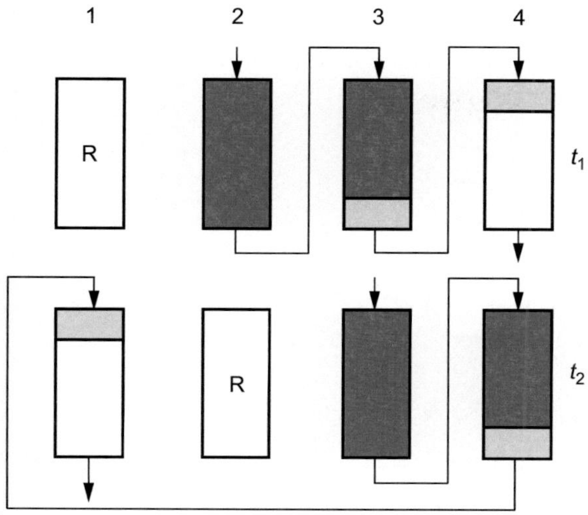

Figure 6.17: Fixed-bed adsorbers in series connection. Legend: R, adsorber out of operation for adsorbent regeneration; dark gray, adsorbent loaded to equilibrium; light gray, MTZ; white, adsorbent free of adsorbate.

and the MTZ is located between adsorbers 3 and 4. Since the MTZ has already left adsorber 2, the adsorbent in this adsorber is fully saturated to the equilibrium. Therefore, adsorber 2 will be the next to go out of operation. Time t_2 represents a later time. In the meantime, the regenerated adsorber 1 has been put in stream again. The MTZ has traveled forward and is now located between adsorbers 4 and 1. The adsorbent in adsorber 3 is fully saturated. Next, adsorber 3 will be put out of operation, and so on. In an ideal case, all adsorbers can be operated until the equilibrium loading is reached for the entire adsorbent bed (maximum utilization of the adsorbent capacity). However, if the MTZ is very long and the number of adsorbers is limited, or if there is more than one MTZ as in mixture adsorbate systems, this maximum utilization might not be completely reached. Nevertheless, the adsorbent capacity is significantly better exploited than in a single adsorber. On the other hand, the cross-sectional area available for water flow through is that of a single adsorber, independent of the number of adsorbers in operation. For a given linear filter velocity, this cross-sectional area limits the volumetric flow rate that can be realized (eq. (6.13)).

In multiple adsorber systems with parallel connection, the total water stream to be treated is split into substreams, which are fed to a number of parallel operating adsorbers. The different adsorbers are put in operation at different start times. Consequently, at a given time, the traveled distances of the MTZs are different in the different adsorbers, and therefore the breakthrough times are also different. The effluents of the different adsorbers with different concentrations are blended to give a total effluent stream. Due to the different adsorber lifetimes, the mixing of the effluents leads to low concentrations in the total effluent stream even if the effluent

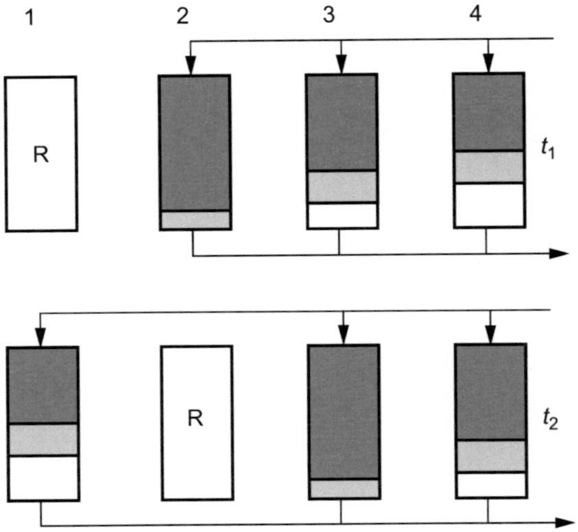

Figure 6.18: Fixed-bed adsorbers in parallel connection. Legend: R, adsorber out of operation for adsorbent regeneration; dark gray, adsorbent loaded to equilibrium; light gray, MTZ; white, adsorbent free of adsorbate.

concentration of the adsorber that was first started is relatively high (Figure 6.18). If the blended effluent concentration becomes too high, the first loaded adsorber is put out of operation for adsorbent regeneration and another adsorber with regenerated adsorbent is put in operation. The scheme in Figure 6.18 shows the location of the MTZs and the degrees of saturation at two different times. The time-shifted operation and the blending of the effluents allow operating all adsorbers in the system longer than in the case of a single-adsorber system. Therefore, the adsorbent capacity is also better exploited than in a single adsorber. Although the effluent concentration in this type of multiple adsorber systems is not zero, the concentration can be minimized and the treatment goal can be met by choosing an appropriate number of adsorbers and an optimum operating time regime. The main advantage of the parallel connection is that the total cross-sectional area increases with an increasing number of adsorbers. This type of a multiple adsorber system is therefore very flexible and can be adapted to different requirements regarding the water volume to be treated. It is in particular suitable for the treatment of large amounts of water.

7 Fixed-bed adsorber design

7.1 Introduction and model classification

Knowledge of the breakthrough behavior of the compounds to be removed is the essential precondition for any fixed-bed adsorber design. To predict the breakthrough behavior of adsorbates in fixed-bed adsorbers, different mathematical tools are available. These tools can be divided into two main groups: scale-up methods and breakthrough curve (BTC) models (Figure 7.1).

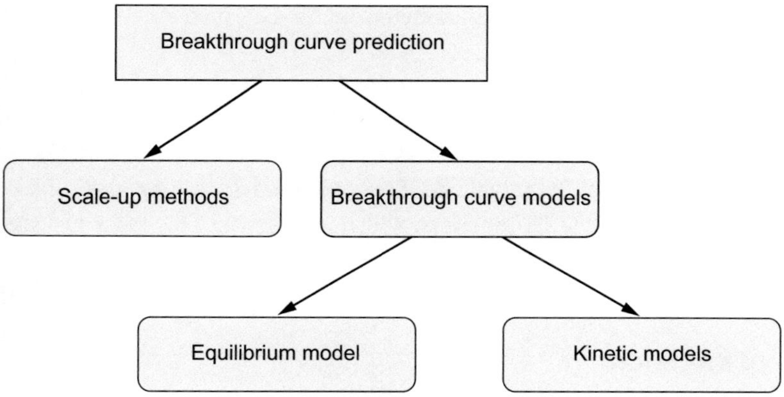

Figure 7.1: Classification of fixed-bed adsorber models.

Scale-up methods require the determination of BTCs in lab-scale experiments as the basis for the prediction of the breakthrough behavior in full-scale adsorbers. The scale-up methods are typically based on one of the following principles:

- Estimation of characteristic parameters of the mass transfer zone (MTZ) from the experimentally determined BTC and application of these parameters to design the full-scale adsorber.
- Application of an experimental setup that guarantees the similarity of mass transfer conditions between small-scale and large-scale adsorbers to determine a normalized (scale-independent) BTC that can be used to describe the adsorption in the large-scale adsorber.

Although the mathematical tools for scale-up are typically based on fundamental relationships between operational parameters, they do not allow a deeper insight into the mechanisms of the adsorption process, because equilibrium relationships and adsorption kinetics are not explicitly, but only indirectly, considered. Scale-up methods are, therefore, only applicable under restrictive conditions, such as specific similarity

https://doi.org/10.1515/9783110715507-007

criteria. A transfer of the lab-scale results to conditions that do not fulfill these criteria is not possible.

The other group of calculation methods includes models that are based on equilibrium relationships and mass transfer equations. Therefore, these models are more flexible in application. In principle, the breakthrough behavior can be predicted from separately determined isotherm parameters and kinetic data by means of these BTC models. However, due to the complexity of the adsorption mechanisms, more or less strong simplifications are also necessary in these models, in particular if multisolute systems are considered. Therefore, it is important to validate each selected model by means of experimental data. However, if the validity of the selected model is once demonstrated for the specific application, it provides the possibility to predict the influence of different process parameters on the BTC on a strict theoretical basis.

The BTC models can be further divided into the equilibrium column model (ECM), which only considers the equilibrium relationships, and models that consider the equilibrium relationships as well as the kinetic equations. The ECM is sometimes also referred to as the local equilibrium model. It only allows determining ideal breakthrough times, whereas complete BTC models, which also account for mass transfer processes, can describe real S-shaped BTCs.

7.2 Scale-up methods

7.2.1 Mass transfer zone model

The mass transfer zone (MTZ) model is a scale-up model that is based on characteristic parameters of the MTZ. The MTZ is that zone within the adsorbent bed in which the adsorption takes place (Chapter 6, Section 6.1). It is also referred to as the adsorption zone. The MTZ model was originally developed for ion exchange processes (Michaels 1952) and later assigned to adsorption processes. It is particularly suitable for single-solute systems. The MTZ model is based on the following assumptions: isothermal adsorption, constant flow velocity, constant initial adsorbate concentration, negligible adsorbate accumulation in the void fraction of the bed, and formation of a constant pattern of the MTZ (Chapter 6, Section 6.4.3). It follows from the constant pattern condition that the MTZ height is independent from the covered distance.

To characterize the BTC, the following parameters are used:

- The height of the MTZ: zone height, h_z.
- The traveling velocity of the MTZ within the adsorber: zone velocity, v_z.
- The time the MTZ needs to travel a distance equal to its own height: zone time, t_z.

The relationship between these three parameters is given by

$$v_z = \frac{h_z}{t_z} \tag{7.1}$$

The zone time, t_z, can be read directly from the experimental BTC. It is given by the difference between the saturation time, t_s, and the breakthrough time, t_b (Figure 7.2)

$$t_z = t_s - t_b \tag{7.2}$$

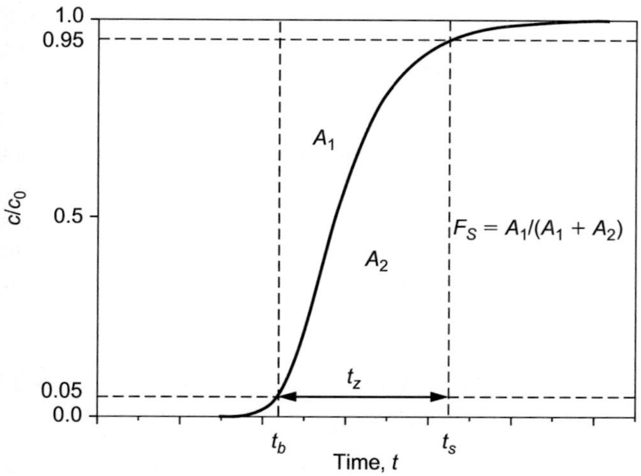

Figure 7.2: Characteristic parameters of the MTZ model.

Due to the asymptotic form of the BTC, the breakthrough time and the saturation time cannot be determined exactly. Therefore, by definition, the times at $c/c_0 = 0.05$ and $c/c_0 = 0.95$ are used as breakthrough time and saturation time, respectively.

Given that the ideal breakthrough time can be approximated by the stoichiometric time (negligible hydraulic residence time), the zone velocity, v_z, can also be expressed by the stoichiometric time, t_{st}, and the adsorber height, h (Chapter 6, Section 6.4.2)

$$v_z = \frac{h}{t_{st}} \tag{7.3}$$

Equalizing eqs. (7.1) and (7.3) provides a relationship that can be used to determine the height of the MTZ

$$h_z = h\frac{t_s - t_b}{t_{st}} = h\frac{t_z}{t_{st}} \tag{7.4}$$

To estimate h_z, the stoichiometric time has to be determined from the BTC. As described in Section 6.4.2, the stoichiometric time is the time at the center of mass of the BTC. In the special case of a symmetric BTC, t_{st} is the time at $c/c_0 = 0.5$. Such symmetric BTCs can be found in cases where film and intraparticle diffusion contribute to the total mass transfer rate to the same extent. In other cases, the center of mass is located at higher concentrations (typical for rate-limiting intraparticle diffusion) or at lower concentrations (typical for rate-limiting film diffusion). In these cases, the stoichiometric time has to be estimated from the BTC by a graphical procedure. After determination of the areas A_1 and A_2 shown in Figure 7.2, a symmetry factor, F_S, can be estimated

$$F_S = \frac{A_1}{A_1 + A_2} \tag{7.5}$$

Given that t_{st} is located at the center of mass of the curve, the area ratio is also equivalent to

$$F_S = \frac{t_{st} - t_b}{t_s - t_b} = \frac{t_{st} - t_b}{t_z} \tag{7.6}$$

and the stoichiometric time is given by

$$t_{st} = t_b + F_S\, t_z \tag{7.7}$$

As can be derived from eq. (7.6), F_S is 0.5 for the special case of a symmetric BTC.

After substituting t_{st} in eq. (7.4) by eq. (7.7), the resulting equation for the zone height, h_z, reads

$$h_z = h\frac{t_s - t_b}{t_b + F_S\, t_z} = h\frac{t_z}{t_b + F_S\, t_z} \tag{7.8}$$

With

$$1 - F_S = \frac{t_s - t_{st}}{t_z} \tag{7.9}$$

the following alternative equation for h_z can be found

$$h_z = h\frac{t_s - t_b}{t_s - (1 - F_S)\, t_z} = h\frac{t_z}{t_s - (1 - F_S)\, t_z} \tag{7.10}$$

The parameters h_z and F_S characterize the shape of the BTC. Their values reflect the impact of the mass transfer rates. The higher the value of h_z, the slower the mass transfer. The value of F_S gives an indication of the dominant mass transfer mechanism ($F_S < 0.5$: dominating intraparticle diffusion; $F_S > 0.5$: dominating film diffusion). Both parameters provide the basis for the scale-up. By using h_z and F_S found from lab-scale experiments, breakthrough times, saturation times, and breakthrough adsorbent loadings for full-scale adsorption processes can be predicted.

The respective equation for the breakthrough time can be derived from eq. (7.8):

$$t_b = \frac{1}{v_z}(h - F_S\, h_z) = \frac{h}{v_z} - F_S\, t_z \tag{7.11}$$

and the equation for the saturation time can be found from eq. (7.10):

$$t_S = \frac{1}{v_z}[h + (1 - F_S)\, h_z] = \frac{h}{v_z} + (1 - F_S)\, t_z \tag{7.12}$$

Equations (7.11) and (7.12) describe the linear dependence of the breakthrough and saturation times on the bed height. The slope represents the reciprocal traveling velocity of the MTZ. It has to be noted that the MTZ model is only valid for constant pattern conditions (Chapter 6, Section 6.4.3). The condition for establishing constant pattern, expressed in terms of the MTZ model, is that the height of the adsorber is larger than the zone height ($h > h_z$). An extrapolation of the lines given by eqs. (7.11) and (7.12) to bed heights lower than h_z would lead to unrealistic breakthrough and saturation times. For t_b, even negative values would be predicted for very low bed heights (Figure 7.3). To verify whether the constant pattern is established, the zone height has to be formally calculated by eq. (7.8) and then compared with the bed height.

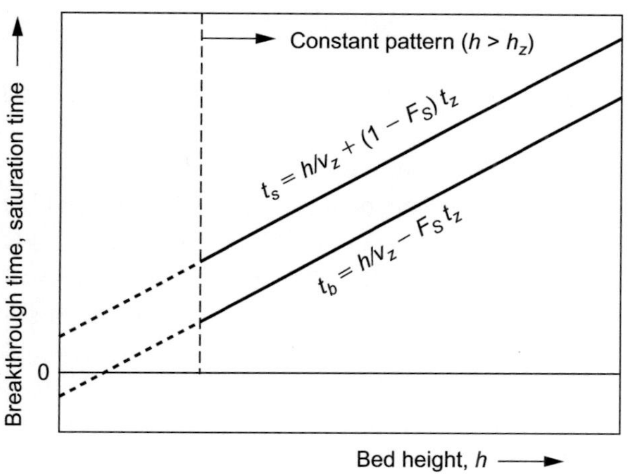

Figure 7.3: Dependence of breakthrough and saturation times on the bed height according to the MTZ model.

For a given adsorber height, the minimum breakthrough time required to establish a constant pattern (MTZ formation time, t_f) can be derived from eq. (7.11) under the condition $h = h_z$

$$t_f = (1 - F_S)\, t_z \tag{7.13}$$

To increase confidence in the estimation of the characteristic MTZ parameters, BTCs should be measured for different bed heights. Then, the parameters h_z, v_z, and F_S can be found by linear regression according to eq. (7.11) or (7.12).

If the equations given previously should be used for scale-up, the possible dependences of the MTZ parameters on the process conditions have to be taken into account. The zone height, h_z, and the symmetry factor, F_S, are only independent of the adsorber dimensions (adsorber height, cross-sectional area) but can be influenced by all factors that have an impact on the mass transfer processes, such as adsorbate concentration, particle diameter, and flow velocity. Therefore, these parameters must be the same in both scales, or, alternatively, the respective dependences have to be determined by experiments.

In contrast to h_z and F_S, the traveling velocity of the MTZ is independent of the mass transfer processes and can be predicted by means of the following equation derived from the integral material balance equation of the fixed-bed adsorber (Chapter 6, Section 6.4.2):

$$v_z \approx \frac{h}{t_{st}} = \frac{v_F \, c_0}{q_0 \, \rho_B} \tag{7.14}$$

If h_z and F_S are known, the amount adsorbed until breakthrough (breakthrough loading, q_b) can be predicted by combining the respective material balance equations with the equations of the MTZ model

$$q_b = q_0 \left(\frac{h - F_S \, h_z}{h} \right) \tag{7.15}$$

where q_0 is the equilibrium adsorbent loading related to the initial concentration, c_0.

An extension of the MTZ model to multicomponent adsorption is only possible under the restrictive condition that the adsorption zones of all components are fully established and do not overlap (Hoppe and Worch 1981a, 1981b). This condition strongly limits the practical applicability of the MTZ model to multicomponent adsorption processes.

7.2.2 Length of unused bed model

The length of unused bed (LUB) model (Collins 1967; Lukchis 1973) is a scale-up model that uses the length of the unused bed at the breakthrough point as a parameter to characterize the breakthrough behavior. The location of the MTZ in a fixed-bed adsorber at the breakthrough point is shown in Figure 7.4.

If the adsorption process is stopped at the breakthrough point, a fraction of the adsorbent capacity remains unused. This fraction is proportional to the distance

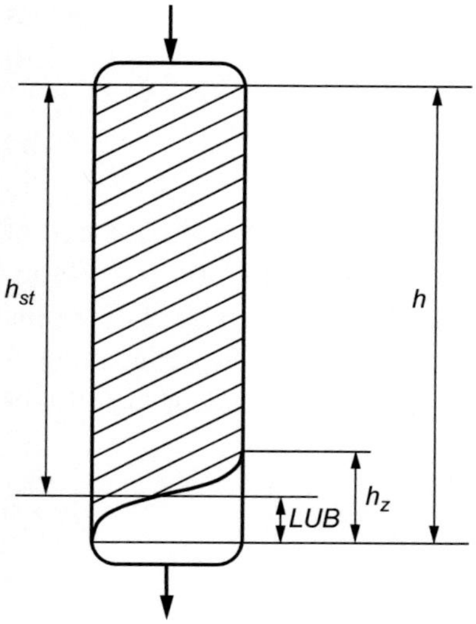

Figure 7.4: Characteristic parameters of the LUB model.

between the location of the stoichiometric front (h_{st}) and the adsorber height, h. Accordingly, the length of the unused bed is given by

$$LUB = h - h_{st} \tag{7.16}$$

The LUB is related to the adsorption rate. The slower the mass transfer processes are, the longer is the LUB.

Since the stoichiometric front travels with the same velocity as the real concentration front, the travel velocity can be expressed by using either the real breakthrough time or the stoichiometric time

$$v_z = \frac{h_{st}}{t_b} = \frac{h}{t_{st}} \tag{7.17}$$

Combining eqs. (7.16) and (7.17) gives

$$LUB = v_z(t_{st} - t_b) = \frac{t_{st} - t_b}{t_{st}} h \tag{7.18}$$

Equation (7.18) allows calculating the LUB on the basis of an experimentally determined BTC. The required stoichiometric time, t_{st}, can be estimated as described for the MTZ model (Section 7.2.1).

For scale-up, at first the desired run time (breakthrough time) for the engineered process has to be defined. For this breakthrough time, the location of the stoichiometric front, h_{st}, can be estimated by eq. (7.17). The zone velocity necessary for the calculation is available from eq. (7.14). The required adsorber height for the desired breakthrough time results from addition of h_{st} and LUB. Finally, the related adsorbent mass can be determined by eq. (6.7) (Chapter 6, Section 6.3).

Rearranging eq. (7.18) gives a relationship that can be used to calculate breakthrough times for different bed heights

$$t_b = \frac{1}{v_z}(h - LUB) \tag{7.19}$$

Generally, the LUB model is subject to the same restrictions as the MTZ model. In particular, the process parameters that have an influence on the LUB (concentration, particle diameter, flow velocity) must be similar in the different scales, or the dependences of the LUB on these parameters must be determined experimentally.

A deeper inspection of the LUB model shows that it is equivalent to the MTZ model. With the equivalence relationship

$$LUB = F_S\, h_z \tag{7.20}$$

both models become identical.

7.2.3 Rapid small-scale column test

The rapid small-scale column test (RSSCT) was developed by Crittenden et al. (1986a, 1987a) as an alternative to time-consuming and expensive pilot-plant studies. The basic idea is to study the breakthrough behavior in small columns that are specifically designed in a manner that guarantees that the operational conditions in the small-scale column reflect exactly the situation in the large-scale fixed-bed adsorber. The respective down-scaling equations were derived from mass transfer models to ensure that the influence of the different mass transfer processes on the breakthrough behavior is identical in both scales. The BTCs determined in such a small-scale column can then be used as the basis for adsorber design. The primary advantages of the RSSCT are as follows:

- In comparison to a pilot study, the RSSCT may be conducted in a much shorter time (days vs. months) and requires a much smaller volume of water.
- Unlike predictive mathematical models, extensive isotherm or kinetic studies are not required to obtain a full-scale performance prediction.

The down-scaling equations for the RSSCT were derived from the dispersed-flow, pore and surface diffusion model (see Table 7.2 in Section 7.4.1). In this model, dimensionless groups are defined that express the relative importance of the mass transfer mechanisms (dispersion, film diffusion, pore diffusion, surface diffusion) and the relative partitioning between the liquid and the solid phase. The definition equations for the dimensionless groups include a number of key variables important for adsorber design, such as length of the fixed bed, interstitial velocity, adsorbent particle radius, bed porosity, particle porosity, and adsorbent density. By setting the dimensionless groups of a small-scale column (SC) equal to those of a large-scale column (LC), relationships among the key variables can be found.

The main down-scaling equation for the RSSCT is

$$\frac{EBCT_{SC}}{EBCT_{LC}} = \left(\frac{d_{P,SC}}{d_{P,LC}}\right)^{2-x} = \frac{t_{SC}}{t_{LC}} \tag{7.21}$$

where $EBCT_{SC}$ and $EBCT_{LC}$ are the empty bed contact times of the small and large columns, respectively; $d_{P,SC}$ and $d_{P,LC}$ are the adsorbent particle diameters in the small and large columns, respectively; and t_{SC} and t_{LC} are the elapsed times in the small and large columns, respectively. The parameter x characterizes the dependence of the intraparticle diffusion coefficients (surface and pore diffusion coefficients) on the particle size. With respect to small and large columns, this dependence can be written as

$$\frac{D_{S,SC}}{D_{S,LC}} = \left(\frac{d_{P,SC}}{d_{P,LC}}\right)^{x} \quad \text{or} \quad \frac{D_{P,SC}}{D_{P,LC}} = \left(\frac{d_{P,SC}}{d_{P,LC}}\right)^{x} \tag{7.22}$$

For down-scaling, at first an appropriate down-scaling factor ($d_{P,LC}/d_{P,SC}$) has to be chosen. This factor defines the particle size that has to be applied in the RSSCT. The smaller-size adsorbent particles are obtained by crushing a representative sample of the adsorbent used in the large-scale adsorber. Based on the down-scaling factor, the other operational parameters for the RSSCT can be determined by using eq. (7.21). With respect to eq. (7.22), two limiting cases have to be considered: constant diffusivity (CD) and proportional diffusivity (PD).

In the case of CD, the diffusion coefficients do not change with particle size ($x = 0$), and eq. (7.21) becomes

$$\frac{EBCT_{SC}}{EBCT_{LC}} = \left(\frac{d_{P,SC}}{d_{P,LC}}\right)^{2} = \frac{t_{SC}}{t_{LC}} \tag{7.23}$$

This equation assures that the extent of spreading of the MTZ caused by intraparticle diffusion in the SC and in the LC is identical in relation to the respective column length. An equal degree of spreading caused by external mass transfer and axial dispersion can be assured if the Reynolds numbers for the SC and the LC are set equal. According to the definition of the Reynolds number

$$Re = \frac{v_F\, d_P}{\varepsilon_B\, \nu} \tag{7.24}$$

where v_F is the filter velocity, ε_B is the bed porosity, and ν is the kinematic viscosity; the following relationship holds:

$$\frac{v_{F,SC}}{v_{F,LC}} = \frac{d_{P,LC}}{d_{P,SC}} \tag{7.25}$$

For both the small-scale and the large-scale adsorber, the height of the adsorber bed is related to the filter velocity and the empty bed contact time by

$$h = EBCT\, v_F \tag{7.26}$$

Combining eqs. (7.23), (7.25), and (7.26) gives the down-scaling equation for the adsorbent bed height

$$h_{SC} = \left(\frac{d_{P,SC}}{d_{P,LC}}\right)^2 EBCT_{LC}\, \frac{d_{P,LC}}{d_{P,SC}}\, v_{F,LC} = \frac{d_{P,SC}}{d_{P,LC}}\, h_{LC} \tag{7.27}$$

In the case of PD, the intraparticle diffusivity is assumed to be proportional to the particle size ($x = 1$). According to eq. (7.21), an identical degree of spreading of the MTZ caused by intraparticle diffusion in relation to the respective column length is given under the condition

$$\frac{EBCT_{SC}}{EBCT_{LC}} = \frac{d_{P,SC}}{d_{P,LC}} = \frac{t_{SC}}{t_{LC}} \tag{7.28}$$

The condition for equal Reynolds numbers is given by eq. (7.25), but in this case, the combination of eqs. (7.25), (7.26), and (7.28) leads to

$$h_{SC} = \frac{d_{P,SC}}{d_{P,LC}}\, EBCT_{LC}\, \frac{d_{P,LC}}{d_{P,SC}}\, v_{F,LC} = h_{LC} \tag{7.29}$$

That means that the bed height in the small-scale adsorber must be the same as in the large-scale adsorber, which would lead to an extremely high pressure drop in the small-scale adsorber. Therefore, Crittenden et al. (1987a) proposed an equation that can be used for reducing the filter velocity as well as the bed height

$$\frac{v_{F,SC}}{v_{F,LC}} = \frac{d_{P,LC}}{d_{P,SC}} \cdot \frac{Re_{SC,min}}{Re_{LC}} \tag{7.30}$$

Here, $Re_{SC,min}$ is the minimum Reynolds number that guarantees the intraparticle diffusion will be rate limiting, and the effects of external mass transfer and dispersion will not be greater in the RSSCT than in the large-scale adsorber. A value of 1 or slightly lower for $Re_{SC,min}$ usually yields good results. If the adsorber length and the resulting pressure drop are unacceptably high, a lower value of $Re_{SC,min}$ has to be used.

To demonstrate the determination of the operational parameters for RSSCT, the results of a down-scaling calculation based on typical large-scale adsorber parameters are shown in Table 7.1. A down-scaling factor ($d_{P,LC}/d_{P,SC}$) of 10 was chosen for this example. Further RSSCT operational parameters can be found after setting the ratio of particle diameter to column diameter to a value that ensures that channeling is avoided (>50 is recommended). In the given example, a ratio of 50 results in a column diameter of 10 mm. If the column diameter is fixed, the cross-sectional area of the filter can be calculated. If the cross-sectional area is known, the SC parameters

Table 7.1: RSSCT design under assumption of constant diffusivity (CD) and proportional diffusivity (PD) based on a down-scaling factor of 10. The large-scale process data for this example were arbitrarily chosen taking into consideration the typical ranges in water treatment practice.

Parameter	Unit	Large-scale column	Small-scale column (CD)	Small-scale column (PD)
d_p	mm	2	0.2	0.2
EBCT	min	20	0.2	2
Re	–	8.5	8.5	1 (Re_{min})
v_F	m/h	6	60	7.06
h	m	2	0.2	0.235

volumetric flow rate, bed volume, and adsorbent mass can be estimated by using the respective equations given in Chapter 6, Section 6.3.

To make the RSCCT and full-scale adsorber breakthrough behavior comparable, the effluent concentration profiles have to be presented by using time-equivalent parameters that normalize the results with respect to bed size. Such normalizing parameters are the bed volumes fed to the adsorber (*BV*) and the specific throughput (volume treated per mass adsorbent, V_{sp}). The definitions of these parameters are also given in Chapter 6, Section 6.3.

It has to be noted that the RSSCT is subject to some limitations:

- The derivation of the down-scaling equation is based on the assumption that the bulk density as well as the bed porosity is the same for the RSSCT and the large-scale adsorber, but crushing of the original adsorbent for use in the RSSCT can change these parameters. The impact of possible differences in bed density and/or bed porosity can be minimized by using the specific throughput to represent the adsorber performance and by calculating the adsorbent mass for the RSSCT on the basis of the bed density of the large-scale adsorber (Crittenden et al. 1991).
- It is not possible to decide from the outset if the CD or the PD approach works better in a considered adsorbate/adsorbent system.
- In large-scale adsorbers with long run times, biological degradation processes, which partially regenerate the adsorbent, may occur. Due to the short duration of the RSSCT, it cannot simulate biodegradation processes and would underestimate the bed life in these cases.
- The impact of dissolved organic matter (DOM) (e.g. carbon fouling) may be different in small-scale and large-scale adsorbers. To this point, different results are reported in the literature.

Due to the problems and uncertainties mentioned previously, RSSCTs can only be used to obtain preliminary information about the adsorber performance. For more design information, at least one pilot-scale experiment is needed to calibrate the RSSCT.

After that, the RSSCT can be used for further studies, such as experiments on the influence of operating conditions or pretreatment processes on the breakthrough behavior of adsorbates.

Special problem: Micropollutant adsorption in the presence of background organic matter

As previously discussed, problems may arise when the RSSCT should be applied to systems consisting of micropollutant (MP) and background organic matter (measured as DOC), because the RSSCT cannot adequately map the effect of carbon fouling in the full-scale adsorber. The term “carbon fouling” is used to describe the permanent reduction of the adsorption capacity for the MP through direct site competition and pore blockage by irreversibly adsorbed background organic matter. Since the carbon fouling is not considered in the RSSCT, MP breakthrough profiles obtained in the test typically overpredict those obtained at the full scale.

Corwin and Summers (2010) have developed a method for adjusting PD-RSSCT breakthrough data of micropollutants to match full-scale data by introducing a fouling index, *FI*. *FI* equals the scaling factor, $d_{P,LC}/d_{P,SC}$, raised to the power of *Y*, where *Y* is the fouling factor

$$FI = \left(\frac{d_{P,LC}}{d_{P,SC}}\right)^{Y} \tag{7.31}$$

The fouling index can be used to normalize the PD-RSSCT throughput, expressed as bed volumes

$$BV_{LC} = \frac{BV_{SC}}{FI} \tag{7.32}$$

If a PD-RSSCT matches the full-scale adsorption capacity perfectly (i.e. no fouling), *Y* would be equal to zero ($FI = 1$). The higher the value of *Y*, the more fouling occurs. However, an independent estimation of *Y* on a theoretical basis is not possible, which limits the applicability of the normalizing method for full-scale prediction purposes.

Kennedy et al. (2017) found that the fouling factor mainly depends on the concentration ratio of MP and DOC, $c_{0,MP}/c_{0,DOC}$; the bed volume until 10% breakthrough in the PD-RSSCT, $BV_{10\%,PD\text{-}RSSCT}$; and the logarithm of the pH-dependent octanol-water partition coefficient, log *D*. The dependence could be expressed by the following empirical correlation:

$$Y = \left(a_0 + a_1\frac{c_{0,MP}}{c_{0,DOC}} + a_2 BV_{10\%}^{PD-RSSCT} + a_3 \log D + a_4\frac{c_{0,MP}}{c_{0,DOC}} BV_{10\%}^{PD-RSSCT} + a_5 BV_{10\%}^{PD-RSSCT} \log D\right)^{a_6} \tag{7.33}$$

with the empirical parameters $a_0 = 2.59 \pm 0.11$, $a_1 = 3.94 \pm 1.48$, $a_2 = -7.87 \times 10^{-6} \pm 1.29 \times 10^{-6}$, $a_3 = -0.402 \pm 0.054$, $a_4 = 4.20 \times 10^{-4} \pm 1.73 \times 10^{-4}$, $a_5 = 2.86 \times 10^{-6} \pm 5.87 \times 10^{-7}$, and $a_6 = -1.47 \pm 0.06$. The reported Y values are in the range between 0.2 and 0.7.

The following trends become apparent:

The fouling factor, Y, increases

- with increasing bed volumes to 10% MP breakthrough because there is more time for fouling to occur
- with decreasing $c_{0,MP}/c_{0,DOC}$ due to the reduced ability of the MP to compete for adsorption sites
- with increasing log D (increasing hydrophobicity), which leads to slower traveling of the MTZ (more time for fouling).

As an alternative to the FI method, Kennedy et al. (2017) developed an empirical relationship between the $BV_{10\%}$ values obtained from the PD-RSSCT, $BV_{10\%}^{PD-RSSCT}$, and those measured in pilot columns, $BV_{10\%}^{full\ scale}$. For 73 pairs of RSSCT and pilot column data points, they found the following linear relationship:

$$\ln BV_{10\%}^{full\ scale} = (0.57 \pm 0.32) + (0.855 \pm 0.029) \ln BV_{10\%}^{PD-RSSCT} \tag{7.34}$$

Kearns et al. (2020) extended this regression approach by including additional micropollutants, adsorbents, and water types. Furthermore, not only PD-RSSCTs but also CD-RSSCTs were considered. For 115 data pairs, the following relationship was found:

$$\ln BV_{10\%}^{full\ scale} = (1.21 \pm 0.41) + (0.80 \pm 0.039) \ln BV_{10\%}^{RSSCT} \tag{7.35}$$

It has to be noted that the equations are restricted to early breakthrough and give no information about the shape of the MP breakthrough curves at higher relative concentrations. Furthermore, conclusions about the relative contributions of the different mass transfer mechanisms cannot be drawn from this empirical approach.

Despite these limitations, the empirical equations help to better adapt the RSSCTs to the practical conditions that are found in drinking water or wastewater treatment.

7.3 Equilibrium column model

The equilibrium column model (ECM) is the simplest BTC prediction model. It assumes an instantaneous establishment of the equilibrium state and neglects the influence of dispersion and adsorption kinetics on the shape of the BTC. Therefore, this model requires only isotherm and basic process data as input. This simplification, however, leads to the restriction that the ECM can only predict concentration steps (ideal BTCs) but not real S-shaped BTCs (see Chapter 6, Section 6.4.2).

Consequently, the BTC prediction is reduced to a prediction of the ideal breakthrough time. Accordingly, appropriate equations for single-solute as well as for multisolute adsorbate systems can be derived from the integral material balance. Although this model does not reflect the real breakthrough behavior, it can be used to identify the maximum service life of a fixed-bed adsorber.

According to eq. (6.21) (Chapter 6), the ideal breakthrough time for a single adsorbate, neglecting the residence time of water, is given by

$$t_b^{id} \approx t_{st} = \frac{q_0 \, m_A}{\dot{V} \, c_0} \tag{7.36}$$

where m_A is the adsorbent mass, $\dot{V}$ is the volumetric flow rate, c_0 is the inlet concentration, and q_0 is the adsorbed amount in equilibrium with the inlet concentration. The latter is available from the isotherm.

The extension of the ECM to multisolute systems will be exemplarily demonstrated for a three-component system. The three ideal BTCs are shown schematically in Figure 7.5. As can be seen, three zones have to be distinguished. In the first zone between $t_{st,1}$ and $t_{st,2}$, only the weakest adsorbable component 1 is present; in the second zone between $t_{st,2}$ and $t_{st,3}$, components 1 and 2 are present; and the third zone ($t > t_{st,3}$) is characterized by the presence of all three components. Zone 3 represents the state of equilibrium in the three-component system. Furthermore, taking into account that the areas above the line $c/c_0 = 1$ represent desorption (displacement by the stronger adsorbable components, see Chapter 6, Section 6.4.2), the following material balance equation for the first (weakest adsorbable) component can be derived:

$$q_{0,1} \, m_A = c_{0,1} \, \dot{V} \, t_{st,1} - (c_{1,1} - c_{1,3}) \, \dot{V} \, (t_{st,2} - t_{st,1}) - (c_{1,2} - c_{1,3}) \, \dot{V} \, (t_{st,3} - t_{st,2}) \tag{7.37}$$

For the second component, the material balance equation reads

$$q_{0,2} \, m_A = c_{0,2} \, \dot{V} \, t_{st,2} - (c_{2,2} - c_{2,3}) \, \dot{V} \, (t_{st,3} - t_{st,2}) \tag{7.38}$$

and for the third (strongest adsorbable) component, the material balance equation is simply given by

$$q_{0,3} \, m_A = c_{0,3} \, \dot{V} \, t_{st,3} \tag{7.39}$$

Analogous to the derivation of eq. (6.38) shown in Section 6.4.2, a set of material balance equations can be written that includes the concentrations and adsorbed amounts in the different plateau zones

$$\frac{q_{1,1}}{c_{1,1}} = \frac{t_{st,1} \, \dot{V}}{m_A} \tag{7.40}$$

$$\frac{q_{1,1} - q_{1,2}}{c_{1,1} - c_{1,2}} = \frac{q_{2,2}}{c_{2,2}} = \frac{t_{st,2} \, \dot{V}}{m_A} \tag{7.41}$$

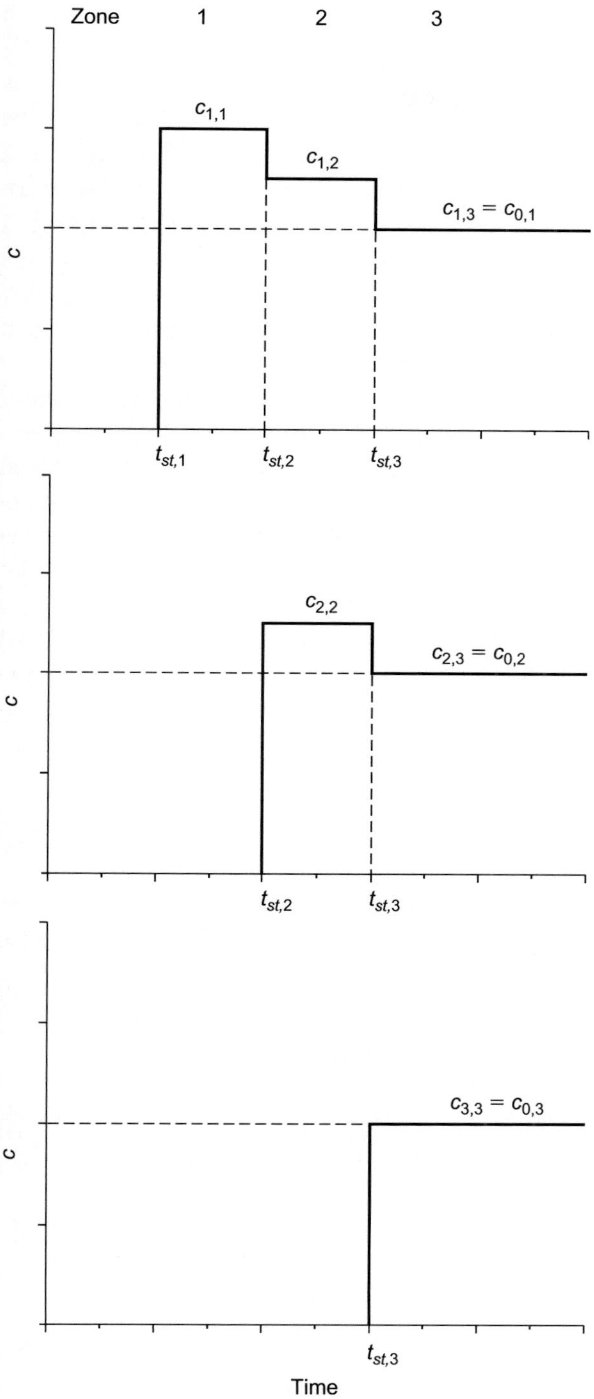

Figure 7.5: Equilibrium column model. Concentration profiles of a three-component adsorbate system.

$$\frac{q_{1,2}-q_{1,3}}{c_{1,2}-c_{1,3}}=\frac{q_{2,2}-q_{2,3}}{c_{2,2}-c_{2,3}}=\frac{q_{3,3}}{c_{3,3}}=\frac{t_{st,3}\,\dot{V}}{m_A} \tag{7.42}$$

Herein, the first subscript denotes the component, and the second subscript denotes the zone. The concentrations in the third zone ($c_{1,3}$, $c_{2,3}$, and $c_{3,3}$) are identical with the initial concentrations $c_{0,1}$, $c_{0,2}$, and $c_{0,3}$. Since equilibrium is assumed for all the zones, the respective equilibrium relationships can be used to find the adsorbed amounts related to the concentrations. For zone 1, the single-solute isotherm of component 1 has to be used, whereas for the zones 2 and 3, where two or three components are present, a competitive adsorption model (e.g. the ideal adsorbed solution theory [IAST]) is required to calculate the equilibrium data (see Chapter 4).

In the given example, the calculation starts with zone 3 where the concentrations $c_{1,3}$, $c_{2,3}$, and $c_{3,3}$ are identical with the known initial concentrations. The corresponding adsorbed amounts $q_{1,3}$, $q_{2,3}$, and $q_{3,3}$ can be found by applying the IAST as shown in Chapter 4, Section 4.5.2 (solution for given equilibrium concentrations). If the single-solute isotherms of the mixture components can be described by the Freundlich isotherm, the following set of equations has to be solved with $N=3$:

$$\sum_{i=1}^{N}\frac{c_{i,N}}{\left(\dfrac{\varphi\, n_i}{K_i}\right)^{1/n_i}}=\sum_{i=1}^{N}z_i=1 \tag{7.43}$$

$$\frac{1}{q_T}=\sum_{i=1}^{N}\frac{z_i}{\varphi\, n_i} \tag{7.44}$$

$$q_{i,N}=z_i\,q_T \tag{7.45}$$

After knowing $c_{3,3}$ and $q_{3,3}$, the stoichiometric time for component 3, $t_{st,3}$, is available from eq. (7.42). The same equation, together with a second relationship, can be used to calculate $q_{1,2}$, $q_{2,2}$, $c_{1,2}$, and $c_{2,2}$ for zone 2. The required second equation is derived from the IAST (see Sections 4.5.1 and 4.5.2 in Chapter 4 and Section 10.2 in Chapter 10) as

$$c_{i,N}=z_{i,N}\,c_{i,N}^{0}=\frac{q_{i,N}}{\sum\limits_{j=1}^{N}q_{j,N}}\left[\frac{n_i\sum\limits_{j=1}^{N}\dfrac{q_{j,N}}{n_j}}{K_i}\right]^{1/n_i} \tag{7.46}$$

with $N=2$ for the second zone. With $c_{2,2}$ and $q_{2,2}$, the stoichiometric time for component 2, $t_{st,2}$, is available from eq. (7.41). To find the concentration and adsorbed amount in the zone 1 ($c_{1,1}$ and $q_{1,1}$), the Freundlich isotherm equation

$$q_{1,1}=K_1\,c_{1,1}^{n_1} \tag{7.47}$$

has to be solved together with eq. (7.41). Finally, the stoichiometric time for component 1, $t_{st,1}$, can be calculated from eq. (7.40).

The ECM can be easily extended to more than three components. The general material balance equation for an N-component system reads

$$\frac{q_{1,N-1}-q_{1,N}}{c_{1,N-1}-c_{1,N}}=\frac{q_{2,N-1}-q_{2,N}}{c_{2,N-1}-c_{2,N}}=\ldots=\frac{q_{N-1,N-1}-q_{N-1,N}}{c_{N-1,N-1}-c_{N-1,N}}=\frac{q_{N,N}}{c_{N,N}}=\frac{t_{st,N}\,\dot{V}}{m_A} \tag{7.48}$$

The solution algorithm (backward from zone N to zone 1) is analogous to that shown previously for the three-component system.

The ECM can also be used to predict BTCs of dissolved organic matter, measured as dissolved organic carbon (DOC). In this case, an adsorption analysis (see Chapter 4, Section 4.7.2) has to be carried out prior to the ECM application. With the isotherm parameters and the concentrations of the fictive components resulting from the adsorption analysis, the ideal BTCs of the DOC fractions can be predicted. Subsequently, the individual BTCs of the fictive components have to be added to get the total DOC BTC. Figure 7.6 shows, as an example, a DOC BTC calculated by the ECM using the results of an adsorption analysis.

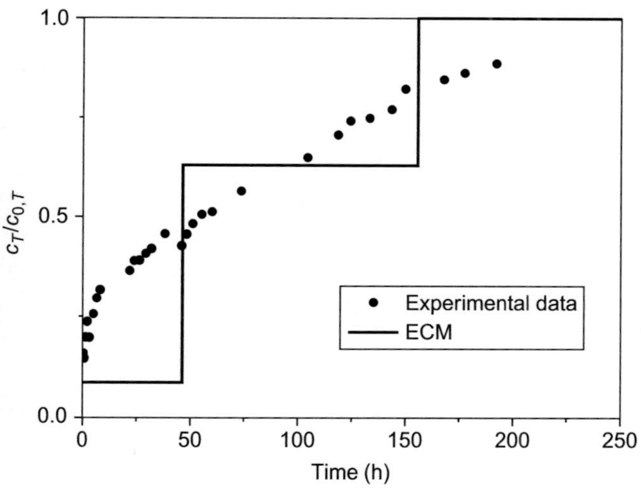

Figure 7.6: Approximate description of a DOC breakthrough curve by the ECM. Experimental data: lab-scale experiment with Elbe River water, 5.2 mg/L DOC (Rabolt 1998).

7.4 Complete breakthrough curve models

7.4.1 Introduction

Here and in the following sections, the term "complete BTC model" is used to indicate that the model considers not only the adsorption equilibrium but also the adsorption kinetics. Consequently, such a complete BTC model can predict real S-shaped BTCs and not only ideal BTCs (ideal breakthrough times) as the ECM does.

In general, a complete BTC model consists of three constituents: (1) the material balance equation, (2) the equilibrium relationship, and (3) a set of equations describing the external and internal mass transfer.

As the material balance equation, the differential material balance (Chapter 6, Section 6.4.3) has to be used. In principle, either the general form or the simplified form, neglecting the dispersion, can be applied. The general form reads

$$v_F \frac{\partial c}{\partial z} + \varepsilon_B \frac{\partial c}{\partial t} + \rho_B \frac{\partial \bar{q}}{\partial t} - D_{ax}\, \varepsilon_B \frac{\partial^2 c}{\partial z^2} = 0 \tag{7.49}$$

In most cases, the dispersion term in the material balance equation is neglected because under the typical flow conditions in engineered adsorption processes, dispersion has no (or only a minor) impact on the breakthrough behavior. Accordingly, eq. (7.49) reduces to

$$v_F \frac{\partial c}{\partial z} + \varepsilon_B \frac{\partial c}{\partial t} + \rho_B \frac{\partial \bar{q}}{\partial t} = 0 \tag{7.50}$$

Models that consider the dispersion are referred to as "dispersed-flow models", whereas models that neglect dispersion are termed "plug-flow models".

In the case of single-solute adsorption, an appropriate isotherm equation has to be used as the equilibrium relationship, whereas in the case of multisolute adsorption, a competitive adsorption model, typically the IAST (Chapter 4, Section 4.5.2), has to be applied. For adsorption from aqueous solutions, the Freundlich isotherm is most frequently used, either to describe single-solute adsorption or as the basic isotherm equation within the IAST. Although in principle possible, other isotherm equations or other competitive adsorption models are rarely applied for modeling engineered adsorption processes in aqueous systems.

The adsorption kinetics (film diffusion, pore diffusion, surface diffusion) can be described by means of the respective mass transfer or diffusion equations presented in Chapter 5.

Table 7.2 lists some BTC models, which differ in the assumptions about dispersion and dominating internal diffusion processes. Although models that include all three mass transfer processes (film, pore, and surface diffusion) were proposed in the literature (e.g. Crittenden et al. 1987b), such complex models are seldom used due to the limited availability of the different internal mass transport parameters. As already discussed in Chapter 5, pore diffusion and surface diffusion act in parallel in the interior of the adsorbent particles, and the related diffusion coefficients, D_P and D_S, cannot be determined independently. Only an approximate estimation of the diffusion coefficients is possible as shown in Chapter 5, Section 5.4.5.

Table 7.2: Breakthrough curve models with different assumptions about the dominating mass transfer processes.

Model	Mass transfer mechanisms			
	Dispersion	Film diffusion	Pore diffusion	Surface diffusion
Dispersed-flow, pore and surface diffusion model	×	×	×	×
Dispersed-flow, pore diffusion model	×	×	×	
Dispersed-flow, homogeneous surface diffusion model	×	×		×
Plug-flow, pore and surface diffusion model		×	×	×
Plug-flow, pore diffusion model		×	×	
Plug-flow, homogeneous surface diffusion model		×		×
Linear driving force model		×	(×)	×

Therefore, models considering only a single internal transport mechanism, in addition to the external film diffusion, are often preferred. In most cases, the internal diffusion is assumed to be dominated by surface diffusion, and possible additional transport by pore diffusion is considered by the effective surface diffusion coefficient. This frequently applied BTC model is referred to as the homogeneous surface diffusion model (HSDM).

As described in Chapter 5, Section 5.4.3, the driving force for surface diffusion is a nonlinear gradient of the adsorbed amount along the radial coordinate of the adsorbent particle. According to this, the radial coordinate, r, has to be considered a third coordinate in the model equations in addition to the axial coordinate of the adsorber, z, and the time, t. To simplify the HSDM, an average loading of the adsorbent particle can be introduced that substitutes the location-depending adsorbent loading. This reduces the number of coordinates (two instead of three) and, consequently, the mathematical effort for solving the set of model equations. This simplified model is known as the linear driving force (LDF) model (see also Section 5.4.6 in Chapter 5). In this model, the driving force for internal diffusion is described by the difference between the adsorbent loading at the external surface and the mean loading of the adsorbent particle. The mathematical form of the simplified internal mass transfer equation is similar to that used for film diffusion.

Due to their frequent application in practice, the HSDM and the LDF model will be discussed in more detail in the following sections.

7.4.2 Homogeneous surface diffusion model

To derive the homogeneous surface diffusion model (HSDM), the material balance for the fixed-bed adsorber has to be combined with the equations for film and surface diffusion and with the equilibrium relationship. In principle, the HSDM can be formulated with and without considering dispersion. Here, the dispersion will be neglected.

Starting with eq. (7.50) and substituting the differential quotient $\partial\bar{q}/\partial t$ by the mass transfer equation for film diffusion (Chapter 5, Section 5.4.2)

$$\frac{\mathrm{d}\bar{q}}{\mathrm{d}t} = \frac{k_F\, a_{VR}}{\rho_B}(c - c_s) \tag{7.51}$$

gives

$$v_F \frac{\partial c}{\partial z} + \varepsilon_B \frac{\partial c}{\partial t} + k_F\, a_{VR}(c - c_s) = 0 \tag{7.52}$$

where a_{VR} is the external surface area related to the reactor volume. For spherical particles, a_{VR} is given by (Chapter 5, Table 5.1)

$$a_{VR} = \frac{3}{r_P}(1 - \varepsilon_B) \tag{7.53}$$

and eq. (7.52) becomes

$$v_F \frac{\partial c}{\partial z} + \varepsilon_B \frac{\partial c}{\partial t} + \frac{3\, k_F(1 - \varepsilon_B)}{r_P}(c - c_s) = 0 \tag{7.54}$$

The rate equation for surface diffusion with a constant diffusion coefficient can be expressed as (Chapter 5, Section 5.4.3)

$$\frac{\partial q}{\partial t} = D_S\left(\frac{\partial^2 q}{\partial r^2} + \frac{2}{r}\frac{\partial q}{\partial r}\right) \tag{7.55}$$

The HSDM will be completed by the equilibrium condition that relates q with c_s at the external surface of the adsorbent particle

$$q(t, z, r = r_P) = \mathrm{f}(c_s(t, z)) \tag{7.56}$$

Equations (7.54–7.56) allow determining the three dependent variables, c, c_s, and q. The initial and boundary conditions for eq. (7.55) are

$$q(t = 0, z, r) = 0 \tag{7.57}$$

$$\left[\frac{\partial q(t, z, r)}{\partial r}\right]_{r=0} = 0 \tag{7.58}$$

$$\left[\frac{\partial q(t,z,r)}{\partial r}\right]_{r=r_P} = \frac{k_F}{\rho_P D_S}\left[c(t,z) - c_s(t,z)\right] \tag{7.59}$$

Equation (7.58) follows from the symmetry of the solid-phase concentration at the center of the adsorbent particle, and eq. (7.59) describes the continuity of the mass transfer through the film and into the particle. The boundary condition for eq. (7.54) is

$$c(t,z=0) = c_0 \tag{7.60}$$

To reduce the number of model parameters and to simplify the mathematical solution, the model equations are usually transformed into their dimensionless forms. To that purpose, dimensionless parameters have to be defined.

From the integral material balance equation (Chapter 6, eq. (6.19)), a solute distribution parameter, D_g, can be derived. This parameter relates the mass of adsorbate in the solid phase to the mass of adsorbate in the liquid phase under equilibrium conditions (i.e. first and second term of the right-hand side of the integral material balance equation)

$$D_g = \frac{q_0\, m_A}{c_0\, \varepsilon_B\, V_R} = \frac{q_0\, \rho_B}{c_0\, \varepsilon_B} \tag{7.61}$$

According to eq. (6.31) (Chapter 6), D_g is related to the throughput ratio, T, which is used in the model as a dimensionless time coordinate

$$T = \frac{t}{t_b^{id}} = \frac{t}{t_{st} + t_r} = \frac{t}{\dfrac{m_A\, q_0}{\dot{V}\, c_0} + \dfrac{V_R\, \varepsilon_B}{\dot{V}}} = \frac{t}{t_r\,(D_g + 1)} \tag{7.62}$$

If the residence time, t_r, is very short in comparison to the stoichiometric time, which is typically the case for strongly adsorbable solutes, eq. (7.62) simplifies to

$$T = \frac{t}{t_b^{id}} \approx \frac{t}{t_{st}} = \frac{t}{\dfrac{m_A\, q_0}{\dot{V}\, c_0}} = \frac{t}{t_r\, D_g} \tag{7.63}$$

The other dimensionless parameters used in the HSDM are listed in Table 7.3. The Stanton number, St^*, and the diffusion modulus, E_d, describe the rate of the respective mass transfer processes (film and surface diffusion) compared to the rate of advection. The Biot number, Bi, characterizes the relative influence of the external and internal mass transfer on the overall mass transfer rate. Since film diffusion and surface diffusion are consecutive processes, the slower process determines the overall kinetics. If Bi increases, the film mass transfer becomes faster as compared to the intraparticle mass transfer. As a rule, intraparticle mass transfer controls the adsorption rate at $Bi > 50$, and film diffusion controls the adsorption rate at $Bi < 0.5$. In the intermediate range, both mechanisms are relevant for the overall adsorption rate.

Table 7.3: Dimensionless parameters used in the HSDM.

Dimensional parameter	Symbol	Definition
Dimensionless concentration	X	$X = \frac{c}{c_0}$
Dimensionless adsorbent loading	Y	$Y = \frac{q}{q_0}$
Dimensionless radial coordinate (within the particle)	R	$R = \frac{r}{r_P}$
Dimensionless axial coordinate (distance from adsorber inlet)	S	$S = \frac{z}{h}$
Solute distribution parameter	D_g	$D_g = \frac{q_0\, m_A}{c_0\, \varepsilon_B\, V_R} = \frac{q_0\, \rho_B}{c_0\, \varepsilon_B}$
Dimensionless time (throughput ratio)	T	$T = \frac{t}{t_b^{id}} = \frac{t}{t_r\,(D_g+1)}$
Stanton number (transport rate ratio: film diffusion/advection)	St^*	$St^* = \frac{k_F\, t_r\,(1-\varepsilon_B)}{\varepsilon_B\, r_P}$
Diffusion modulus (transport rate ratio: surface diffusion/advection)	E_d	$E_d = \frac{D_S\, D_g\, t_r}{r_P^2}$
Biot number (transport rate ratio: film diffusion/surface diffusion)	Bi	$Bi = \frac{(1-\varepsilon_B)\, r_P\, c_0\, k_F}{\rho_B\, q_0\, D_S} = \frac{St^*}{E_d}$

By using the dimensionless parameters, the equations of the HSDM can be written as

$$\frac{\partial X}{\partial S} + \frac{1}{(D_g+1)}\frac{\partial X}{\partial T} + 3\,St^*(X - X_s) = 0 \tag{7.64}$$

$$\frac{\partial Y}{\partial T} = E_d\,\frac{D_g+1}{D_g}\left(\frac{\partial^2 Y}{\partial R^2} + \frac{2}{R}\frac{\partial Y}{\partial R}\right) \tag{7.65}$$

$$Y(T,S,R=1) = \mathrm{f}(X_s(T,S)) \tag{7.66}$$

The initial and boundary conditions for eq. (7.65) are

$$Y(T=0,S,R) = 0 \tag{7.67}$$

$$\left[\frac{\partial Y(T,S,R)}{\partial R}\right]_{R=0} = 0 \tag{7.68}$$

$$\left[\frac{\partial Y(T,S,R)}{\partial R}\right]_{R=1} = Bi\,[X(T,S) - X_s(T,S)] \tag{7.69}$$

and the boundary condition for eq. (7.64) is

$$X(T, S=0)=1 \tag{7.70}$$

For the frequently used Freundlich isotherm, the explicit form of eq. (7.66) reads (see Chapter 3, Section 3.7)

$$Y(T, S, R=1)=(X_s)^n \tag{7.71}$$

To solve the set of eqs. (7.64–7.66) under consideration of the initial and boundary conditions, numerical methods have to be applied.

The HSDM can also be extended to multisolute systems. For that purpose, it is necessary to consider the competitive adsorption effects in the BTC model. As a rule, the consideration of competitive effects is restricted to the equilibrium, whereas the impact of competition on the adsorption kinetics is assumed to be negligible (Chapter 5, Section 5.4.8). Consequently, the isotherm equation has to be substituted by an appropriate competitive adsorption model, typically the IAST. The mass transfer equations, which have to be formulated separately for each component, are the same as used in the case of single-solute adsorption.

As in the case of single-solute adsorption, it is appropriate to write the model equations in their dimensionless forms. An essential precondition for the numerical solution is that the same normalized coordinates are valid for each component in the multisolute system. This requirement is fulfilled for the radial and axial coordinates because their normalization is independent of the adsorbate data. In contrast, the dimensionless time (throughput ratio, T) is defined on the basis of the stoichiometric time or the solute distribution parameter. Both are related to the equilibrium data, q_0 and c_0, and therefore, the different components have individual values. That would lead to different dimensionless time coordinates for the different components. To overcome this problem, a fictive reference value (t_{st} or D_g) can be defined that is valid for all components – for instance, by using mean values for concentration and loading.

7.4.3 Constant pattern approach to the HSDM

Hand et al. (1984) developed a method that allows describing BTCs by simple polynomials. This method, based on the HSDM and usually abbreviated as CPHSDM, is applicable to single-solute adsorption under constant pattern conditions.

It can be seen from the HSDM (Section 7.4.2) that after introducing dimensionless measures, only five parameters are necessary to describe all the factors that influence the BTC. These parameters are the distribution coefficient, D_g; the Biot number, Bi; the Stanton number, St^*; the diffusion modulus, E_d; and the Freundlich exponent, n.

The parameters D_g, Bi, St^*, and E_d are defined in Table 7.3. Given that E_d can be expressed by the ratio St^*/Bi and D_g is included in the dimensionless time, T, by

$$T = \frac{t}{t_b^{id}} = \frac{t}{t_r\,(D_g + 1)} \tag{7.72}$$

only three independent parameters (n, Bi, St^*) are necessary to describe all solutions to the HSDM in the form $X = f(T)$. If it is further assumed that constant pattern is established after a minimum bed contact time, only the solution for this minimum contact time has to be determined. For all longer contact times, the solutions can be found by simple parallel translation. Hand et al. (1984) have estimated the minimum Stanton number, St^*_{min}, that is necessary to establish constant pattern by model calculations with the exact HSDM. They have found that the minimum Stanton number depends on n and Bi, and the relationship between these parameters can be expressed by the linear equation

$$St^*_{min} = A_0\,Bi + A_1 \tag{7.73}$$

where A_0 and A_1 are empirical parameters that depend on n. Furthermore, the numerically calculated BTCs for the minimum Stanton number can be approximated by the empirical polynomial

$$T(n, Bi, St^*_{min}) = A_0 + A_1\,X^{A_2} + \frac{A_3}{1.01 - X^{A_4}} \tag{7.74}$$

Tables with the parameters for eqs. (7.73) and (7.74) are given in the Appendix (Tables 10.5 and 10.6). Based on these equations, a BTC prediction can be carried out in the following manner. At first, the minimum Stanton number for the given Bi and n has to be estimated by using eq. (7.73). After that, the BTC for this minimum Stanton number and for the given Bi and n can be calculated by eq. (7.74). The minimum residence time related to St^*_{min} can be found after rearranging the definition equation for the Stanton number given in Table 7.3

$$t_{r,min} = \frac{St^*_{min}\,r_P\,\varepsilon_B}{k_F(1 - \varepsilon_B)} \tag{7.75}$$

The transformation of the throughput ratio into the real time is possible by means of D_g

$$t_{min} = t_{r,min}(D_g + 1)\,T \tag{7.76}$$

and the parallel translation can be carried out by using the residence time for the process conditions of interest, t_r,

$$t = t_{min} + (t_r - t_{r,min})\,(D_g + 1) \tag{7.77}$$

As can be derived from eq. (7.77), the residence time in the considered adsorber must be greater than the minimum residence time necessary for establishing constant pattern. Otherwise, the time, t, would be shorter than t_{min}, and the precondition for the calculation method (constant pattern) is not fulfilled.

To illustrate the quality of the simplified BTC prediction method, results of CPHSDM calculations for different Biot numbers are compared with exact HSDM results in Figure 7.7. In the graph, the throughput is expressed as bed volumes (see eq. (6.17) in Chapter 6). For the model calculations, practice-oriented process data were used (Table 7.4). Since the film mass transfer coefficient is determined by the filter velocity, it cannot be freely chosen. The variation of Bi was therefore done by changing only the surface diffusion coefficient. For the lower Biot number

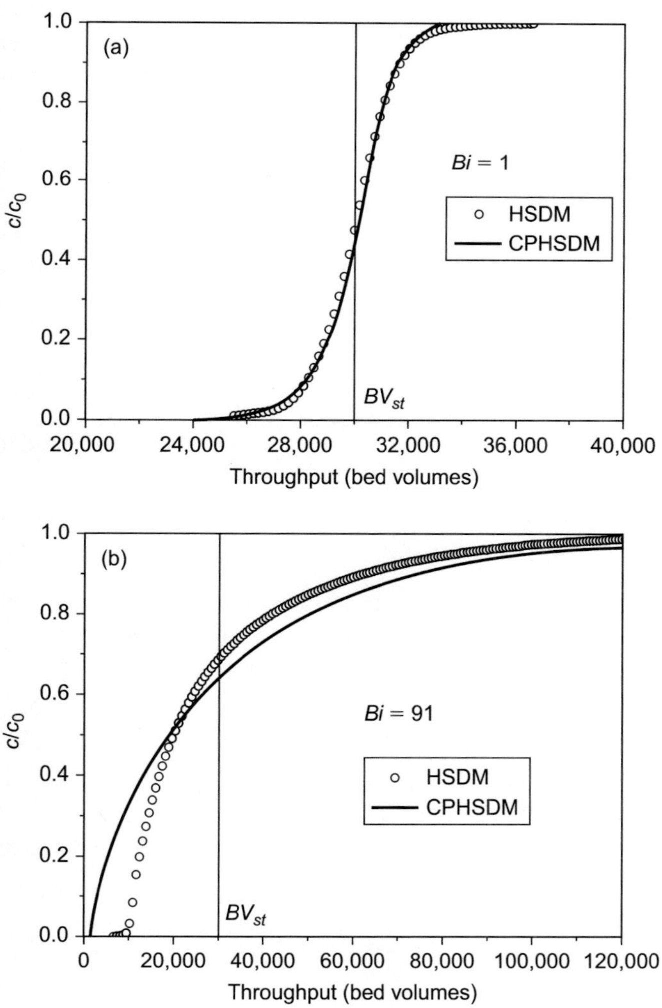

Figure 7.7: Comparison of the constant pattern solution (CPHSDM) with the exact HSDM solution for (a) $Bi = 1$ and (b) $Bi = 91$ (for other data, see Table 7.4).

Table 7.4: Process parameters used for comparative model calculations with the HSDM, CPHSDM, and LDF model.

Parameter	Unit	Value	
Adsorbent bed height, h	m	2	
Adsorber diameter, d_R	m	2	
Bed density, ρ_B	kg/m^3	500	
Bed porosity, ε_B	–	0.4	
Adsorbent particle diameter, d_P	mm	2	
Inlet concentration, c_0	mg/L	1	
Adsorbent mass, m_A	kg	3,140	
Empty bed contact time, $EBCT$	min	20	
Superficial filter velocity, v_F	m/h	6	
Freundlich coefficient, K	$(mg/g)/(mg/L)^n$	60	
Freundlich exponent, n	–	0.3	
Film mass transfer coefficient, k_F	m/s	$1.37 \cdot 10^{-5}$	
Volumetric film mass transfer coefficient, $k_F a_{VR}$	1/s	$2.46 \cdot 10^{-2}$	
Intraparticle mass transfer coefficient, k_S^*	1/s	$4.0 \cdot 10^{-6}$	$4.40 \cdot 10^{-8}$
Surface diffusion coefficient, D_S	m^2/s	$2.67 \cdot 10^{-13}$	$2.93 \cdot 10^{-15}$
Biot number, Bi	–	1	91

(Figure 7.7a), the constant pattern condition is fulfilled ($t_{r,min} = 0.036$ h, $t_r = 0.133$ h) and the predicted BTC matches the exact HSDM solution. However, with increasing Biot number, the minimum Stanton number and also the minimum residence time increases. In the example shown in Figure 7.7b ($Bi = 91$), the constant pattern condition is not fulfilled ($t_{r,min} = 0.44$ h, $t_r = 0.133$ h) despite the relatively long adsorber, and, consequently, the CPHSDM solution differs considerably from the exact solution. Under this condition, the CPHSDM predicts an earlier breakthrough. This can be explained by the fact that the MTZ is still in the formation phase and is therefore shorter than the fully established constant pattern MTZ for which the BTC was calculated. A formal parallel translation of the broader constant pattern BTC to earlier times according to eq. (7.77) must therefore lead to a breakthrough time that is shorter than the real breakthrough time.

7.4.4 Linear driving force model

The linear driving force (LDF) model is an alternative to the HSDM. The characteristic of the LDF model consists of a simplified description of the intraparticle diffusion (Chapter 5, Section 5.4.6). Instead of Fick's law (eq. (7.55)), a mass transfer equation with linear driving force is used, analogous to the LDF approach for film diffusion. That makes the solution easier, which is an advantage, in particular, in the case of complex multisolute systems.

Under the further assumptions of negligible dispersion and validity of the Freundlich isotherm, the following set of model equations can be derived:

$$v_F \frac{\partial c}{\partial z} + \varepsilon_B \frac{\partial c}{\partial t} + \rho_B \frac{\partial \bar{q}}{\partial t} = 0 \tag{7.78}$$

$$c(t=0,z)=0, \quad \bar{q}(t=0,z)=0 \tag{7.79}$$

$$c(t,z=0)=c_0 \tag{7.80}$$

$$\frac{\mathrm{d}\bar{q}}{\mathrm{d}t} = \frac{k_F\, a_{VR}}{\rho_B}(c-c_s) \tag{7.81}$$

$$\frac{\mathrm{d}\bar{q}}{\mathrm{d}t} = k_S^*(q_s - \bar{q}) \tag{7.82}$$

$$q_s = K\,(c_s)^n \tag{7.83}$$

where k_F is the film mass transfer coefficient, a_{VR} is the area available for mass transfer related to the reactor volume, and k_S^* is the intraparticle mass transfer coefficient as defined in Chapter 5, Section 5.4.6. The concentration and adsorbent loading at the external particle surface, c_s and q_s, are assumed to be in the state of equilibrium, expressed by the Freundlich isotherm (eq. (7.83)). The volume-related surface area, a_{VR}, can be estimated for spherical particles by eq. (7.53).

According to Glueckauf's approach (Chapter 5, Section 5.4.6), the intraparticle mass transfer coefficient, k_S^*, is related to the surface diffusion coefficient, D_S, by

$$k_S^* = \frac{15\,D_S}{r_P^2} \tag{7.84}$$

where r_P is the particle radius. In principle, the modified mass transfer coefficient, k_S^*(1/s), which has the character of a volumetric mass transfer coefficient, can be further separated into a mass transfer coefficient, k_S (m/s), and a volume-related surface area, a_{VR} (m^2/m^3), as shown in Chapter 5 (Table 5.2). However, such a separation of k_S^* is not necessary for the practical application of the LDF model.

For systems where pore diffusion contributes to the overall mass transfer, eq. (7.84) can be extended to

$$k_S^* = \frac{15 D_{S,eff}}{r_P^2} = \frac{15 D_S}{r_P^2} + \frac{15\,D_P}{r_P^2}\frac{c_0}{\rho_P\, q_0} \tag{7.85}$$

where $D_{S,eff}$ is the effective surface diffusion coefficient, D_P is the pore diffusion coefficient, and ρ_P is the particle density (see Section 5.4.5). In the limiting case $D_S = 0$, a formal relationship between k_S^* and the pore diffusion coefficient results from eq. (7.85).

Table 7.5: Dimensionless parameters used in the LDF model.

Dimensionless parameter	Symbol	Definition
Dimensionless concentration	X	$X = \frac{c}{c_0}$
Dimensionless adsorbent loading	Y	$Y = \frac{q}{q_0}$
Dimensionless distance	S	$S = \frac{z}{h}$
Dimensionless time (throughput ratio)	T	$T = \frac{t}{t_b^{id}} \approx \frac{t}{t_{st}}$
Dimensionless mass transfer coefficient (film diffusion)	N_F	$N_F = \frac{k_F\, a_{VR}\, c_0\, t_{st}}{\rho_B\, q_0}$
Dimensionless mass transfer coefficient (intraparticle diffusion)	N_S	$N_S = k_S^*\, t_{st}$

To further simplify the model equations, dimensionless parameters are introduced (Table 7.5). By using these dimensionless parameters, the set of equations becomes

$$\frac{\partial X}{\partial S} + \frac{\partial \bar{Y}}{\partial T} = 0 \tag{7.86}$$

$$X(T=0,S)=0, \quad \bar{Y}(T=0,S)=0 \tag{7.87}$$

$$X(T,S=0)=1 \tag{7.88}$$

$$\frac{d\bar{Y}}{dT} = N_F(X - X_s) \tag{7.89}$$

$$\frac{d\bar{Y}}{dT} = N_S(Y_s - \bar{Y}) \tag{7.90}$$

$$Y_s = (X_s)^n \tag{7.91}$$

Equation (7.86) is derived from eq. (7.78) under the additional assumption that the accumulation in the liquid phase is very small in comparison to the accumulation in the solid phase, and, consequently, the second term in eq. (7.78) can be neglected. Under this condition, the definition of the throughput ratio, T, simplifies to

$$T = \frac{t}{t_{st}} \tag{7.92}$$

because the residence time is much shorter than the stoichiometric time and $t_b^{id} \approx t_{st}$ (see also eq. (7.63)).

An alternative derivation of eq. (7.86) is possible by using the definition of T based on the solute distribution parameter (Table 7.3). That leads to the material balance in the form

$$\frac{D_g + 1}{D_g}\frac{\partial X}{\partial S} + \frac{1}{D_g}\frac{\partial X}{\partial T} + \frac{\partial Y}{\partial T} = 0 \tag{7.93}$$

from which, under the assumption of negligible accumulation in the liquid phase ($D_g \gg 1$), eq. (7.86) results.

From the dimensionless mass transfer coefficients, conclusions regarding the rate-limiting mechanisms can be drawn. Due to the definition of the parameters N_F and N_S, which include the solute distribution between the liquid and the solid phase, their absolute values are directly comparable. If $N_F = N_S$, both mechanisms contribute to the overall kinetics in equal parts. If the N_F value is higher than the N_S value, then film diffusion is faster and surface diffusion determines the adsorption rate. At ratios $N_F/N_S > 10$, it can be assumed that only surface diffusion controls the adsorption kinetics. Conversely, if N_S is greater than N_F, then surface diffusion is faster and film diffusion determines the adsorption rate. At ratios $N_F/N_S < 0.1$, it can be assumed that the adsorption kinetics is controlled only by film diffusion.

Figure 7.8 shows the influence of the N_F/N_S ratio on the shape of the BTC. In general, the higher the N values are, the faster is the mass transfer and the steeper are the BTCs. If intraparticle diffusion determines the adsorption kinetics, as it is

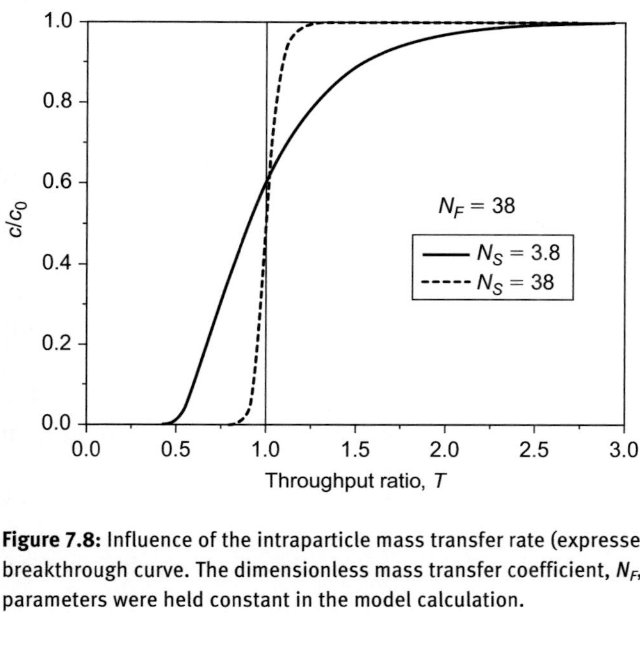

Figure 7.8: Influence of the intraparticle mass transfer rate (expressed by N_S) on the shape of the breakthrough curve. The dimensionless mass transfer coefficient, N_F, and all other process parameters were held constant in the model calculation.

often the case in practice, the BTC becomes asymmetrical with a long tailing in the upper part.

N_F and N_S can be related to the dimensionless mass transfer parameters used in the HSDM. By comparing the definition equations, the following relationships can be found:

$$N_F = 3\,St^* \qquad N_S = 15\,E_d \qquad \frac{N_F}{N_S} = \frac{1}{5}\,Bi \tag{7.94}$$

To illustrate the quality of BTC prediction, Figure 7.9 shows an example for the application of the LDF model to a single-solute system. Here, the calculation results agree well with the experimental BTCs measured in laboratory experiments with activated carbon of different particle sizes. This example also depicts the strong influence of the particle size on the adsorption kinetics and the steepness of the BTC.

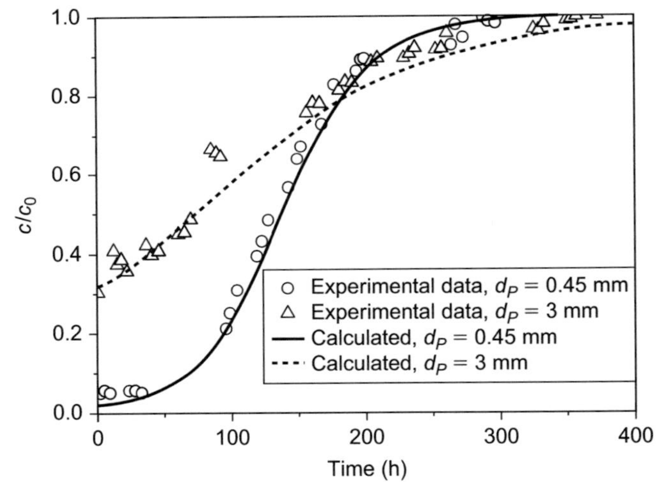

Figure 7.9: Application of the LDF model to the single-solute system 2,4-dinitrophenol/activated carbon F300. Experimental data from Heese (1996).

The LDF model can easily be extended to competitive adsorption processes by introducing the IAST. In this case, the relationship between the concentrations and adsorbed amounts at the external surface is described by the set of equations given in Section 4.5.2. As in the HSDM (Section 7.4.2), a fictive stoichiometric time as reference has to be introduced to assure that the dimensionless time axis is defined on the same basis for all components.

Figure 7.10 shows an example for the application of the LDF model to a bisolute system. In this example, the BTCs of phenol and 4-chlorophenol were predicted by using only single-solute isotherm parameters and mass transfer coefficients estimated by the methods described in Section 7.5. In such full predictions, frequently

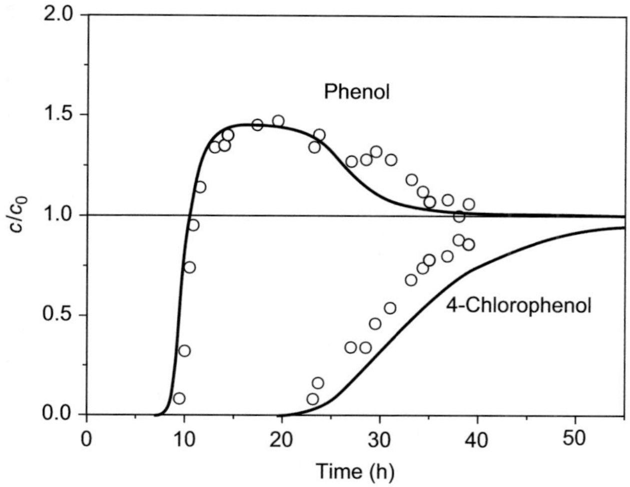

Figure 7.10: Application of the LDF model to the system phenol/4-chlorophenol/activated carbon WL2. Laboratory experiment with c_0(phenol) = 9.62 mmol/L, c_0(4-chlorophenol) = 4,74 mmol/L, m_A = 50 g, v_F = 6.2 m/h.

slight differences between the predicted and experimental BTCs are found, mainly caused by IAST prediction errors.

The LDF model can also be applied to complex multicomponent systems. An example will be presented in Section 7.5.

7.4.5 Comparison of HSDM and LDF model

As shown before, the LDF model is a simplified version of the HSDM. Therefore, it is interesting to evaluate under which conditions both models give comparable results. The essential difference between the models lies in the different mathematical description of the surface diffusion, in particular in the description of its driving force. Whereas the HSDM is based on Fick's law with a nonlinear solid-phase concentration gradient, the LDF model simplifies the gradient to a linear driving force and makes it independent of the radial coordinate. On the other hand, there is no difference in the mathematical description of the film diffusion, which is included in both models. Therefore, it can be expected that differences between the models will appear in particular under conditions where the surface diffusion dominates the overall adsorption rate. This effect can be illustrated by model calculations where only the Biot number is varied and all other operating condition are held constant. Figure 7.11 shows the results of BTC predictions by both models for Biot numbers of 1 and 91. For the model calculations, the same practice-oriented process data were used as for the HSDM/CPHSDM comparison (Table 7.4). To make both models comparable, the Glueckauf

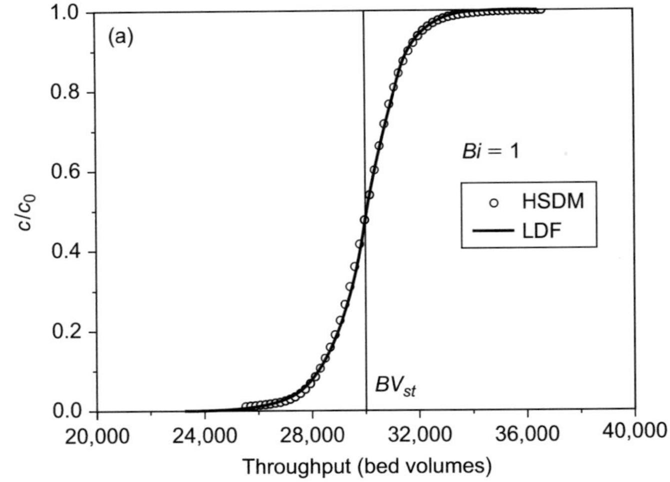

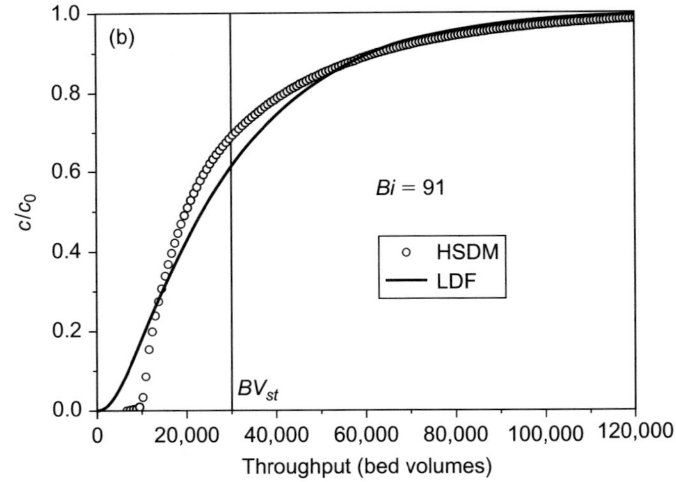

Figure 7.11: Comparison of LDF and HSDM solutions for (a) $Bi = 1$ and (b) $Bi = 91$ (for other data, see Table 7.4).

approach (eq. (7.84)) was used to find k_S^* values equivalent to the surface diffusion coefficient, D_S. The respective values are also given in Table 7.4.

According to the discussions in Sections 7.4.2 and 7.4.4, a Biot number of 1 ($N_F/N_S = 0.2$) indicates that both film diffusion and surface diffusion are relevant, with a slightly higher impact of film diffusion. As can be seen from Figure 7.11a, there is no significant difference in the calculated BTCs under this condition. As expected, the differences between the calculated BTCs become larger with increasing Biot number, as demonstrated in Figure 7.11b for $Bi = 91$ ($N_F/N_S = 18.2$). Under this condition, the film diffusion is so fast that its influence on the overall adsorption rate is negligibly small and only surface diffusion limits the adsorption rate. It has to be noted that the second example describes an extreme situation. For lower Biot numbers, the LDF model is a good approximation to the exact HSDM. A detailed study on the differences between LDF, CPHSDM, and HSDM solutions under consideration of broad ranges of Biot and Stanton numbers was published by Sperlich et al. (2008).

A better approximation to the HSDM results, in particular for high Biot numbers, can be achieved if the intraparticle mass transfer coefficient is not considered a constant but a parameter that depends on the adsorbed amount (see also Chapter 5, Section 5.4.6). This dependence can be described by

$$k_S^* = k_S^*(0)\,\exp(\omega\,\bar{q}) \tag{7.95}$$

where $k_S^*(0)$ is the intrinsic mass transfer coefficient and ω is an empirical parameter that describes the strength of the influence of the adsorbed amount, $\bar{q}$, on the intraparticle mass transfer. Under the realistic assumption that the intraparticle mass transfer resistance increases with increasing loading, the parameter ω must be negative. Taking the data of the example shown in Figure 7.11b and introducing eq. (7.95) into the LDF model, a strongly improved agreement of the LDF and HSDM results can be achieved even for the case $Bi = 91$ (Figure 7.12). In the given example, the constant mass transfer coefficient $k_S^* = 4.4 \cdot 10^{-8}$ 1/s (see Table 7.4) was replaced by eq. (7.95) with the parameters $k_S^*(0) = 9 \cdot 10^{-8}$ 1/s and $\omega = -0.025$.

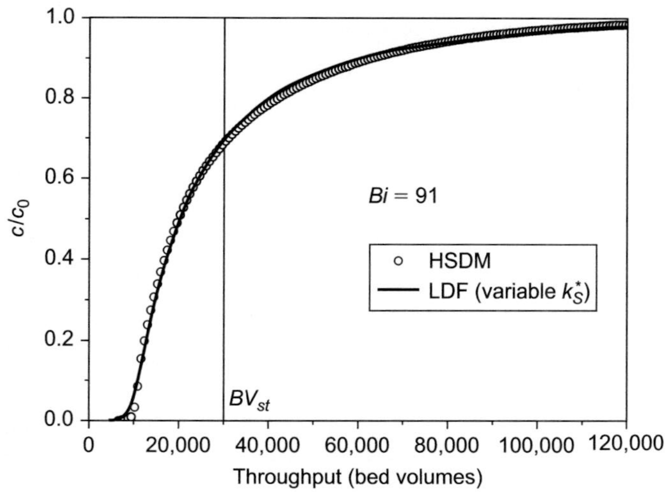

Figure 7.12: Improvement of the LDF approach by introducing a variable mass transfer coefficient according to eq. (7.95) ($Bi = 91$, $k_S^*(0) = 9 \cdot 10^{-8}$ 1/s and $\omega = -0.025$, for other data, see Table 7.4).

Generally, the application of the LDF model with variable k_S^* becomes more difficult because two parameters, $k_S^*(0)$ and ω, are needed to describe the intraparticle mass transfer. Therefore, it has to be decided on a case-by-case basis whether the effort is worth it. This modified LDF approach has proven particularly useful in cases where the intraparticle mass transfer is very slow compared to the film diffusion and the BTCs show a pronounced tailing.

7.4.6 Simplified breakthrough curve models with analytical solutions

For single-solute adsorption, simple BTC models with analytical solutions can be derived under the following conditions: (a) only a single mechanism determines the adsorption rate, (b) constant pattern has been established, and (c) the separation factor, R^*, as defined in Section 3.7, is constant.

Under the condition that dispersion as well as accumulation in the void volume can be neglected, the dimensionless material balance reads (Section 7.4.4)

$$\frac{\partial X}{\partial S} + \frac{\partial \bar{Y}}{\partial T} = 0 \tag{7.96}$$

where X is the dimensionless concentration, $\bar{Y}$ is the dimensionless mean adsorbent loading, S is the dimensionless distance, and T is the dimensionless time (throughput ratio).

In the simplified models that will be discussed here, the material balance equation is combined either with the rate equation for dominating film diffusion or with the rate equation for dominating surface diffusion (LDF approximation). To enable an analytical integration, the material balance has to be transformed at first in a manner that only derivatives with respect to one variable remain. This can be done by introducing a new coordinate system that moves with the concentration front. The new origin is then the center of the concentration front. The transformed time, t^*, in this new coordinate system is defined as

$$t^* = t - \frac{z}{v_c} \tag{7.97}$$

where z is the distance from the adsorber inlet and v_c is the velocity of a concentration point in the MTZ. Under constant pattern conditions, v_c has the same value for all concentration points and equals

$$v_c = \frac{h}{t_{st}} \tag{7.98}$$

where h is the adsorber height and t_{st} is the stoichiometric time.

Introducing eq. (7.98) into eq. (7.97) and additionally substituting the transformed time (t^*), the time (t), and the distance (z) according to

$$T^* = \frac{t^*}{t_{st}} \qquad T = \frac{t}{t_{st}} \qquad S = \frac{z}{h} \tag{7.99}$$

gives the dimensionless transformation equation

$$T^* = T - S \tag{7.100}$$

where T is the throughput ratio and T^* is the transformed throughput ratio. At the adsorber outlet ($z = h$), the transformed throughput ratio is

$$T^* = T - 1 \tag{7.101}$$

By using eq. (7.100), the dimensionless material balance equation can be written as

$$\frac{\partial X}{\partial T^*} = \frac{\partial \bar{Y}}{\partial T^*} \tag{7.102}$$

The right-hand side of eq. (7.102) is equal to the dimensionless form of the rate equation, which is

$$\frac{\partial \bar{Y}}{\partial T^*} = N_F(X - X_S) \tag{7.103}$$

for film diffusion and

$$\frac{\partial \bar{Y}}{\partial T^*} = N_S(Y_s - \bar{Y}) \tag{7.104}$$

for surface diffusion (LDF approximation). The dimensionless rate parameters, N_F and N_S, are defined in the same manner as given in Table 7.5. The dimensionless liquid-phase concentration, X, and the dimensionless mean loading, $\bar{Y}$, are related by the constant pattern material balance (Chapter 6, eq. (6.54))

$$X = \bar{Y} \tag{7.105}$$

The relationship between the equilibrium concentration and adsorbent loading, X_s and Y_s, can be expressed by means of the separation factor, R^* (Section 3.7)

$$R^* = \frac{X_S(1 - Y_S)}{Y_S(1 - X_S)} \tag{7.106}$$

To obtain a simple BTC model that can be solved analytically, R^* must be constant. This precondition is only fulfilled in the case of the Langmuir isotherm that reads in the dimensionless form

$$Y_S = \frac{X_S}{R^* + (1 - R^*)\,X_S} \tag{7.107}$$

with

$$R^* = \frac{1}{1 + b\,c_0} \tag{7.108}$$

where b is the parameter of the Langmuir isotherm and c_0 is inlet concentration. For other isotherms, an average value of R^* has to be estimated for the concentration range of interest as described in Section 3.7.

Figure 7.13 gives graphical representations of the driving forces for the limiting cases film diffusion (eq. (7.103)) and surface diffusion (eq. (7.104)).

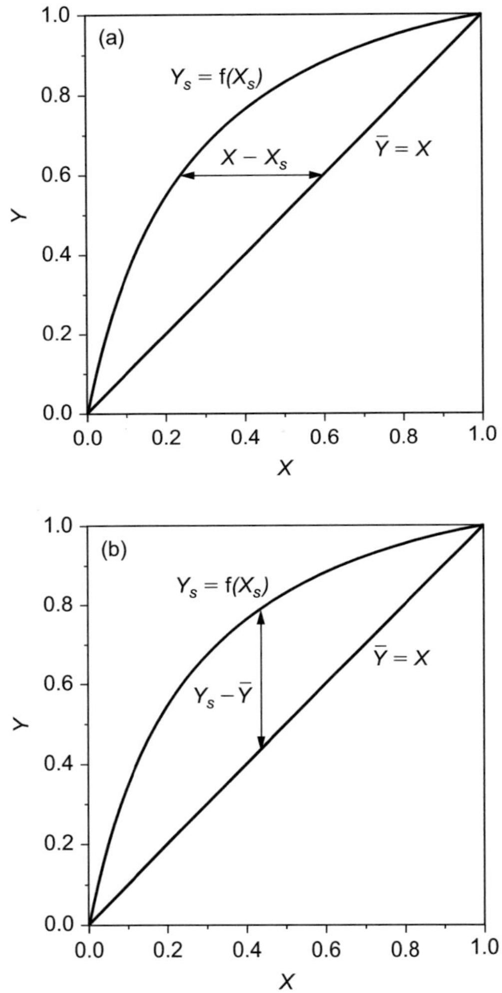

Figure 7.13: Driving forces in the case of dominating film diffusion (a) and dominating intraparticle diffusion (b).

The differential equation that has to be integrated can be derived from the material balance (eq. (7.102)), the adsorption isotherm (eq. (7.107)), and the mass transfer equation for the respective rate-limiting transport mechanism (eq. (7.103) or (7.104)).

To find the integration constants, the following condition can be used (Figure 7.14):

$$\int_{T=0}^{T(X=1)} (1-X)\,\mathrm{d}T = 1 \tag{7.109}$$

Equation (7.109) is the dimensionless form of the integral material balance for a real BTC (Chapter 6, eq. (6.26)).

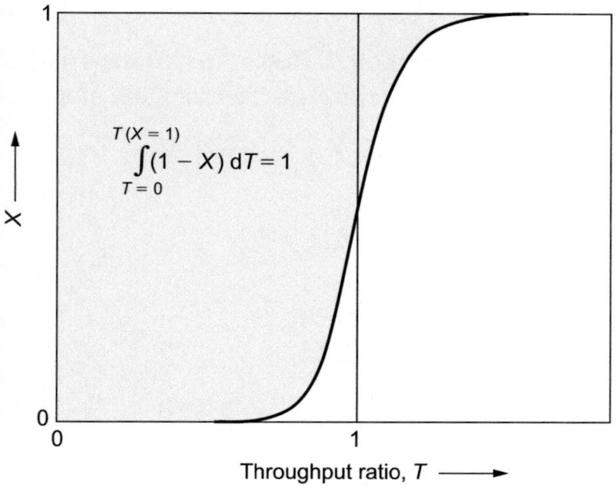

Figure 7.14: Graphical presentation of the material balance given by eq. (7.109).

Below, solutions are shown for favorable isotherms ($R^* < 1$) and either film or surface diffusion as the dominating transport mechanism.

Film diffusion ($R^* < 1$)

Combining eqs. (7.102) and (7.103) gives

$$\frac{\partial X}{\partial T^*} = N_F(X - X_S) \tag{7.110}$$

and rearranging eq. (7.107) yields

$$X_S = \frac{R^* \, Y_S}{1-(1-R^*)\,Y_S} \tag{7.111}$$

If the adsorption rate is limited by film diffusion, the loading is balanced over the particle ($Y_S = \bar{Y}$). Together with the material balance equation for constant pattern condition (eq. (7.105)), the following equation results:

$$Y_S = \bar{Y} = X \tag{7.112}$$

Therefore, Y_s in eq. (7.111) can be substituted by X, and eq. (7.110) can be transformed to

$$\frac{\partial X}{\partial T^*} = N_F \left(X - \frac{R^* X}{1 - (1 - R^*) X} \right) \tag{7.113}$$

The solution to eq. (7.113) is (Michaels 1952)

$$\frac{1}{1 - R^*} \ln X - \frac{R^*}{1 - R^*} \ln(1 - X) = N_F(T - 1) + \delta_{IF} \tag{7.114}$$

The value of the integration constant δ_{IF} depends on R^*. Table 7.6 lists some values of δ_{IF} together with the minimum values of N_F that assure the formation of a constant pattern.

Table 7.6: Integration constants in eqs. (7.114) and (7.118) and the minimum mass transfer coefficients necessary for constant pattern formation (Vermeulen et al. 1973).

R^*	δ_{IF}	δ_{IS}	$N_{F,min}$ or $N_{S,min}$
0	−1.00	1.00	5
0.2	−1.10	1.05	10
0.5	−1.17	1.14	25
0.8	−0.69	1.17	140

Figure 7.15 shows typical BTCs for different R^* values and dominating film diffusion. In general, the BTCs for dominating film diffusion show a leakage (or fronting), that means, they are flatter at lower relative concentrations, c/c_0, than at higher relative concentrations. This effect is more pronounced the lower R^* is, which means the stronger the curvature of the isotherm is.

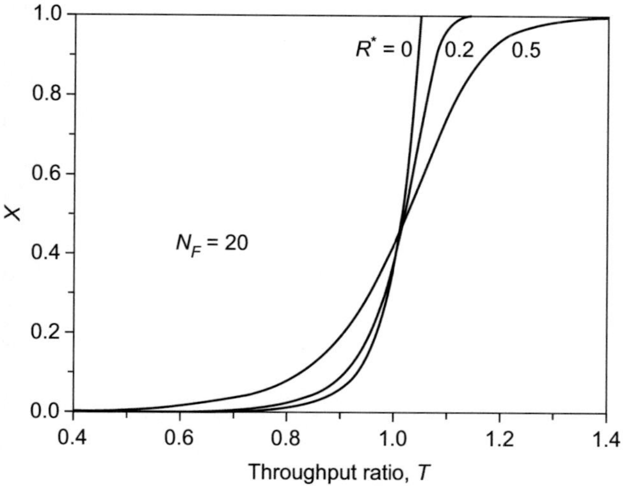

Figure 7.15: Influence of the separation factor, R^*, on the shape of the breakthrough curve in the case of film diffusion controlled mass transfer.

Surface diffusion ($R^* < 1$)

Taking into consideration the material balance (eq. (7.102)), the rate equation (LDF approximation, eq. (7.104)) can be expressed as

$$\frac{\partial X}{\partial T^*} = N_S(Y_s - \bar{Y}) \tag{7.115}$$

If surface diffusion dominates the overall adsorption rate, the concentration gradient within the film can be neglected ($X_s = X$). Under constant pattern conditions (eq. (7.105)), the following relationship holds:

$$X_S = X = \bar{Y} \tag{7.116}$$

Introducing eq. (7.116) into eq. (7.115) and substituting Y_s by the isotherm equation (eq. (7.107)) yields

$$\frac{\partial X}{\partial T^*} = N_S\left(\frac{X}{R^* + (1-R^*)X} - X\right) \tag{7.117}$$

Integration gives (Glueckauf and Coates 1947)

$$\frac{R^*}{1-R^*}\ln X - \frac{1}{1-R^*}\ln(1-X) = N_S(T-1) + \delta_{IS} \tag{7.118}$$

The integration constants and the minimum N_S values required for constant pattern formation are given in Table 7.6.

Figure 7.16 shows typical BTCs for different R^* values and dominating intraparticle diffusion. In contrast to the situation under dominating film diffusion, the BTCs for

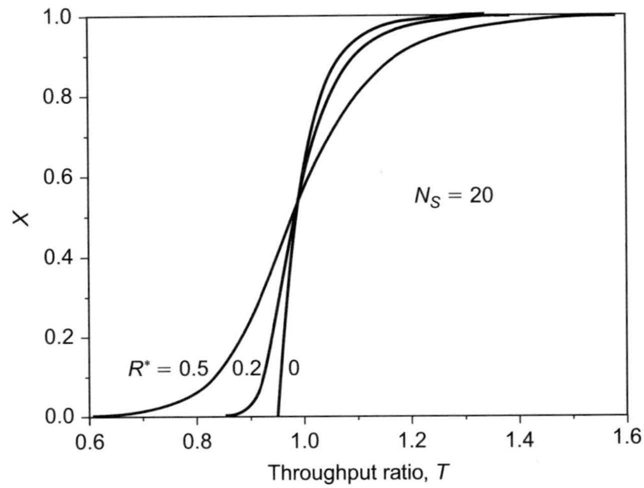

Figure 7.16: Influence of the separation factor, R^*, on the shape of the breakthrough curve in the case of intraparticle diffusion controlled mass transfer.

dominating intraparticle diffusion show a tailing, that means, they are flatter at higher relative concentrations, c/c_0, than at lower relative concentrations. This effect is more pronounced the lower R^* is, which means the stronger the curvature of the isotherm is.

7.5 Determination of model parameters

7.5.1 General considerations

To predict breakthrough curves (BTCs) by means of complete BTC models, the characteristic equilibrium and kinetic parameters must be known. In general, the number of parameters required for calculation increases with increasing complexity of the BTC model. On the other hand, the acceptance of a BTC model in practice depends on the easy availability of the required parameters. Therefore, in view of model selection for practical purposes, a compromise has to be found between the accuracy of process description and the effort necessary for model parameter estimation. Since BTC prediction without equilibrium data is impossible, simplifications can only be made in the description of dispersion and diffusion processes. The HSDM and the LDF model are examples of such a compromise. In both models, the dispersion is neglected and the internal diffusion is assumed to be dominated by only one mechanism. In this case, the number of considered mechanisms, and therefore also the number of the needed coefficients, is reduced to the half in comparison to a model that includes all possible mechanisms (see Table 7.2).

In the following sections, methods for parameter estimation are presented for different cases: single-solute adsorption, competitive adsorption of defined adsorbate mixtures, and adsorption of unknown multicomponent mixtures (e.g. background dissolved organic matter, DOM).

The important case of micropollutant adsorption in the presence of DOM is excluded here and will be discussed separately in Section 7.6.1 because the complexity of the system requires specific modeling approaches and parameter estimation methods.

7.5.2 Single-solute adsorption

For single-solute adsorption, the single-solute isotherm parameters of the adsorbate as well as its mass transfer coefficients are needed for BTC prediction. The isotherm parameters have to be determined by isotherm measurements as described in Chapter 3. If the HSDM or the LDF model is used as the BTC model, the film mass transfer coefficient (k_F) and the respective coefficients for intraparticle mass transfer (D_S or k_S^*) are needed.

Film mass transfer coefficients

As already discussed in Chapter 5, the film mass transfer coefficient strongly depends on the hydrodynamic conditions within the reactor. Therefore, the kinetic measurement has to be carried out under exactly the same hydrodynamic conditions as exist in the adsorber to be designed. For this reason, film mass transfer coefficients that are intended to be used for fixed-bed adsorber modeling cannot be determined by using one of the batch reactors described in Chapter 5.

An appropriate method to determine film mass transfer coefficients that are applicable for BTC predictions is to measure the concentration in the effluent of a short fixed-bed adsorber (Weber and Liu 1980; Cornel and Fettig 1982), which is operated under the same flow conditions as exist in the full-scale adsorber. This method is based on the facts that the initial part of a BTC is mainly determined by film diffusion and that the residence time in a short fixed-bed adsorber does not allow a full establishment of the BTC. Accordingly, an instantaneous breakthrough can be expected with a characteristic breakthrough concentration that is constant over a certain time span and depends on the film mass transfer rate. However, since the void fraction of the adsorbent bed is typically filled with adsorbate-free water before the experiment starts, the time-independent concentration occurs only after this water is displaced. Figure 7.17 shows schematically the adsorber effluent concentration as a function of time resulting from such an experimental setup. The constant concentration exists only as long as the film diffusion determines the overall mass transfer rate and the concentration at the external particle surface, c_s, is negligible (initial phase of the adsorption process). Later, the effluent concentration increases with time.

From the plateau concentration, the film mass transfer coefficient can be estimated by means of the film diffusion model. Taking eq. (7.52), which combines the

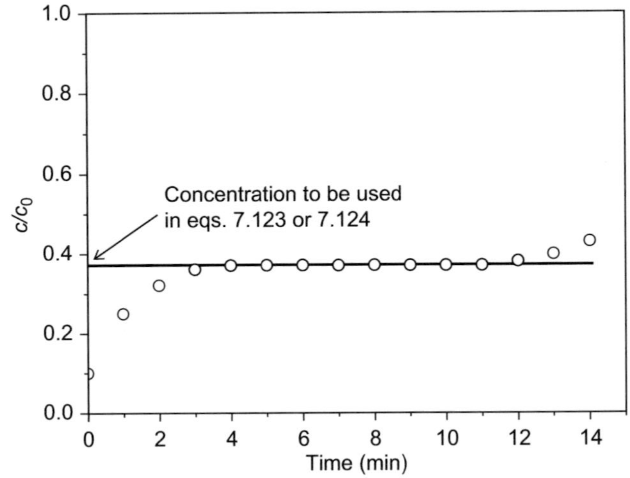

Figure 7.17: Constant effluent concentration of a short fixed-bed adsorber to be used for determination of the film mass transfer coefficient by eq. (7.123) or (7.124).

material balance of the fixed-bed adsorber with the mass transfer equation for film diffusion, and neglecting the adsorbate accumulation in the void fraction gives

$$v_F \frac{\partial c}{\partial z} + k_F\, a_{VR}(c - c_s) = 0 \tag{7.119}$$

At the beginning of the adsorption process, the bulk concentration is much higher than the concentration at the external surface of the adsorbent ($c \gg c_s$). Therefore, eq. (7.119) reduces to

$$-v_F \frac{\partial c}{\partial z} = k_F\, a_{VR}\, c \tag{7.120}$$

Integration of eq. (7.120) according to

$$-\int_{c=c_0}^{c} \frac{\mathrm{d}c}{c} = \int_{z=0}^{z=h} \frac{k_F\, a_{VR}}{v_F}\, \mathrm{d}z \tag{7.121}$$

gives

$$-\ln \frac{c}{c_0} = \frac{k_F\, a_{VR}\, h}{v_F} \tag{7.122}$$

and

$$k_F\, a_{VR} = -\frac{v_F}{h} \ln \frac{c}{c_0} \tag{7.123}$$

or, alternatively,

$$k_F\,a_{VR} = -\frac{\dot{V}}{V_R}\ln\frac{c}{c_0} = -\frac{\dot{V}\rho_B}{m_A}\ln\frac{c}{c_0} \tag{7.124}$$

where v_F is the (superficial) filter velocity, h is the height of the adsorbent layer, $\dot{V}$ is the volumetric flow rate, V_R is the volume of the reactor filled with adsorbent, ρ_B is the bed density, and m_A is the adsorbent mass.

The plateau concentration c/c_0 has to be determined from the experimental BTC as shown in Figure 7.17, and then eq. (7.123) or eq. (7.124) has to be used to estimate $k_F\,a_{VR}$. In principle, it is also possible to change the filter velocity during the experiment. In this case, different plateau concentrations occur successively, from which the mass transfer coefficients as a function of the flow velocity can be determined. Although it is not necessary to separate $k_F\,a_{VR}$ for further application in the BTC model, it is, in principle, possible after estimating a_{VR} from eq. (7.53).

An alternative way to estimate film mass transfer coefficients without laborious experiments consists in the application of empirical correlations. In the literature, a number of different correlations can be found that relate the film mass transfer coefficient to hydrodynamic conditions, adsorbent characteristics, and adsorbate properties. A selection of frequently used equations is listed in Table 7.7.

Table 7.7: Correlations for estimating film mass transfer coefficients (adapted from Smith and Weber 1989, extended).

Authors	**Correlation**	**Validity range**
Williamson et al. (1963)	$Sh = 2.4\ \varepsilon_B\,Re^{0.34}\,Sc^{0.42}$	$0.08 < Re < 125$; $150 < Sc < 1{,}300$
Wilson and Geankoplis (1966)	$Sh = 1.09\ \varepsilon_B^{-2/3}\,Re^{1/3}\,Sc^{1/3}$	$0.0016 < \varepsilon_B\,Re < 55$; $950 < Sc < 70{,}000$
Kataoka et al. (1972)	$Sh = 1.85\,[(1-\varepsilon_B)/\varepsilon_B]^{1/3}\,Re^{1/3}\,Sc^{1/3}$	$Re\,[\varepsilon_B/(1-\varepsilon_B)] < 100$
Dwivedi and Upadhyay (1977)	$Sh = (1/\varepsilon_B)\,[0.765\,(\varepsilon_B\,Re)^{0.18} + 0.365\,(\varepsilon_B\,Re)^{0.614}]\,Sc^{1/3}$	$0.01 < Re < 15{,}000$
Gnielinski (1978)	$Sh = \left[2 + \left(Sh_L^2 + Sh_T^2\right)^{0.5}\right]\,[1 + 1.5\,(1-\varepsilon_B)]$ $Sh_L = 0.644\,Re^{1/2}Sc^{1/3}$ $Sh_T = \dfrac{0.037\,Re^{0.8}\,Sc}{1 + 2.443\,Re^{-0.1}\,(Sc^{2/3} - 1)}$	$Re\,Sc > 500$; $Sc < 12{,}000$
Ohashi et al. (1981)	$Sh = 2 + 1.58\ Re^{0.4}\ Sc^{1/3}$ $Sh = 2 + 1.21\ Re^{0.5}\ Sc^{1/3}$ $Sh = 2 + 0.59\ Re^{0.6}\ Sc^{1/3}$	$0.001 < Re < 5.8$ $5.8 < Re < 500$ $Re > 500$
Vermeulen et al. (1973)	$k_F\,a_{VR} = \dfrac{2.62\,(D_L\,v_F)^{0.5}}{d_P^{1.5}}$	$\varepsilon_B \approx 0.4$

Except for the equation proposed by Vermeulen et al. (1973), all equations have the general form

$$Sh = f(Re, Sc) \tag{7.125}$$

where Sh is the Sherwood number, Re is the Reynolds number, and Sc is the Schmidt number. The dimensionless numbers are defined as follows:

$$Sh = \frac{k_F\, d_P}{D_L} \tag{7.126}$$

$$Re = \frac{v_F\, d_P}{\varepsilon_B\, \nu} \tag{7.127}$$

$$Sc = \frac{\nu}{D_L} \tag{7.128}$$

where d_P is the particle diameter and ν is the kinematic viscosity. For kinematic viscosity of water see Table 10.7 in the Appendix (Chapter 10, Section 10.6).

At first, the Reynolds number and the Schmidt number have to be calculated for the given process conditions. Then, the Sherwood number can be found from one of the correlations, and finally, k_F can be calculated from the Sherwood number. To find the volumetric mass transfer coefficient, $k_F\, a_{VR}$, the volume-related surface area, a_{VR}, has to be calculated by eq. (7.53).

To apply the empirical correlations, the aqueous-phase diffusivity, D_L, must be known. If D_L is not available from databases, it can be estimated from one of the empirical equations given in Table 7.8. It has to be noted that, due to the empirical nature of the equations, it is absolutely necessary to use the units given in the table.

Table 7.8: Correlations for estimating aqueous-phase diffusivities, D_L.

Authors	**Correlation**	**Nomenclature and units**
Worch (1993)	$D_L = \frac{3.595 \cdot 10^{-14}\, T}{\eta\, M^{0.53}}$	T – temperature, K η – dynamic viscosity (solvent), Pa·s M – molecular weight (solute), g/mol
Hayduk and Laudie (1974)	$D_L = \frac{5.04 \cdot 10^{-12}}{\eta^{1.14}\, V_b^{0.589}}$	η – dynamic viscosity (solvent), Pa·s V_b – molar volume at boiling point (solute), cm^3/mol
Wilke and Chang (1955)	$D_L = 7.4 \cdot 10^{-15} \frac{(\Phi\, M_{solv})^{0.5}\, T}{\eta\, V_b^{0.6}}$	M_{solv} – molecular weight (solvent), g/mol Φ – association factor (2.6 for water) T – temperature, K η – dynamic viscosity (solvent), Pa·s V_b – molar volume at boiling point (solute), cm^3/mol
Polson (1950)	$D_L = 2.74 \cdot 10^{-9}\, M^{-1/3}$ for $M > 1{,}000$ g/mol	M – molecular weight (solute), g/mol

Except for the first equation, the empirical correlations are subject to several restrictions. Two of the equations include the molar volume of the solute at the boiling point, which is often not available and has to be estimated by further empirical correlations. The Polson equation, proposed especially for high-molecular-weight compounds, ignores the influence of temperature, which is included in the other equations directly or indirectly (via temperature-dependent viscosity). For dynamic viscosity of water, see Table 10.7 in the Appendix (Chapter 10, Section 10.6).

For low-molecular-weight compounds, the results of the different correlations are usually in good agreement as shown exemplarily in Table 7.9. Although developed on the basis of a data set with molecular weights ranging from 56 to 404 g/mol, the first equation in Table 7.8 can also be applied to adsorbates with higher molecular weights. For instance, for $M = 2{,}000$ g/mol and $T = 298.15$ K, a diffusivity of $2.1 \cdot 10^{-10}$ m²/s is found, which is nearly the same as results from the Polson equation for high-molecular-weight compounds ($2.2 \cdot 10^{-10}$ m²/s).

Table 7.9: Comparison of different correlations for estimating aqueous-phase diffusivities.

Compound	Molecular weight (g/mol)	Temperature (°C)	D_L (10^{-9} m²/s)		
			Worch	Wilke-Chang	Hayduk-Laudie
4-Nitrophenol	139.1	20	0.77	0.76	0.73
2,4-Dichlorophenol	163.0	20	0.71	0.76	0.73
1,4-Dichlorobenzene	147.0	22	0.79	0.85	0.81
Trichloroethylene	131.4	22	0.84	1.02	0.97

Intraparticle mass transfer coefficients

Generally, the intraparticle mass transfer coefficient, k_S^*, as well as the respective diffusion coefficient, D_S (or also D_P), can be estimated from experimentally determined kinetic curves by application of the respective kinetic model. Since the intraparticle transport coefficients are independent of the stirrer or flow velocity, the coefficients are transferable to other conditions. Accordingly, batch experiments as described in Chapter 5 can be used for their evaluation.

Alternatively, to avoid time-consuming experiments, the empirical equation given in Section 5.4.6

$$k_S^* = 0.00129\sqrt{\frac{D_L\, c_0}{q_0\, r_P^2}} \tag{7.129}$$

can be applied for an approximate estimation of k_S^*. The empirical factor in eq. (7.129) is valid under the condition that the following units are used: m²/s for the diffusion coefficient, D_L; mg/L for the inlet concentration, c_0; mg/g for the related equilibrium loading, q_0; and m for the adsorbent particle radius, r_P. The unit of the resulting mass transfer coefficient is 1/s. Equation (7.129) can also be used to estimate surface diffusion coefficients needed for the HSDM. Combining eq. (7.129) with Glueckauf's approach (eq. (7.84)) leads to

$$D_S = 8.6 \cdot 10^{-5} r_P \sqrt{\frac{D_L c_0}{q_0}} \tag{7.130}$$

The units that have to be used in the empirical eq. (7.130) are the same as in eq. (7.129), and the unit of the resulting D_S is m^2/s.

To apply eq. (7.129) or (7.130), the aqueous-phase diffusivity, D_L, has to be known. It can be estimated, if necessary, from one of the empirical equations given in Table 7.8.

7.5.3 Competitive adsorption in defined multisolute systems

In multicomponent BTC models, the adsorbate competition is typically considered only with regard to the description of the equilibrium, whereas the mass transfer is assumed to be uninfluenced by competitive effects (Chapter 5, Section 5.4.8). Therefore, the respective single-solute mass transfer or diffusion coefficients are used to predict the BTCs for the mixture components, and, consequently, the estimation methods for single-solute adsorption described previously can be used.

To describe the equilibrium in multisolute systems, the IAST can be applied. Since the IAST is a predictive model (Chapter 4, Section 4.5), only the single-solute isotherm parameters of the mixture components are necessary for the BTC calculation.

In this respect, the estimation of the model parameters for the components of a defined multisolute system does not differ from parameter estimation for single solutes.

7.5.4 Competitive adsorption in complex systems of unknown composition

Natural organic matter (NOM) in drinking water treatment and effluent organic matter (EfOM) in treated wastewater, usually measured as dissolved organic carbon (DOC), are typical examples of complex adsorbate mixtures of unknown composition. As a precondition for any BTC prediction, the unknown mixture has to be formally transformed into a known mixture system. This can be suitably done by the fictive component approach (adsorption analysis) as described in Chapter 4, Section 4.7.2. From the adsorption analysis, a set of concentrations and Freundlich parameters (c_0, n, and K) are found for each fictive component, which allows applying the IAST within the BTC model.

In contrast, the estimation of mass transfer parameters is more complicated because the mixture constituents are fictive components. To find appropriate film mass transfer coefficients, at first the aqueous-phase diffusivity, D_L, can be estimated by using the equation given in Table 7.8

$$D_L = \frac{3.595 \cdot 10^{-14}\, T}{\eta\, M^{0.53}} \tag{7.131}$$

and the received D_L can then be used in one of the equations given in Table 7.7. Although the molecular weight of the background organic matter (NOM or EfOM) generally shows a broad distribution, the main fraction seems to fall into the range of approximately 500 g/mol to 1,500 g/mol. Therefore, values on this order of magnitude should be used for D_L calculation. Table 7.10 compares mass transfer coefficients, $k_F\, a_{VR}$, calculated for 500 g/mol and 1,500 g/mol by using the Wilson-Geankoplis correlation (Table 7.7) together with eqs. (7.131) and (7.53). Despite the large difference in the molecular weights, the difference between the predicted film mass transfer coefficients is relatively small. Furthermore, taking into account that BTCs under practical conditions are not very sensitive to minor $k_F\, a_{VR}$ changes, this estimation method based on arbitrarily assumed molecular weight is an acceptable approach.

Table 7.10: Influence of the molecular weight on calculated aqueous-phase diffusion coefficients, D_L, and volumetric film mass transfer coefficients, $k_F\, a_V$. The calculation was carried out by using eqs. (7.131) and (7.53) together with the Wilson-Geankoplis correlation given in Table 7.7. Process conditions: $v_F = 6$ m/h, $\varepsilon_B = 0.4$, $d_P = 2$ mm, and temperature: 10 °C.

Molecular weight, M (g/mol)	Aqueous-phase diffusion coefficient, D_L (m²/s)	Film mass transfer coefficient, $k_F\, a_{VR}$ (1/s)
500	$2.9 \cdot 10^{-10}$	$1.6 \cdot 10^{-2}$
1,500	$1.6 \cdot 10^{-10}$	$1.1 \cdot 10^{-2}$

In contrast to the $k_F\, a_{VR}$ estimation, finding appropriate values for k_S^* or D_S is more difficult. In principle, two different approaches can be taken. The first is to find an average value of the mass transfer parameter, which is used for all fictive components. This value can be estimated by BTC fitting, by separate kinetic experiments, or by using the following empirical equation (Chapter 5, Section 5.4.6):

$$k_S^* = a + b\frac{c_0(\text{DOC})}{r_P^2} \tag{7.132}$$

where c_0(DOC) is the total concentration of all adsorbable DOC fractions. The empirical parameters, found from experiments with natural waters from different sources, are $a = 3 \cdot 10^{-6}$ 1/s and $b = 3.215 \cdot 10^{-14}$ (m²·L)/(mg·s). If combined with Glueckauf's approach (eq. (7.84)), eq. (7.132) can also be used to estimate surface diffusion coefficients for the HSDM.

An alternative approach is to use different mass transfer coefficients for the different DOC fractions. According to the surface diffusion mechanism (migration in the adsorbed state along the pore walls), the weakest adsorbable component should have the highest value of k_S^* or D_S. Graduated mass transfer coefficients can be found from BTC fitting, which requires at least one lab-scale experiment. An average starting value can be derived from eq. (7.132). As an example, Figure 7.18 shows the BTC for NOM-containing river water calculated with the LDF model using different sets of the mass transfer coefficient, k_S^*, for the two adsorbable fractions. Here, the average value of k_S^* found from eq. (7.132) was $6 \cdot 10^{-6}$ 1/s.

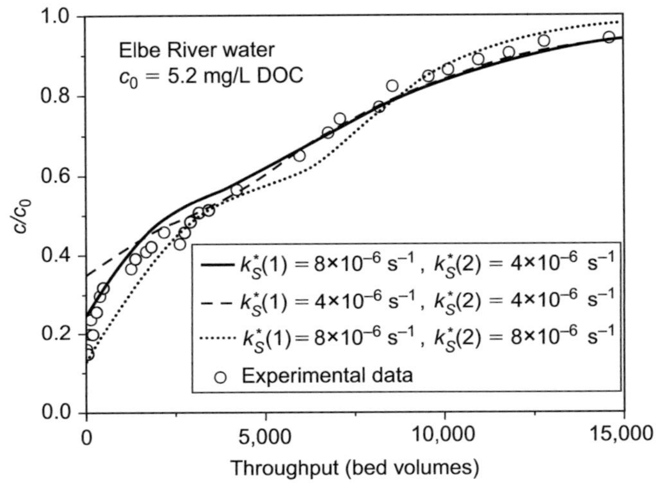

Figure 7.18: DOC breakthrough curve calculation for river water by using different sets of the intraparticle mass transfer coefficients of the fictive components.

This example shows the slight improvement of the calculation results if graduated mass transfer coefficients are used. On the other hand, this example also makes clear that the differences between the results for the different parameter sets are not very large. Therefore, if only an approximate prediction of the breakthrough behavior is required, it is not necessary to make great demands on the accuracy of the mass transfer parameter estimation.

7.6 Special applications of breakthrough curve models

7.6.1 Micropollutant adsorption in the presence of background organic matter

As pointed out in Chapter 4 (Section 4.7.1), natural organic matter (NOM) is present in all natural waters and is therefore of relevance for adsorption processes used in drinking water treatment. In particular, NOM strongly influences the micropollutant (MP) adsorption, and therefore has to be considered in the respective BTC models. The same is true for effluent organic matter (EfOM) in wastewater treatment. In the following discussion, we want to use the general term "dissolved organic matter (DOM)" to designate the background organic matter in the modeling approaches.

In principle, the DOM impact on MP adsorption can be explained by two different mechanisms. The first mechanism is a direct competitive adsorption of the DOM fractions and the MP within the micropore system, referred to as site competition. If only site competition is relevant, the equilibrium data for the MP/DOM system can be predicted by using the IAST and its modifications (tracer model (TRM), equivalent background compound model (EBCM)) as shown in Chapter 4.

The second possible mechanism is pore blockage. Here, it is assumed that larger DOM molecules accumulate in the pore system (in particular in the mesopores) and hamper the transport of both the MP and the small DOM molecules to the adsorption sites located in the micropores. This mechanism is also referred to as intraparticle pore blockage. In a number of studies, it was shown that the external mass transfer may also be affected by blocking (referred to as external surface pore blockage or surface blockage); however, this mechanism in not well understood. Changes in the film thickness or in the viscosity of the boundary layer are discussed as possible explanations for the hampered external mass transfer.

In general, pore blockage does not only affect the adsorption equilibrium of the MP but also results in a decrease of its adsorption rate. To account for this effect, lower mass transfer and diffusion coefficients as in the respective single-solute system or – more realistic – variable coefficients, which decrease with the DOM accumulation, have to be applied in the BTC model.

The extent of pore blockage depends on the molecular weight distribution of the DOM and the pore-size distribution of the activated carbon. The following limiting cases can be distinguished (Pelekani and Snoeyink 1999):

- When the pores are large enough to admit the MP but too small to admit DOM, pore blockage is the dominant competition mechanism.
- When the pores are large enough to admit both the MP and DOM, direct site competition becomes the important mechanism.

Furthermore, it has to be noted that the accumulation of DOM is a function of the operation time of the fixed-bed adsorber. Therefore, it can be expected that the impact of pore blockage becomes the more pronounced the longer the adsorber is preloaded with DOM.

Preloading effects can be expected if the DOM enters the adsorber earlier than the MP (e.g. if the MP occurs in the raw water only after a certain operation time of the adsorber). However, preloading occurs also if the MP is present in the DOM-containing water from the beginning of the operation. The reason for this is that most of the DOM is not as effectively adsorbed as the MP and therefore travels faster through the adsorber and preloads the fresh granular activated carbon (GAC) layers.

Depending on the assumptions about the dominating mechanism, different models can be used to describe the MP/DOM competition in fixed-bed adsorbers. In the following paragraphs, two modeling approaches will be shown. The first considers only site competition; the second considers site competition as well as pore blockage.

Case 1: Only site competition

The simplest way to describe MP/DOM adsorption in a fixed-bed adsorber is to ignore possible impacts of DOM on the MP mass transfer and to consider only the direct site competition. For BTC modeling under the assumption of site competition as the dominating mechanism, multisolute BTC models that include the IAST can be used – for instance, the multisolute HSDM as well as the multisolute LDF model.

The DOM and MP input data needed for the IAST equilibrium calculation within the BTC models ($c_{0,i}$, n_i, K_i) have to be determined prior to the BTC calculation by one of the methods described in Chapter 4 (Section 4.7.3). As discussed in Chapter 4, the simple application of the IAST to a system consisting of a MP and fictive DOC fractions, characterized by an adsorption analysis, typically leads to an overestimation of the competition effect of DOM on the MP adsorption. As a result, the predicted MP breakthrough time would be shorter than in reality.

To overcome this problem and to improve the BTC prediction, the tracer model (TRM) can be used to find modified isotherm parameters of the MP, which have to be applied together with the parameters of the DOC fractions as input data for the BTC prediction. As an alternative, the equivalent background compound model (EBCM) can be used to describe the competitive adsorption equilibrium. In this case, the multisolute system is reduced to a bisolute system consisting of the MP and a fictive equivalent background compound (EBC), which represents that fraction of DOC that competes with the MP. The methods for estimating the corrected MP isotherm parameters of the TRM or, alternatively, the EBC parameters are described in detail in Section 4.7.3.

Accordingly, the input equilibrium parameters for the BTC calculation are either the isotherm parameters of the DOC fractions together with the corrected isotherm parameters of the MP (if using the TRM) or the isotherm parameters of the EBC and the MP (if using the EBCM). It has to be noted that only the TRM allows predicting in parallel the MP BTC and the DOC BTC (as the sum of the BTCs of the DOC fractions). In contrast, the EBCM cannot describe the DOC breakthrough behavior because the EBC is a single component.

The mass transfer parameters required for the BTC prediction can be determined as described in Section 7.5 for single solutes and pure DOM systems, respectively. The latter is only necessary if fictive DOM fractions are considered (TRM). In the case of the EBCM, the EBC mass transfer parameters can be assumed to be the same as for the MP.

This follows from the EBC concept, which postulates that the MP and the EBC have comparable adsorption properties.

In Figure 7.19, an atrazine BTC determined in a lab-scale adsorber is compared with BTCs calculated by using the multisolute LDF model based on the IAST. The background DOM was characterized by an adsorption analysis, and the mass transfer parameters were estimated by the methods discussed previously. If using the original isotherm parameters of atrazine as input data, the predicted breakthrough is too early. This is in accordance with the typical overestimation of the competition effect by the conventional IAST (see Chapter 4, Section 4.7.3). In contrast, the modification of the Freundlich parameters by the TRM improves considerably the prediction of the MP breakthrough in the presence of DOM.

In this context, it is interesting to evaluate the result of using a pseudo single-solute isotherm for the MP adsorption in the presence of DOM within a single-solute BTC model. Here, the isotherm of atrazine in the presence of DOM was formally described by the Freundlich isotherm equation. As can be seen in Figure 7.19, this pseudo single-solute approach provides better results than the BTC model with the conventional IAST but cannot describe the BTC as well as the multisolute BTC model in combination with the TRM.

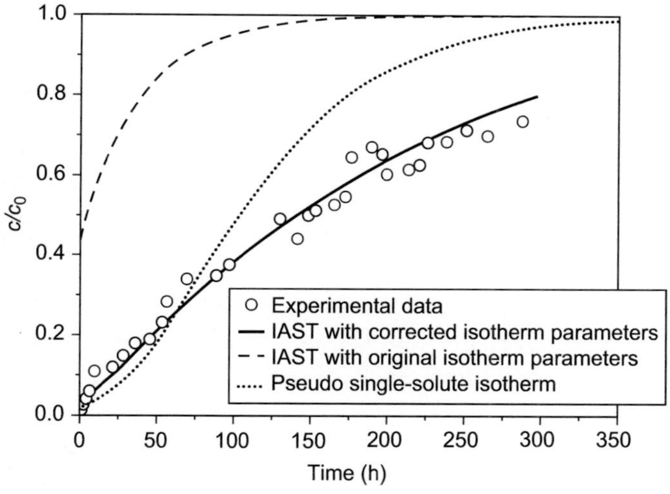

Figure 7.19: Experimental and calculated breakthrough curves of atrazine adsorbed from Elbe River water onto activated carbon F300. Comparison of the results of different BTC models: pseudo single-solute LDF model, multisolute LDF model with IAST and use of uncorrected atrazine isotherm parameters, multisolute LDF with IAST and use of atrazine isotherm parameters corrected by the TRM. Experimental data from Rabolt (1998).

Case 2: Site competition and pore blockage

A competitive adsorption model for fixed-bed adsorption, which accounts for site competition as well as for pore blockage (COMPSORB-GAC), was presented by Schideman et al. (2006a, 2006b) based on the earlier work of Li et al. (2003) and Ding et al. (2006). The basic concept is analogous to the EBCM but with additional consideration of the pore-blocking effects. In this approach, the HSDM is used as the basic BTC model, and all single-solute isotherms are described by the Freundlich equation.

In the COMPSORB-GAC model, the complex MP/DOM system is described by only three components: (1) the MP or trace compound (TRC), (2) a strongly competing DOM compound (SCC), and (3) a pore-blocking DOM compound (PBC). The assumptions regarding the behavior of these three compounds are as follows:

- The PBC comprises the larger-sized fractions of DOM. The PBC is only adsorbed in mesopores and larger micropores. The PBC adsorption is not affected by the other two compounds, and vice versa.
- SCC and TRC are smaller and are adsorbed in the smaller micropores where they compete for the adsorption sites. The SCC corresponds to the EBC in the EBCM.
- Due to the much higher concentration of the The SCC, the adsorption capacity for the SCC is independent of the TRC.
- The SCC reduces the capacity for the TRC. This direct competition can be described by the IAST.
- The PBC reduces the film mass transfer rate of both the SCC and TRC (external surface pore blockage); k_F is therefore assumed to be a function of the mean surface loading with the PBC.
- The PBC also reduces the intraparticle mass transfer (intraparticle pore blockage) because the large molecules accumulate in the pores and impede TRC diffusion. This effect will occur only after a critical surface loading is reached. After this critical point, D_S is a function of the mean surface loading with the PBC.

With these assumptions, the adsorption behavior of the three components in a fixed-bed can be calculated by a sequential application of a pseudo single-solute HSDM, following the algorithm given by Schideman et al. (2006a).

1. Application of the conventional HSDM with constant equilibrium and kinetic parameters to predict the concentration profile of the PBC, which is independent of the other compounds.
2. Application of a modified HSDM with variable kinetic parameters to predict the concentration profile of the SCC, which is adsorbed independently of the TRC. The variable kinetic parameters depend on the local PBC surface loadings, which were computed previously.
3. Application of a modified HSDM with variable equilibrium and kinetic parameters to predict the concentration profile of the TRC. The variable kinetic parameters depend on the previously computed local surface loadings of the PBC.

For the TRC only, the initial concentration is known and the isotherm parameters can be determined from a single-solute isotherm test. The other compounds are fictive compounds. Their initial concentrations and single-solute isotherm parameters have to be determined prior to the application of the BTC model. This can be done by a special algorithm based on three experimental isotherms: the TRC isotherm in the presence of DOM, the TRC isotherm in the absence of DOM, and the DOC isotherm.

The initial concentration of the fictive SCC is available from the TRC isotherm in the presence of DOM by using the EBCM with SCC isotherm parameters equal to those of the TRC (one-parameter fitting, see Chapter 4, Section 4.7.3). To allow the application of mass concentrations within the IAST, the molecular weight of the SCC is assumed to be the same as that for the TRC.

The initial concentration of the PBC can be found from a DOC mass balance as the difference between the total DOC concentration and the concentration of the SCC. Here, the DOC is assumed to be 50% of the DOM mass.

Other points of the PBC isotherm can be found by subtracting the SCC concentration from the respective DOC isotherm concentration for each carbon dose used in the DOC isotherm determination. The required SCC concentration for a given carbon dose has to be calculated from the SCC isotherm (K and n equal to that of TRC) and the material balance equation. Finally, the related PBC isotherm parameters are determined by fitting the isotherm data.

Figure 7.20 shows the algorithm that has to be applied for estimating the initial concentrations and the isotherm parameters of the three model components. It has to be noted that a nonadsorbable DOM fraction is not considered in the COMPSORB-GAC model.

As mentioned previously, the TRC adsorption is assumed to be influenced by the SCC. To predict the competitive adsorption of SCC and TRC, a simplified IAST approach can be used. Since the values of the Freundlich exponents of the SCC and TRC are the same (according to the EBC approach), and given that it can be further assumed that the surface loading of the SCC (the competing DOM fraction) dominates over the TRC ($q_{SCC} \gg q_{TRC}$), the IAST equations of the bisolute system SCC/TRC reduce to two simple isotherm equations (see also Chapter 10, Section 10.3)

$$q_{SCC} = K_{SCC}(c_{SCC})^{n_{SCC}} \tag{7.133}$$

$$q_{TRC} = \left(\frac{K_{TRC}\, n_{SCC}}{n_{TRC}}\right)^{1/n_{TRC}} [q_{SCC}(z,t)]^{(1-1/n_{TRC})} c_{TRC} \tag{7.134}$$

Accordingly, the bisolute system is decoupled, and a pseudo single-solute HSDM can be applied at first to the SCC and then, after the profile of the SCC ($q_{SCC}[z,t]$) is computed, also to the TRC.

The change of the film mass transfer parameter for both TRC and SCC with increasing preloading with PBC is described by the empirical correlation

$$\frac{k_F(t,z)}{k_{F,0}} = kx_{F,min} + (1 - kx_{F,min})\exp[-\alpha\, \bar{q}_{PBC}(t,z)] \tag{7.135}$$

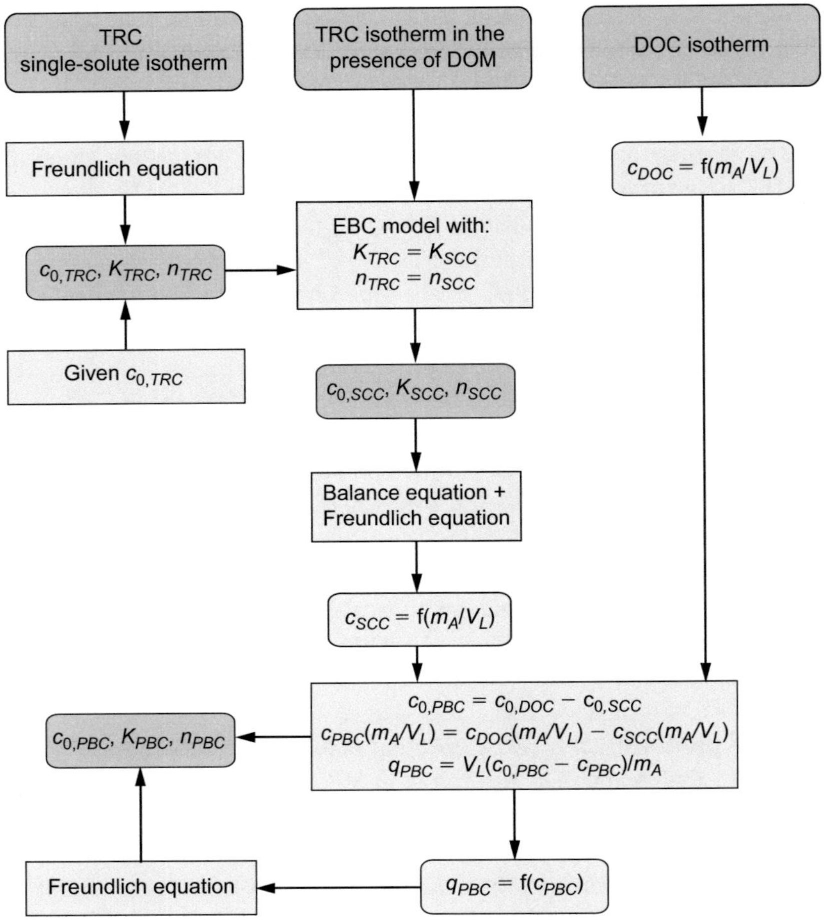

Figure 7.20: Algorithm for estimating the initial concentrations and the isotherm parameters of the three components of the COMPSORB-GAC model (TRC: trace compound, SCC: strongly competing compound, PBC: pore-blocking compound).

where $k_{F,0}$ is the initial film mass transfer coefficient without DOM preloading, $kx_{F,min}$ is the minimum mass film mass transfer coefficient expressed as a fraction of its initial value, α is an empirical parameter that describes the decrease of the mass transfer parameter with preloading, and $\bar{q}_{PBC}(t,z)$ is the mean surface loading of the PBC at a given time and location. As can be derived from eq. (7.135), the ratio $k_F/k_{F,0}$ remains constant ($kx_{F,min}$) after a certain preloading has been reached.

Table 7.11: Suitable methods to determine the rate parameters needed for the COMSORB-GAC model as proposed by Schideman et al. (2006b).

Parameter	Experimental setup	Calculation method	Remarks
$k_{F,0}$ (TRC) $D_{S,0}$ (TRC)	BTC determination in an SBA, TRC in DOM-containing water	HSDM fit, $k_{F,0}$: earliest breakthrough data $D_{S,0}$: late data	In the initial part of the BTC, film diffusion is the dominating mechanism, whereas surface diffusion dominates in the later part of the BTC. Under SBA conditions, the preloading effect during the experiment is assumed to be negligible.
$k_{F,0}$ (PBC) $D_{S,0}$ (PBC)	DOC breakthrough curve determination in an SBA with larger EBCTs (>1 min)	HSDM fit	Larger EBCTs as for TRC parameter determination are necessary because otherwise the breakthrough concentrations approach the inlet concentration too fast and the parameters cannot be determined exactly. Under the given conditions, the SCC breakthrough is negligible and DOC can be set equal to the PBC concentration.
$kx_{F,min}$ and α	1. Preloading of the activated carbon to different extents in a batch reactor 2. SBA tests with TRC and differently preloaded activated carbon	1. Calculation of the PBC surface loading as difference between DOC surface loading and SCC surface loading 2. HSDM fit of the earliest SBA breakthrough data gives k_F for different PBC loadings 3. Fitting the data by using eq. (7.135)	The DOC surface loading can be found from the DOC removal. The SCC surface loading can be calculated by applying the HSDM for batch systems.
q_{cr} and β	1. Preloading of the activated carbon to different extents in a batch reactor 2. Batch kinetic tests with TRC and differently preloaded activated carbon	1. Calculation of the PBC surface loading as difference between DOC surface loading and SCC surface loading 2. HSDM fit of batch kinetic curves gives D_S for different PBC loadings 3. Fitting the data by using eq. (7.136)	The DOC surface loading can be found from the DOC removal. The SCC surface loading is assumed to be the equilibrium loading, which can be calculated from the SCC isotherm parameters and the carbon dose.

For the surface diffusion coefficient, D_S, it was found that a decrease only occurs after a certain level of preloading has been reached. This surface loading is referred to as the critical loading, q_{cr}. The respective empirical equation is

$$\frac{D_S(t,z)}{D_{S,0}} = \exp[-\beta\,(\bar{q}_{PBC}(t,z) - q_{cr})] \quad \text{for} \quad \bar{q}_{PBC} > q_{cr} \tag{7.136}$$

where $D_{S,0}$ is the initial surface diffusion coefficient without preloading and β is an empirical parameter.

To predict the TRC BTC by applying the algorithm given previously, the initial film mass transfer coefficients as well as the initial surface diffusion coefficients for all three components are required. Furthermore, the parameters of eqs. (7.135) and (7.136) ($kx_{F,\min}$, α, q_{cr}, β), which quantify the decrease of the kinetic parameters, have to be known.

To determine these parameters, a number of different batch and short bed adsorber (SBA) tests have to be carried out. The SBA is in particular appropriate for determining the film mass transfer coefficients from the initial part of the BTC (see also Section 7.5) but can also be used to determine D_S from the upper part of the BTC because these late breakthrough data are nearly uninfluenced by film mass transfer. In other cases, it can be advantageous to apply batch kinetic tests to determine D_S. Table 7.11 gives an overview of the methods that are suitable for determining all the parameters needed to describe the adsorption rate of the model components together with some background information. As already mentioned, the same isotherm and kinetic parameters are used for the SCC and the TRC. For more details, the original literature should be consulted.

The COMPSORB-GAC model requires a high experimental and computational effort as can be seen from Figure 7.20 and Table 7.11. It should, therefore, be checked for each practical objective whether a simpler BTC model or a rapid small-scale column test is sufficient.

7.6.2 Biologically active carbon filters

If fixed-bed adsorbers are exposed to raw waters, which contain microorganisms, accumulation and growth of these microorganisms can take place on the surface of the adsorbent particles. Such a situation can be found, for instance, in GAC adsorbers used for drinking water treatment. Activated carbon with its rough surface and its adsorption properties is a favored medium for accumulation of microorganisms. Under these conditions, the GAC filter acts not only as adsorber but also as bioreactor, and biodegradation of NOM fractions and micropollutants contributes to the net removal of organics.

The effect of NOM degradation is reflected in the shape of the BTC. The DOC BTC does not end at the inlet concentration level but at a lower steady-state concentration. Figure 7.21 shows a typical DOC BTC of a biologically active GAC adsorber in

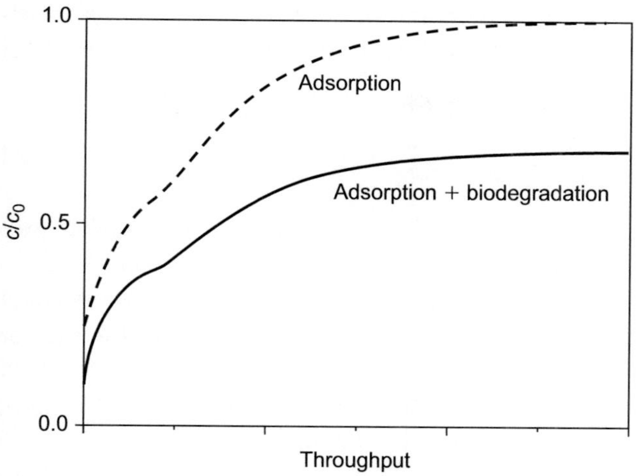

Figure 7.21: NOM breakthrough behavior in the case of adsorption with and without biodegradation (schematic).

comparison with a BTC of an adsorber without biological activity. The lower steady-state concentration of the BTC is a result of the mass loss due to degradation of NOM fractions. The value of the steady-state concentration depends on the content of degradable NOM fractions, the degradation rates, and the empty bed contact time. A total degradation cannot be expected, because NOM always contains a certain amount of nondegradable components or slowly degradable components, which will be not degraded within the given contact time.

Degradable micropollutants are also removed in the biologically active carbon filter. In contrast to NOM, the steady-state concentration can even be zero if the degradation rate is fast enough.

Nondegradable micropollutants may be indirectly affected by the biological processes. The removal of competing NOM fractions by biodegradation can lead to weaker competition and consequently to a stronger adsorption of the nondegradable micropollutants.

Under conventional conditions, the effect of biological NOM reduction is not strongly pronounced, because NOM in its original form contains mainly larger molecules, which are not, or only slowly, degradable. However, the NOM degradation can be enhanced by preozonation. Ozonation prior to adsorption breaks down the NOM molecules, which makes them more assimilable and microbially oxidizable. As a consequence, the biomass concentration within the adsorbent bed increases, and NOM is degraded to a higher degree. Therefore, a lower steady-state concentration at the adsorber outlet can be expected. The combination of ozonation and GAC application is referred to as biological activated carbon (BAC) process or biologically enhanced activated carbon process.

If biodegradation should be involved in a fixed-bed adsorber model, the differential mass balance equation (eq. (7.50)) has to be expanded by a reaction term (R_T) that describes the biodegradation rate

$$v_F \frac{\partial c}{\partial z} + \varepsilon_B \frac{\partial c}{\partial t} + \rho_B \frac{\partial \bar{q}}{\partial t} + R_T = 0 \tag{7.137}$$

It has to be noted that dispersion is neglected here. As shown in the previous sections, BTC models that consider only adsorption are already of high complexity, in particular if they include different transport mechanisms and multisolute adsorption. This complexity further increases if biodegradation is additionally integrated into the model. Therefore, as in the case of pure adsorption, a compromise has to be found between the model complexity and the exactness of BTC prediction.

The reaction term, R_T, can be described by different approaches. Here, only some selected examples with relatively simple structure will be given. For more sophisticated models, special literature has to be consulted.

First-order rate law

If we assume that degradation takes place only in the liquid phase and follows a first-order rate law, the respective rate equation reads

$$\frac{\partial c}{\partial t} = -\lambda_l \, c \tag{7.138}$$

where λ_l is the liquid-phase degradation rate constant. The reaction term in eq. (7.137) is then

$$R_T = \varepsilon_B \, \lambda_l \, c \tag{7.139}$$

If the degradation is assumed to occur not only in the liquid phase but also in the adsorbed phase, an additional rate equation for the adsorbed phase has to be formulated

$$\frac{\partial \bar{q}}{\partial t} = -\lambda_s \, \bar{q} \tag{7.140}$$

where $\bar{q}$ is the mean adsorbent loading and λ_s is the solid-phase degradation rate constant. The reaction term to be introduced in eq. (7.137) is then given by

$$R_T = \varepsilon_B \, \lambda_l \, c + \rho_B \, \lambda_s \, \bar{q} \tag{7.141}$$

or

$$R_T = \varepsilon_B \left(\lambda_l + \frac{\rho_B}{\varepsilon_B} \, \lambda_s \, \frac{\bar{q}}{c} \right) c \tag{7.142}$$

Monod equation

A more sophisticated approach is to express the reaction term by means of the well-known Monod equation, which describes the biomass growth (e.g. De Wilde et al. 2009). Considering additionally the decay of biomass, the net growth can be expressed by

$$\frac{\mathrm{d}c_{BM}}{\mathrm{d}t} = \mu_{max} \frac{c}{K_s + c} c_{BM} - k_{decay}\, c_{BM} \tag{7.143}$$

where c_{BM} is the biomass concentration, μ_{max} is the maximum growth rate of the microorganisms, c is the concentration of the degraded substance (substrate), K_s is the half saturation constant, and k_{decay} is the decay constant of the biomass. Introducing a yield coefficient, Y_C, which is defined as the ratio of biomass production and substrate consumption, allows describing the change of the substrate concentration with time as

$$\frac{\mathrm{d}c}{\mathrm{d}t} = -\frac{1}{Y_C}\left(\mu_{max} \frac{c}{K_s + c}\right) c_{BM} \tag{7.144}$$

and the reaction term reads

$$R_T = \frac{\varepsilon_B}{Y_C}\left(\mu_{max} \frac{c}{K_s + c}\right) c_{BM} \tag{7.145}$$

If the substrate concentration, c, is much lower than the half saturation constant, K_s, eq. (7.144) simplifies to

$$\frac{\mathrm{d}c}{\mathrm{d}t} = -\left(\frac{\mu_{max} c_{BM}}{Y_C K_s}\right) c = -\left(\mu^*_{max} c_{BM}\right) c \tag{7.146}$$

with

$$\mu^*_{max} = \frac{\mu_{max}}{Y_C K_s} \tag{7.147}$$

and the reaction term reads

$$R_T = \varepsilon_B \left(\mu^*_{max} c_{BM}\right) c \tag{7.148}$$

which becomes identical with eq. (7.139) under the condition that μ^*_{max} and c_{BM} are assumed to be constant.

Further model extensions

The extended form of the material balance equation (eq. (7.137)) provides the general basis for considering biodegradation in a BTC model. However, often further model extensions are necessary to describe the breakthrough behavior under real conditions. Below, some important aspects will be noted without going into detail.

As mentioned previously, NOM degradation is the main effect in the BAC process. Since NOM is a multicomponent system consisting of fractions with different adsorbabilities and degradabilities, a fictive component approach is necessary to characterize its behavior in a GAC adsorber. Here, a compromise in view of the appropriate number of fictive components has to be found. On the one hand, the fictive components should represent all important combinations of adsorbability and degradability. On the other hand, the complexity of the system and the problems connected with the parameter estimation confine the number of fictive components. If the fictive components are defined, the reaction terms given previously can be used to describe their biodegradation.

Another model extension becomes necessary if the impact of the biofilm on the mass transfer from the liquid to the activated carbon particles is not negligible. In the case of a relatively thick and dense biofilm, the mass transfer can be considerably hampered, and the biofilm diffusion has to be introduced into the BTC model as an additional mass transfer mechanism. In connection with this, it could also be necessary to consider the detachment mechanism of the biofilm.

8 Desorption and reactivation

8.1 Introduction

The operating time of an adsorption unit is limited by the capacity of the adsorbent. When the adsorbent capacity is exhausted, the adsorbent has to be removed from the reactor and has to be replaced by fresh or regenerated adsorbent material. Since engineered adsorbents are typically highly refined and very expensive products, their regenerability and the regeneration costs are important factors in view of the economic efficiency of the entire adsorption process. This aspect becomes more important the higher the adsorbent costs are and the faster the adsorbent capacity is exhausted. Furthermore, if an adsorbent should be used for the recovery of valuable solutes from water, regenerability is an essential property.

Low-cost adsorbents – for instance, waste products – are typically not regenerated. Instead of that, the loaded adsorbent materials are disposed of by landfill or incineration. The same is true for powdered activated carbon (PAC). The main reasons for using PAC as a one-way adsorbent are the relatively low costs of PAC as compared to granular activated carbon (GAC), difficulties in separating PAC from associated suspended solids, and the small particle size, which complicates handling during regeneration.

During regeneration, the adsorbates are desorbed from the surface and transferred into the adjacent phase. Since desorption is the reversal of the adsorption process, all conditions that lead to a decrease of adsorption increase the amount of adsorbate that can be desorbed. Depending on the phase in which the desorbed substances are transferred, a distinction can be made between desorption into the gas phase and desorption into the liquid phase.

Thermal desorption or desorption by steam are processes where the adsorbed substances are transferred into the gas phase, whereas during desorption by solvent extraction or by pH shift, the adsorbed species are transferred into a liquid phase. If valuable substances should be recovered from the desorbate, for instance, in process wastewater treatment, an additional phase separation process is necessary to remove the desorbed substance from the receiving gas or liquid phase. In principle, extraction can be carried out not only with conventional solvents but also with supercritical fluids, in particular with supercritical CO_2.

Regeneration of adsorbents loaded with organic adsorbates is also possible by biodegradation. This process is referred to as bioregeneration. Bioregeneration can take place during the operation time of the adsorber simultaneously to adsorption. In this case, microorganisms occurring in the treated water are immobilized and form a biofilm on the adsorbent surface. As a consequence, the adsorber acts partially as a bioreactor where the immobilized microorganisms degrade organic substances and therefore extend the lifetime of the adsorber (see also Section 7.6.2 in

https://doi.org/10.1515/9783110715507-008

Chapter 7). Although, in principle, bioregeneration can also be operated as a separate regeneration step, this offline regeneration is seldom used.

The restoration of the adsorption capacity of activated carbons, in particular GAC, is typically carried out by reactivation. Although the terms "regeneration" and "reactivation" are sometimes used synonymously, it is necessary to distinguish between them because they describe different processes. The term "regeneration" describes the removal of the adsorbed substances from the adsorbent surface by desorption without irreversible transformation of the adsorbent surface. In contrast, the term "reactivation" is used for a specific thermal treatment of loaded activated carbon. Here, the adsorbed substances are removed not only by thermal desorption but also by thermal destruction and subsequent burning off of the carbonaceous residues. During reactivation, the adsorbent material takes part in the burn-off reaction to a certain extent, which leads to a loss of adsorbent mass and a change in the pore structure. The operational conditions in reactivation are similar to those in activated carbon production by gas activation (Chapter 2, Section 2.2.1). Reactivation is typically applied to activated carbons that were used for treatment of waters with complex composition such as drinking water or wastewater.

The selection of an appropriate regeneration/reactivation process depends on a number of factors, including

- the type of the adsorbent;
- the character of the treated water, in particular number and nature of the accumulated adsorbates;
- further treatment objectives, additional to the restoration of the adsorbent capacity (e.g. recovery of valuable substances); and
- economic efficiency.

In the following sections, an overview of the most important regeneration and reactivation processes and their application fields is given.

8.2 Physicochemical regeneration processes

8.2.1 Desorption into the gas phase

Desorption into the gas phase can be realized by thermal desorption or by desorption with steam. In both cases, the adsorbate is transferred from the adsorbed state to an adjacent gas phase. This type of desorption is particularly appropriate for volatile adsorbates. It is also part of the reactivation process (Section 8.3).

Thermal desorption

In thermal desorption, the temperature dependence of the adsorption process (Chapter 3, Section 3.5) is utilized. Desorption is carried out by heating the adsorbent up to a temperature of approximately 400 °C. As a result, volatile substances are removed from the adsorbent surface. Additionally, the residual water is evaporated. The desorbed substances are then condensed together with the water vapor and can be recovered from the condensate phase.

The kinetics of desorption is determined either by the intrinsic desorption step or by pore diffusion. Pore diffusion is rate limiting in the case of low activation energy for desorption, high temperature, and low pore diffusion coefficient. In this case, a pore diffusion model has to be used to describe the overall desorption rate. More frequently, desorption itself is rate limiting, particularly in the case of low temperature, high activation energy, and high pore diffusion coefficient. Under these conditions, the desorption rate is proportional to the substance amount that is still adsorbed at the considered time, t, and the rate equation reads (Seewald and Jüntgen 1977)

$$\frac{\mathrm{d}q_{des}}{\mathrm{d}t} = k_{des}\, q(t) = k_{des}(q_0 - q_{des}) \tag{8.1}$$

where q_{des} is the amount of desorbed substance per unit adsorbent mass, $q(t)$ is the substance amount still adsorbed at time t, k_{des} is the desorption rate constant, and q_0 is the initially adsorbed amount. The temperature dependence of the rate constant, k_{des}, can be described by the Arrhenius equation

$$k_{des} = k_A \exp\left(-\frac{E_{A,des}}{R\,T}\right) \tag{8.2}$$

where k_A is the preexponential factor (or frequency factor), $E_{A,des}$ is the activation energy of the desorption process, R is the gas constant, and T is the absolute temperature.

For a constant heating rate, v_H,

$$v_H = \frac{\mathrm{d}T}{\mathrm{d}t} = \text{constant} \tag{8.3}$$

the following equation for the temperature dependence of desorption can be derived:

$$\frac{\mathrm{d}q_{des}}{\mathrm{d}T} = \frac{k_A}{v_H}\exp\left(-\frac{E_{A,des}}{R\,T}\right)(q_0 - q_{des}) \tag{8.4}$$

The activation energy as well as the preexponential factor are functions of the adsorbent loading and have to be determined in experiments. For activated carbons, the required activation energy of desorption increases with increasing molecule size of the adsorbate and decreasing adsorbent loading. This can be explained by the differences in the binding strengths. As already discussed in Chapter 2 (Section 2.2.1), larger molecules are more strongly bound to the surface than smaller molecules. Furthermore, in the case of energetically heterogeneous adsorbents, the sites with the highest adsorption energy are occupied at first, which leads to a relatively stronger binding at low surface coverage.

In general, thermal desorption is appropriate for heat-resistant adsorbents and volatile adsorbates. Thus, this desorption method is restricted to special applications of activated carbon where a volatile adsorbate should be recycled – for instance, in process wastewater treatment. Furthermore, as already mentioned, thermal desorption is generally the first step in reactivation of activated carbon (Section 8.3).

Desorption with steam

Desorption with superheated steam is known as a common technique in solvent recovery from waste air. Here, the desorption effect is based on a combination of temperature increase and partial pressure reduction under simultaneous utilization of the steam volatility of the adsorbates. In principle, this process can also be used to desorb organic solvents that were adsorbed from the aqueous phase – for instance, during process water treatment or groundwater remediation. For instance, halogenated hydrocarbons can be efficiently desorbed from activated carbons and polymeric adsorbents. On the other hand, there are some problems connected with the application of steam desorption. Since after their application the adsorbents still contain high amounts of water, they must be dried prior to the desorption process. Otherwise, a high steam demand and a high condensate amount have to be accepted. The condensate has to be treated by an appropriate separation process to recover the organic compounds from water. If the desorbed organic substances are not completely miscible with water, which is typically the case for many solvents, a simple decanter can be used for phase separation. However, a considerable amount of the solvent remains in the aqueous phase and cannot be easily recovered. In conclusion, it has to be stated that steam desorption plays only a minor role as a regeneration method for adsorbents used in water treatment.

8.2.2 Desorption into the liquid phase

In principle, an adsorbate can be removed from the adsorbent and transferred into a liquid phase if the adsorption from this liquid is weaker than the adsorption from the original aqueous solution. This desorption liquid can be another solvent in which the adsorbate is more soluble than in water (extractive desorption), but it can also be an aqueous solution in which an adsorption-influencing property (concentration, temperature, pH) has been changed in comparison to the original adsorbate solution. Under these adsorption-influencing properties, the pH in particular is of practical relevance for desorption because it determines, for instance, the adsorption strength of weak acids and bases on activated carbons and of ions on oxidic adsorbents. For such adsorption processes, desorption by pH shift is a technical option.

Theoretical basics

As already mentioned, desorption into the liquid phase is possible if the loaded adsorbent is brought into contact with a liquid from which the adsorbate is adsorbed to a lesser extent than from the original aqueous solution. That means that the respective isotherm is shifted to lower adsorbent loadings as compared to the original isotherm. Figure 8.1 shows the situation for a batch system. In the diagrams, the isotherms valid for adsorption and desorption are shown together with the operating lines for different volumes of the liquid used for desorption. The operating lines can be found from the material balance equation as shown in Chapter 3 (Section 3.6). During adsorption, the adsorbent is loaded with the adsorbate according to isotherm A. The equilibrium after the adsorption stage is characterized by c_0 and q_0. The desorption process starts with the adsorbent loading, q_0, and proceeds along the operating line until the new equilibrium state, valid for the desorption liquid, is reached (isotherm B). The equilibrium state after desorption is characterized by the concentration in the regeneration liquid, c_R, and the corresponding adsorbent loading, q_R. Depending on the volume of the regeneration liquid, desorption can result in concentrations higher or lower than the concentration in the primarily treated solution, c_0. However, the higher the concentration is in the desorbate solution, the higher the adsorbed amount is that remains on the adsorbent. Consequently, in a batch system, it is not possible to attain a very high concentration in the desorbate solution and a high degree of regeneration at the same time. This may be a problem if the desorption is carried out with the objective of adsorbate recycling where high concentrations in the desorbate solution are desired. As can be seen from a comparison of Figures 8.1a and 8.1b, this conflict is of minor significance if the difference between the isotherms is very large. Here, small volumes of the regeneration solution can be used, which permits high desorbate concentrations without residual loadings that are too high.

In the case of fixed-bed adsorption, the desorption process can also be carried out directly in the adsorber. Here, the adsorbate-free desorption solution is continuously percolated through the adsorbent bed. If the concentration of the desorbed adsorbate is measured at the column outlet, typical elution curves, as shown in Figure 8.2, can be found. The throughput is given here as t/t_r where t_r is the retention time (Chapter 6, Section 6.3). Since the void volume of the adsorbent bed is still filled with adsorbate solution after the end of the adsorption stage, a throughput volume equal to the void volume is needed to replace the original adsorbate solution by the desorption liquid. Therefore, in the beginning, the effluent concentration is constant and equal to the adsorbate feed concentration. Later, if the desorbate has reached the adsorber outlet, the concentration strongly increases and reaches

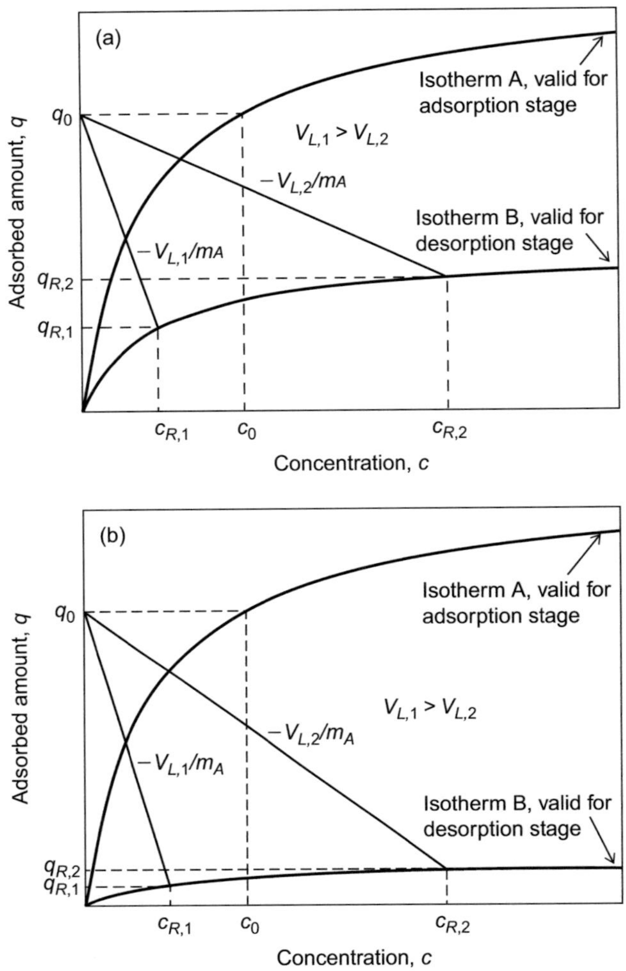

Figure 8.1: Desorption operating lines for weak (a) and strong (b) differences between the adsorption and desorption isotherms.

a maximum. With increasing time, the desorbed amount decreases, and therefore the concentration also decreases again. The shape of the elution curve and the value of the maximum concentration depend on different factors, in particular on the equilibrium conditions, the desorption rate, and the flow velocity. Since after the maximum the concentration decreases with increasing throughput, an optimum throughput with respect to desorbate concentration and degree of regeneration has to be found.

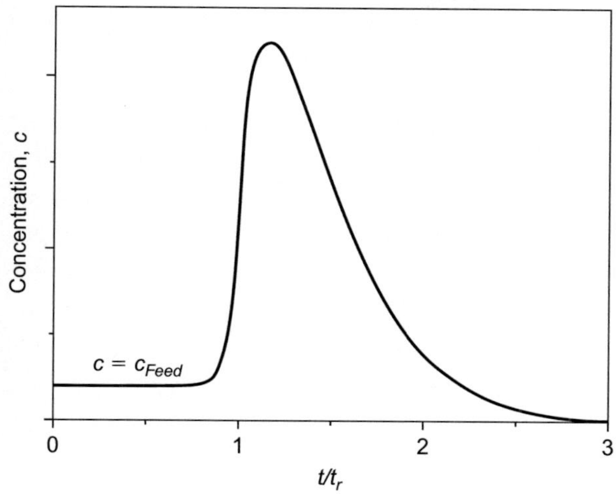

Figure 8.2: Typical elution curve as can be found for fixed-bed desorption.

As shown by Sutikno and Himmelstein (1983), desorption curves can be described with a model based on the material balance equation for fixed-bed adsorbers (Chapter 6, Section 6.4.3)

$$v_F \frac{\partial c}{\partial z} + \varepsilon_B \frac{\partial c}{\partial t} + \rho_B \frac{\partial \bar{q}}{\partial t} = 0 \tag{8.5}$$

where v_F is the filter velocity of the desorption liquid. The desorption rate is expressed by the mass transfer equation

$$\rho_B \frac{\partial \bar{q}}{\partial t} = -k_{des}(c_{eq} - c) \tag{8.6}$$

where k_{des} is the desorption rate constant, c_{eq} is the concentration of the desorbed component in equilibrium with the actual adsorbent loading, and c is the concentration of the desorbed component in the liquid phase. The isotherm completes the set of equations that has to be solved in order to describe the elution curve. A simplified model can be derived under the assumption that the equilibrium will be established spontaneously at all points of the column (equilibrium model, Chinn and King 1999).

Extraction

In general, an adsorbate is better adsorbed from aqueous solution the lower its affinity to water is. If a loaded adsorbent is brought into contact with a solvent in which the adsorbate is more soluble than in water, the state of equilibrium is shifted into the direction of desorption. This effect can be utilized for the regeneration of adsorbents. This extractive desorption is frequently applied to regenerate polymeric adsorbents loaded with organic adsorbates. Here, often alcohols (methanol, isopropanol) are used as extracting agents. Activated carbons can also be regenerated in this manner, for instance, with acetone, alcohols, or dimethylformamide.

If the regeneration is carried out within the fixed-bed adsorber, typical elution curves as shown in Figure 8.2 are found. Since the eluate consists of a mixture of the extraction solvent and the desorbed adsorbate, a distillation is necessary to separate the components. Residues of the extracting agent in the adsorber have to be stripped with steam. Therefore, an additional process stage is necessary to separate the solvent from the condensate. Figure 8.3 shows exemplarily a process scheme for adsorbent regeneration by extraction. The high effort needed for this type of regeneration restricts its application to special cases, in particular recycling processes.

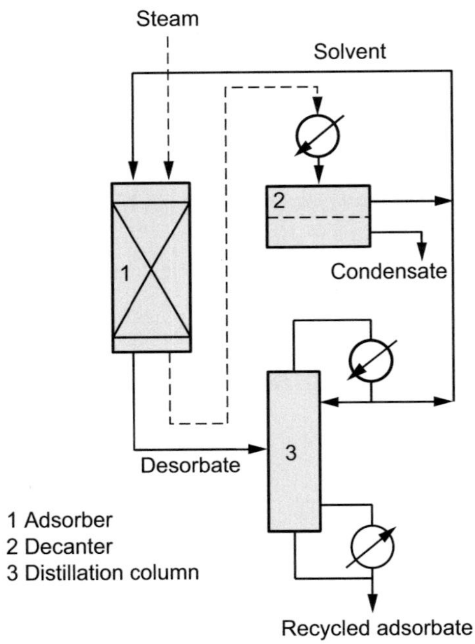

Figure 8.3: Process scheme of an adsorption unit with extractive desorption.

Desorption by pH shift

If the adsorption is pH-dependent, this dependence can be utilized for desorption by pH shift. To promote desorption, the pH has to be changed in the direction where the adsorption strength decreases.

In the case of activated carbons or polymeric adsorbents, a pH effect is typically found for weak acids and bases. This effect mainly results from the change of polarity of the adsorbates due to deprotonation (acids) or protonation (bases). On these adsorbents, neutral species are more strongly adsorbed than ionized species. Therefore, weak acids adsorbed in neutral form can be removed from the adsorbent by increasing the pH (transformation into the ionic form), and conversely, weak bases adsorbed in neutral form can be desorbed by decreasing the pH. If, as in the case of activated carbons, surface groups exist that can be protonated or deprotonated, the surface charge is positive at low pH values and negative at high pH values. Therefore, the adsorbate-related pH effect is increased or weakened by an additional adsorbent-related effect (see also Chapter 4, Section 4.6).

In the case of oxidic adsorbents, which are preferentially used to adsorb ionic species, the protonation/deprotonation of surface OH groups is the main reason for the pH dependence of adsorption. Anions are preferentially adsorbed at low pH values where the surface is positively charged, whereas cations are preferentially adsorbed at high pH values where the surface is negatively charged. Consequently, to enable desorption, the pH has to be shifted to high values if anions should be desorbed and to low values if cations should be desorbed. In practice, oxidic adsorbents (e.g. aluminum oxide or ferric hydroxide) are frequently used to remove phosphate or arsenate from water. In these cases, desorption can be carried out with strong bases.

Exemplarily, desorption of phosphate from granular ferric hydroxide (GFH) by sodium hydroxide solution is shown in Figure 8.4. The throughput is given here in bed volumes (BVs) (Chapter 6, Section 6.3). Whereas the loading was carried out with an element-related concentration of about 16 mg/L P, the mean concentration in the first four bed volumes of the desorbate solution is higher than 2 g/L P which corresponds to a concentration factor of more than 100. However, with increasing desorption time (increasing bed volumes), the concentration decreases. In practice,

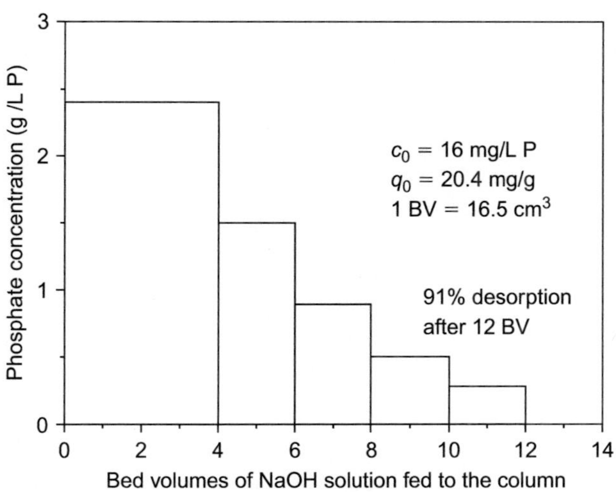

Figure 8.4: Desorbate concentration as a function of regeneration solution throughput. Example: fixed-bed desorption of phosphate from ferric hydroxide by sodium hydroxide solution.

optimum conditions with respect to desorbate concentration, volumetric flow rate, and throughput have to be found by experiments.

At this point, it has to be noted that the calcium ions that occur in the waters to be treated can disturb the desorption process. Some of the calcium ions are co-adsorbed with the oxyanions to be removed (phosphate or arsenate) resulting in surface complex formation and local surface precipitation (Antelo et al. 2015). Precipitates are also formed with other anions, such as carbonate or silicate (Kumar et al. 2018). The increased pH in the desorption step enhances the precipitation effect. The formed solid calcium compounds shield the surface against the access of the regeneration solution and impede the complete restoration of the adsorption capacity. As a consequence, the adsorption capacity decreases with an increasing number of regeneration cycles.

This negative effect can be avoided by an acidic conditioning (acid wash) prior to the desorption step (Kunaschk et al. 2015, Kumar et al. 2018). Reducing the pH by acid addition exchanges the co-adsorbed calcium ions and dissolves the surface precipitates, which makes the pores and adsorption sites accessible for the desorption solution. For GFH, a pH of 2.5 was found to be an optimum value for the removal of the blocking compounds. Lower pH values should be avoided to protect the adsorbent material from destruction. Figure 8.5 shows the efficiency of the acid-alkaline regeneration of phosphate-loaded GFH in comparison with the conventional alkaline regeneration.

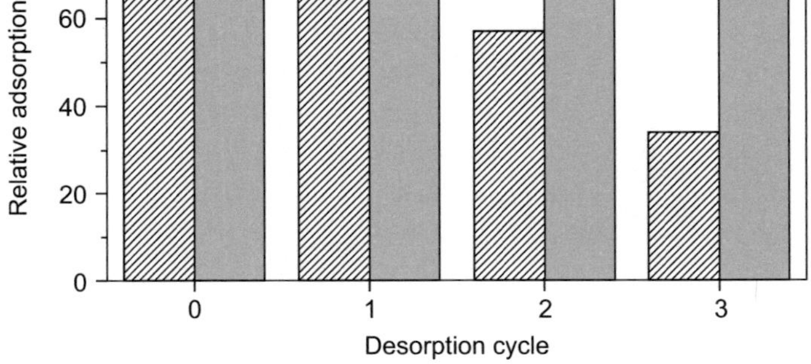

Figure 8.5: Development of the phosphate adsorption capacity of GFH over several regeneration cycles by using the acid-alkaline regeneration and the conventional alkaline regeneration (Data from Kunaschk 2020).

8.3 Reactivation

For activated carbon, the application of the regeneration processes described previously is limited to special applications – for instance, if single solutes should be recovered from process wastewater. However, activated carbon is mainly used for drinking water treatment, wastewater treatment, or groundwater remediation where the adsorbent is loaded with a multitude of different substances. In these cases, recovery of adsorbates is not possible and also not wanted, and therefore restoration of the adsorption capacity remains the only objective of the adsorbent treatment. This objective can be efficiently achieved by reactivation.

Reactivation is a thermal process that typically occurs in four stages (Sontheimer et al. 1988):

- Thermal desorption of adsorbed compounds
- Thermal decomposition of adsorbed compounds followed by desorption of the products
- Carbonization of nondesorbed products that were formed during thermal decomposition or chemisorbed during the adsorption step
- Surface reactions between carbonaceous residuals and water vapor or oxidizing gases to form gaseous products

Each of these stages comprises several simultaneous steps and is associated with a particular temperature range. The reaction conditions during reactivation are similar to those used for the manufacturing of activated carbons by gas activation, but the objectives are slightly different. In gas activation, a part of the solid carbon material is transformed into gaseous products by reactions with the activation gases in order to receive an optimized pore structure. In contrast, during reactivation, the adsorbed species should be removed by the previously mentioned reactions with no or only minor altering of the original pore structure. Since the reaction rates of the adsorbates are much higher than the reaction rate of carbon in the temperature range between 700 °C and 900 °C, a selective burn-off of the adsorbed material is, in principle, possible. However, due to the complexity of the processes and the variation in the composition of the adsorbed material, it is not easy to find the optimum process conditions, in particular the optimum residence time of the carbon in the reactor. Reactivation is therefore often carried out by the manufacturer of the activated carbon based on long-term experience. Nevertheless, an impact on the carbon structure connected with weight loss and changes in the capacity cannot be completely avoided. Problems may occur if the residence time in the reactor falls considerably below or exceeds the optimum residence time. If the reactivation is insufficient, the adsorbed material is not completely removed and the capacity decreases. In case of overactivation, the micropore walls can be burned out to form mesopores with the consequence

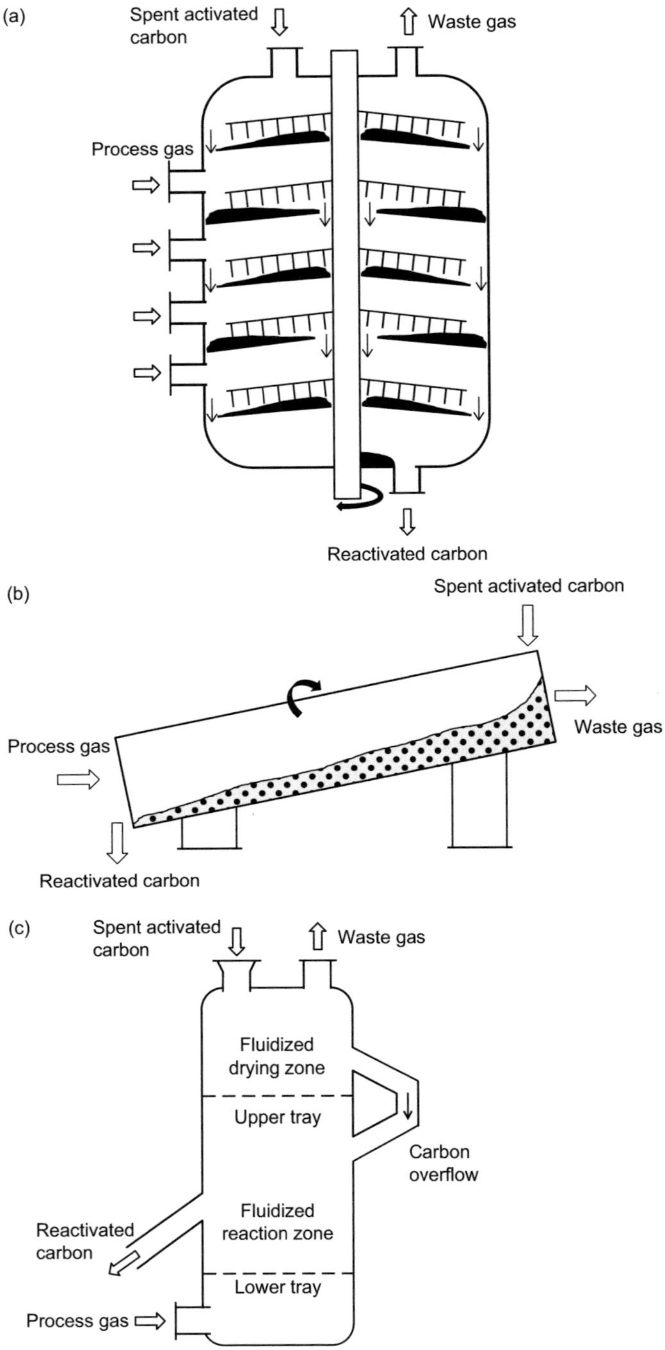

Figure 8.6: Reactor types appropriate for reactivation: (a) multiple hearth furnace, (b) rotary kiln, and (c) fluidized-bed reactor.

that the adsorption capacity for larger molecules increases, but the adsorption capacity for smaller molecules decreases.

For reactivation, different reactors are in use. Due to the similarities of the gas activation and reactivation processes, the same reactor types can be applied. The multiple hearth furnace is the reactor that is most commonly used for activated carbon reactivation. Other reactor types are rotary kiln and fluidized-bed reactor (Figure 8.6).

9 Geosorption processes in water treatment

9.1 Introduction

Under certain conditions, natural attenuation processes can be utilized to support engineered water treatment processes. Typical examples of such natural or seminatural processes are bank filtration or groundwater recharge by infiltration. The general principles of these processes have already been discussed in Chapter 1 (Section 1.3).

Bank filtration can be applied as a pretreatment step in drinking water treatment in cases where polluted surface water has to be used as a raw water source. During the subsurface transport from the river bed to the extraction wells of the waterworks, different attenuation processes take place, in particular filtration, biodegradation, and sorption.

During groundwater recharge, water is infiltrated into the subsurface and flows through the vadose zone (unsaturated zone) to the aquifer. In principle, treated or untreated surface water as well as treated wastewater can be infiltrated. The utilization of the attenuation potential during infiltration of wastewater effluent is also known as soil-aquifer treatment (SAT). During infiltration, the same attenuation processes take place as during bank filtration.

All these processes have in common that they are based on subsurface water transport, connected with different attenuation mechanisms. Typically, the water is transported in a preferred direction (e.g. from the river bed to the extraction well in the case of bank filtration). Therefore, as a first approximation, the transport process can be considered to be a one-dimensional transport. Under this precondition, it is possible to simulate the water and solute transport as well as the relevant attenuation processes by laboratory-scale column experiments in which the column is filled with soil or aquifer material from the considered bank filtration or infiltration site.

In the following sections, an introduction to the modeling of one-dimensional solute transport in porous media will be given with particular emphasis on sorption. Sorption as an attenuation process is, in particular, relevant for substances that are persistent or only poorly degradable. In the latter case, biodegradation and sorption act in parallel, and the sorption model has to be extended to a combined sorption and biodegradation model. A simple model for this case will also be considered in this chapter.

The theoretical considerations in the next sections are restricted to saturated conditions where the void space between the solid particles is completely filled with water. This is the typical situation during bank filtration. In the case of infiltration, the conditions are different because the water travels at first through the unsaturated (vadose) zone before it reaches the aquifer. In the vadose zone, the situation is generally more complicated because a gas phase exists in addition to the solid phase

https://doi.org/10.1515/9783110715507-009

and the liquid phase. Therefore, besides sorption and degradation, liquid/gas transfer is also possible. Furthermore, in contrast to the saturated zone, the water velocity during infiltration is often not constant. Under the following simplifying conditions, however, model approaches for the saturated zone can also be applied to the vadose zone:

- The liquid/gas mass transfer can be neglected (nonvolatile solutes).
- The typically fluctuating velocity is expressed by a constant average value.
- The effective porosity is replaced by the volumetric water content, which is defined as the ratio of the water volume in the void space and the total volume (i.e. solid volume, water volume, and air volume).

Prior to the model discussion, it is necessary to recall some important terms. As already mentioned in Chapter 2, natural adsorbents are referred to as geosorbents. Many geosorbents such as soil or aquifer material are of complex composition. With regard to the binding capacity for neutral organic solutes, the organic fractions of the geosorbents are the most important constituents. Since the interaction between organic solutes and the organic fractions of the solids cannot be exactly specified as adsorption or absorption, the more general term "sorption" is preferred in such processes. According to the term "geosorbents", sorption onto these materials is also referred to as "geosorption".

The main objective of geosorption modeling is to predict the breakthrough behavior of solutes. The sorption parameters required for prediction can be estimated from laboratory column experiments by fitting the experimental breakthrough curves (BTCs) with an appropriate BTC model. In this respect, the methodology is comparable to that of modeling the engineered adsorption in fixed-bed adsorbers.

Generally, the sorption of solutes during the one-dimensional subsurface transport or in the respective lab-scale experiments shows a multitude of analogies to the engineered adsorption in fixed-bed adsorbers, but there are also some important differences. These differences have to be considered in model development and lead to differences in the resulting BTC models.

Due to the strong differences in the flow velocities between engineered fixed-bed adsorption and subsurface transport, the role of dispersion is quite different. The flow velocity of the water in the subsurface is much slower than in engineered adsorbers. Typical flow velocities in riverbank filtration are approximately 1 m/day, whereas the filter velocities in fixed-bed adsorbers are up to 15 m/h. Due to the slow velocities in the natural processes, dispersion cannot be neglected in the BTC models as is typically done in models for engineered systems. Dispersion is even the main factor for the BTC spreading during subsurface transport. In engineered adsorbers, the BTC spreading is mainly caused by slow adsorption kinetics (slow mass transfer), whereas in natural systems the impact of sorption kinetics is negligible or contributes only to a small extent to the BTC spreading.

Another major difference consists in the typical isotherm form. The adsorbate isotherms for engineered adsorbents are typically nonlinear and often follow the Freundlich isotherm equation with exponents, n, much lower than 1. In contrast, in geosorption, the sorbate isotherms are mostly linear. Nonlinear isotherms are seldom found, and if they occur, the n values typically do not differ much from 1. Therefore, the assumption of a linear isotherm is a good approximation in most cases. That makes the BTC models much simpler and allows for finding analytical solutions.

The existence of linear isotherms also simplifies the modeling of multisolute sorption. It can be shown by means of the ideal adsorbed solution theory (IAST) (Chapter 4, Section 4.5.4) that the sorption of a considered sorbate is not influenced by other sorbates if all components of the mixture exhibit a linear isotherm (Schreiber and Worch 2000). This finding was also corroborated by experimental results. This means that all sorbates in the multisolute system sorb independently of each other as long as their single-solute isotherms are linear. Consequently, the BTCs of all components can be calculated separately by means of a single-solute model, and no specific competitive sorption model is necessary.

In this chapter, only a brief introduction to the principles of geosorption modeling will be given. For more detailed information, special literature (e.g. Bear and Cheng 2010) should be consulted.

9.2 Experimental determination of geosorption data

In principle, the same experimental methods as shown for engineered adsorbents can be used to find characteristic sorption data, in particular batch isotherm measurements and lab-scale column experiments. However, a limitation of applicability exists for batch isotherm measurements, resulting from the generally weaker sorption in geosorption systems in comparison to engineered systems. As shown in Chapter 3 (Section 3.2), batch isotherm measurements are based on the material balance equation

$$q_{eq} = \frac{V_L}{m_A}(c_0 - c_{eq}) \tag{9.1}$$

where q_{eq} is the equilibrium loading (sorbed amount), V_L is the volume of the solution, m_A is the mass of the sorbent, c_0 is the initial concentration, and c_{eq} is the residual concentration after equilibration. To eliminate the impact of analytical errors and to obtain accurate equilibrium data, the difference between initial and equilibrium concentrations should not be too small. That requires the addition of appropriate amounts of sorbent. However, in the case of geosorption, it is not always possible to fulfill this requirement, in particular if the sorbate shows only weak sorption. In the case of weak sorption, the maximum sorbent amounts that can be added to the liquid volume in the experiments are often not high enough to obtain observable concentration differences. Consequently, the determination of equilibrium data by batch experiments is restricted to systems with stronger sorption.

As an alternative, sorption measurements can be carried out in column experiments (Figure 9.1). This method is comparable to the BTC measurement for engineered adsorption systems as described in Chapter 6 (Section 6.2). With this kind of experimental setup, not only can equilibrium data be determined, but characteristic parameters for dispersion and, if relevant, for sorption kinetics can also be obtained. Furthermore, this experimental setup better reflects the practical conditions during subsurface solute transport than batch experiments do. On the other hand, the experiments are more complex, in particular for systems with nonlinear sorption behavior. In this case, a number of BTC measurements with different sorbate concentrations have to be carried out to find the parameters of the nonlinear isotherm. In contrast, in systems with a linear isotherm, in principle only one measurement is necessary to find the sorption coefficient. In the case of missing information about the isotherm linearity in the given sorbate/sorbent system, a validity check of the linear isotherm assumption is recommended. On the other hand, it is known from experience that the deviation from the linear course of the isotherm, if existing, is typically small, and neglecting this deviation causes only minor errors.

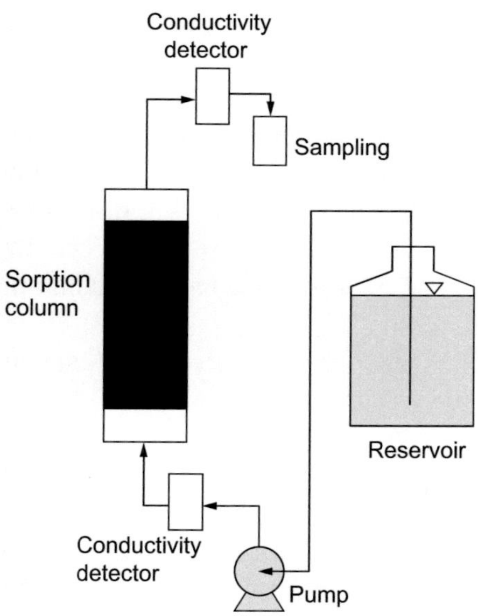

Figure 9.1: Experimental setup for column experiments.

The conductivity detectors in the experimental setup (Figure 9.1) are necessary for determining the BTC of a conservative tracer (e.g. chloride or bromide) in addition to the BTC of the sorbate. From the tracer BTC, characteristic column data such as residence time and bed porosity can be derived (Section 9.4).

9.3 The advection-dispersion equation and the retardation concept

Models that describe the solute transport in the subsurface or during column experiments can be derived from the differential material balance equation, which is, particularly in the hydrogeological literature, also referred to as the advection-dispersion equation (ADE). For the sake of simplification, the following discussion will be restricted to the conditions of column experiments. In principle, the same model equations can be used to describe large-scale subsurface transport. However, for scale-up it has to be taken into account that some parameters are scale-dependent (e.g. dispersivity) and that soil and aquifer typically exhibit a heterogeneous structure.

For the one-dimensional transport taking into consideration sorption, the ADE reads

$$v_F \frac{\partial c}{\partial z} + \varepsilon_B \frac{\partial c}{\partial t} + \rho_B \frac{\partial \bar{q}}{\partial t} = D_{ax}\, \varepsilon_B \frac{\partial^2 c}{\partial z^2} \tag{9.2}$$

where v_F is the filter velocity (Darcy velocity, superficial velocity), c is the concentration, z is the distance, t is the time, ε_B is the bulk porosity, ρ_B is the bulk density, $\bar{q}$ is the sorbed amount, and D_{ax} is the axial (longitudinal) dispersion coefficient. Here, it is assumed that the filter velocity, the bulk density, the porosity, and the dispersion coefficient are constant over time and space.

The four terms in eq. (9.2) describe, from left to right, the processes of advection, accumulation in the void volume, sorption onto the solid material, and dispersion. Obviously, eq. (9.2) is the same as the material balance equation for engineered fixed-bed adsorbers (eq. (6.46)). However, as explained in Section 9.1, dispersion cannot be neglected here.

In the following discussion, the mean loading, $\bar{q}$, is set equal to the equilibrium loading, q (assumption of local equilibrium).

Dividing eq. (9.2) by ε_B and introducing the mean pore water velocity (interstitial velocity), v_W,

$$v_W = \frac{v_F}{\varepsilon_B} \tag{9.3}$$

gives

$$v_W \frac{\partial c}{\partial z} + \frac{\partial c}{\partial t} + \frac{\rho_B}{\varepsilon_B} \frac{\partial q}{\partial t} = D_{ax} \frac{\partial^2 c}{\partial z^2} \tag{9.4}$$

The derivatives with respect to time can be combined after applying the chain rule

$$\frac{\partial q}{\partial t} = \frac{\partial q}{\partial c} \frac{\partial c}{\partial t} \tag{9.5}$$

$$v_W \frac{\partial c}{\partial z} + \left(1 + \frac{\rho_B}{\varepsilon_B} \frac{\partial q}{\partial c}\right) \frac{\partial c}{\partial t} = D_{ax} \frac{\partial^2 c}{\partial z^2} \tag{9.6}$$

The term within the brackets is referred to as the retardation factor, R_d,

$$R_d = 1 + \frac{\rho_B}{\varepsilon_B} \frac{\partial q}{\partial c} \tag{9.7}$$

The retardation factor indicates how strong the solute is retarded by sorption. Introducing R_d into eq. (9.6) gives

$$v_W \frac{\partial c}{\partial z} + R_d \frac{\partial c}{\partial t} = D_{ax} \frac{\partial^2 c}{\partial z^2} \tag{9.8}$$

For the most frequent case of the linear isotherm, R_d is constant for a given system and directly related to the linear sorption coefficient, K_d, which is also referred to as the distribution coefficient

$$q = K_d\, c \tag{9.9}$$

$$\frac{\partial q}{\partial c} = K_d \tag{9.10}$$

$$R_d = 1 + \frac{\rho_B}{\varepsilon_B} K_d \tag{9.11}$$

In contrast, for nonlinear isotherms, R_d depends on the concentration. For instance, for the Freundlich isotherm

$$q = K\, c^n \tag{9.12}$$

the following equations hold:

$$\frac{\partial q}{\partial c} = K\, n\, c^{n-1} \tag{9.13}$$

$$R_d = 1 + \frac{\rho_B}{\varepsilon_B} K\, n\, c^{n-1} \tag{9.14}$$

As can be seen from eq. (9.11) or (9.14), R_d becomes 1 for nonsorbable species ($K_d = 0$ or $K = 0$).

The physical meaning of R_d can also be derived from eq. (6.50) developed in Chapter 6, Section 6.4.3. This equation describes the velocity of a concentration point of the mass transfer zone of the solute

$$v_c = \frac{v_F}{\varepsilon_B + \rho_B \dfrac{\partial q}{\partial c}} \tag{9.15}$$

Introducing K_d for the case of a linear isotherm and rearranging the equation to obtain a velocity ratio gives

$$\varepsilon_B + \rho_B K_d = \frac{v_F}{v_c} \tag{9.16}$$

After dividing both sides by ε_B and introducing the pore water velocity (eq. (9.3)), the following expression is obtained:

$$1 + \frac{\rho_B}{\varepsilon_B} K_d = \frac{v_w}{v_c} \tag{9.17}$$

Comparison of eqs. (9.17) and (9.11) shows that R_d can also be interpreted as the ratio of the pore water velocity and the travel velocity of the retarded sorbate.

$$R_d = \frac{v_w}{v_c} \tag{9.18}$$

If no sorption takes place, the solute travels with the same velocity as the water and R_d is 1 as already derived from eq. (9.11). If sorption takes place, the concentration front of the sorbed compound travels with a velocity slower than the pore water velocity; that is, the compound is retarded, and R_d becomes greater than 1. Thus, the retardation factor is a parameter that shows how strong the solute is retarded in comparison to the pore water velocity or the transport velocity of nonsorbable species.

By means of eq. (9.11), the retardation coefficient, R_d, can be converted into the sorption coefficient, K_d, and vice versa. For this conversion, the characteristic system parameters bulk porosity and bulk density, ε_B and ρ_B, have to be known.

9.4 Determination of the retardation factor from experimental breakthrough curves

A simplified method for determining the retardation factor from a measured BTC can be derived on the basis of some general definitions and relationships. Given that each real BTC can be approximated by a concentration step at its center of mass (ideal BTC, see Chapter 6, Section 6.4.2), the velocity of the retarded solute, v_c, can be expressed as

$$v_c = \frac{h}{t_{b,c}^{id}} \tag{9.19}$$

where h is the height of the sorbent bed within the column and $t_{b,c}^{id}$ is the ideal breakthrough time of the retarded compound. In the case of a symmetrical S-shaped BTC, the ideal breakthrough time is the time where the relative concentration of the real BTC equals 0.5.

The pore water velocity is related to the residence time, t_r, of water (or of a non-sorbed solute) in the column by

$$v_W = \frac{h}{t_r} \tag{9.20}$$

Accordingly, the retardation factor, R_d, can be expressed as the ratio of the ideal breakthrough time of the retarded solute and the residence time of water

$$R_d = \frac{v_W}{v_c} = \frac{t_{b,c}^{id}}{t_r} \tag{9.21}$$

If using the relative time, t/t_r, instead of the absolute time, t, as the time axis in the BTC diagram, the retardation factor, R_d, can be read directly from the BTC as relative time, t/t_r, at $c/c_0 = 0.5$ (Figure 9.2).

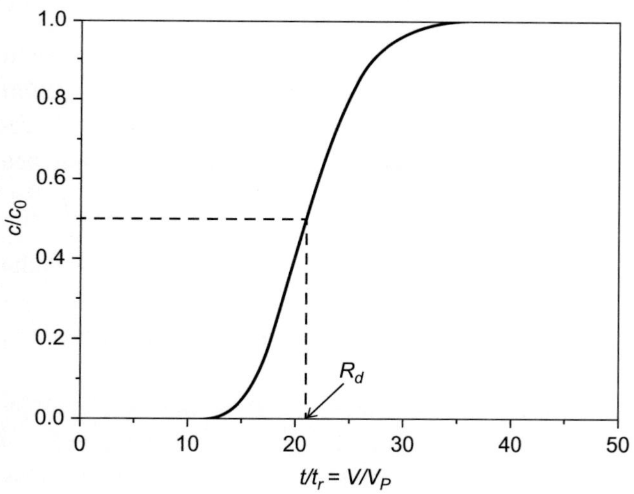

Figure 9.2: Principle of the simplified R_d determination.

In the hydrogeological literature, often the number of pore volumes fed to the column is used as the abscissa instead of the relative time. Given that the filter velocity is defined as

$$v_F = \frac{\dot{V}}{A_R} \tag{9.22}$$

the pore water velocity can be expressed taking into consideration eq. (9.3) as

$$v_W = \frac{v_F}{\varepsilon_B} = \frac{\dot{V}}{A_R\,\varepsilon_B} \tag{9.23}$$

where $\dot{V}$ is the volumetric flow rate and A_R is the cross-sectional area of the column.

Taking the definition of the residence time and expanding the fraction by the term $A_R\,\varepsilon_B$ leads to

$$t_r = \frac{h}{v_W} = \frac{h\,A_R\,\varepsilon_B}{v_W\,A_R\,\varepsilon_B} = \frac{V_P}{\dot{V}} \tag{9.24}$$

As can be seen from eq. (9.24), the residence time can be expressed as the ratio of the pore volume and the volumetric flow rate. It has to be noted that the term "pore volume" is used here according to the hydrogeological literature. The meaning is different from that in the engineered adsorption literature where the term "pore volume" means the internal pore volume (V_{pore}). In eq. (9.24), the term "pore volume" means the volume of pores that contain mobile water (water that is free to move through the packed bed). V_P, therefore, corresponds to the external pores (free volume between the sorbent particles, V_L, in Chapter 6) rather than to the internal pores.

Since the run time of the column can be expressed as the ratio of the volume fed to the column and the volumetric flow rate

$$t = \frac{V}{\dot{V}} \tag{9.25}$$

the relative time is identical to the ratio V/V_P, which is the number of pore volumes fed to the column

$$\frac{t}{t_r} = \frac{V}{V_P} \tag{9.26}$$

Therefore, if plotting the BTC in the form c/c_0 over V/V_P, the retardation factor can also be read directly from the curve at $c/c_0 = 0.5$.

As a precondition for plotting the BTC by using a relative time axis, the residence time or the pore volume must be known. These characteristic parameters of the column can be estimated from the BTC of a conservative tracer. A conservative tracer is a solute that is not sorbed and also not removed by other processes. In practice, solutions of salts such as chlorides or bromides are often used as conservative tracers. The ion concentration of these solutes can be easily determined by measuring the electrical conductivity (see Figure 9.1). Under the typically fulfilled condition that the BTC of such a conservative tracer is symmetrical and given that the residence time of the water is equal to the ideal breakthrough time of the conservative tracer, the residence time can be read from the tracer BTC at the point $c/c_0 = 0.5$. Knowing the residence time, the pore volume as well as the pore water velocity can be calculated from the volumetric flow rate and the bed height by using eq. (9.24). The porosity is then available from eq. (9.23). The required cross-sectional area can be found from the column diameter.

The methods for determining R_d and the characteristic column parameters t_r, V_P, and ε_B described in this section are all based on the assumption that the shape of the BTC is symmetrical. Only under this condition can the ideal breakthrough times of the tracer or the retarded solute be exactly determined from the BTC at $c/c_0 = 0.5$. In contrast to the tracer BTCs where this condition is fulfilled in most cases, the BTCs of retarded substances often show a tailing, for instance, caused by slow sorption kinetics. In this case, the ideal breakthrough time is shifted to relative concentrations $c/c_0 > 0.5$ but cannot be located exactly. To overcome this problem, a complete BTC model, which allows determining the sorption parameters by curve fitting, has to be used.

9.5 Breakthrough curve modeling

9.5.1 Introduction and model classification

The application of a BTC model to estimate characteristic sorption parameters instead of using the simple approach described in Section 9.4 provides a number of benefits: (a) the application of a BTC model is not restricted to symmetrical BTCs; (b) curve fitting considers a multitude of experimental points and is therefore more accurate than the simple one-point method; and (c) not only equilibrium parameters (R_d or K_d) but also characteristic parameters of dispersion and sorption kinetics can be determined by means of the BTC model.

In this section, at first only BTC models that consider dispersion and sorption will be discussed. Models that additionally include biodegradation will be presented separately in Section 9.6.

The available BTC models differ in the isotherm type and the consideration of sorption kinetics. In view of the isotherm, a distinction can be made between models that assume a linear isotherm and models that include the case of nonlinear isotherms. As already mentioned in Section 9.1, sorption kinetics plays a minor role in geosorption processes. Therefore, BTC models are often based on the assumption that local equilibrium exists in any cross section of the column (local equilibrium model, LEM). In this case, the spreading of the traveling concentration front, which can be experimentally observed as the spreading of the measured BTC, is assumed to be a result of dispersion only. For the LEM, an analytical solution is available.

On the other hand, models are also available that include sorption kinetics. There are different ways to consider sorption kinetics within the BTC model. In principle, the BTC models described in Section 7.4 (Chapter 7) for engineered adsorption can also be used for geosorption modeling. The only difference is that dispersion cannot be neglected as is typically done in the case of engineered adsorption processes. In particular, the homogeneous surface diffusion model (Crittenden et al. 1986b; Yiacoumi and Tien 1994; Yiacoumi and Rao 1996) as well as the linear driving force (LDF) model (Worch 2004) have been shown to be applicable to geosorption processes.

Two other types of BTC models are in common use, the two-region model and the two-site model. The two-region model is based on the assumption of slow mass transfer processes between mobile and immobile regions in the sorbent layer (van Genuchten and Wierenga 1976). In this model, the mass transfer between the different regions is formally described by a first-order mass transfer equation using the first-order mass transfer coefficient as the fitting parameter. In several studies (e.g. Young and Ball 1995; Maraqa 2001), it was shown that this mass transfer parameter depends strongly on column run conditions. In particular, the first-order mass transfer coefficient was found to be dependent on the column length, which is atypical for a mass transfer coefficient. Therefore, this model has to be considered empirical rather than theoretically founded. Nevertheless, two-region models were successfully used in several studies.

Another approach assumes the existence of two different types of sorption sites (Cameron and Klute 1977). At the first type of site, sorption takes place instantaneously. At the other type, the sorption process is slow, and kinetics has to be considered. As for the two-region model, a formal first-order mass transfer approach is used to describe the sorption kinetics.

The two-region model and the two-site model have in common that the transferability of the mass transfer coefficients to other conditions is limited due to their empirical character. A further problem results from the fact that the fractions of the different regions or sites cannot be estimated independently and must be estimated by curve fitting together with the respective kinetic parameters.

As an alternative to the models mentioned previously, a simple approach can be derived from the LDF model that allows extending the LEM to cases where sorption kinetics plays a certain role. The main advantage of this extended LEM is that, in principle, the same analytical solution as for the LEM can be used. On the other hand, this simplified approach is only applicable to systems with linear isotherms.

Exemplarily, the LEM, the LDF model, and the extended LEM will be described in more detail subsequently.

9.5.2 Local equilibrium model

The local equilibrium model (LEM) is based on the assumption that the sorption rate is infinitely fast and that, consequently, sorption equilibrium is established instantaneously at any point in the sorbent bed. Accordingly, dispersion is considered the only reason for the spreading of the BTC. Usually, as a further simplification, the isotherm is assumed to be linear. As already mentioned, this is a realistic assumption for most geosorption processes.

The LEM is based on the material balance as given in eq. (9.8). Dividing this equation by R_d gives

$$\frac{v_w}{R_d}\frac{\partial c}{\partial z}+\frac{\partial c}{\partial t}=\frac{D_{ax}}{R_d}\frac{\partial^2 c}{\partial z^2} \tag{9.27}$$

After introducing the velocity of the retarded solute, v_c, and the retarded dispersion coefficient, D^*_{ax},

$$v_c=\frac{v_w}{R_d} \tag{9.28}$$

$$D^*_{ax}=\frac{D_{ax}}{R_d} \tag{9.29}$$

Equation (9.27) can be written as

$$\frac{\partial c}{\partial t}=D^*_{ax}\frac{\partial^2 c}{\partial z^2}-v_c\frac{\partial c}{\partial z} \tag{9.30}$$

The analytical solution to eq. (9.30) is

$$c(z,t)=\frac{c_0}{2}\left(\operatorname{erfc}\left(\frac{z-v_c\,t}{2\sqrt{D^*_{ax}\,t}}\right)+\exp\left(\frac{v_c\,z}{D^*_{ax}}\right)\operatorname{erfc}\left(\frac{z+v_c\,t}{2\sqrt{D^*_{ax}\,t}}\right)\right) \tag{9.31}$$

where c is the concentration, c_0 is the inlet concentration, z is the transport length, and t is the time (Ogata and Banks 1961). The operator erfc is the complementary error function. It is related to the error function of the same variable (x) by

$$\operatorname{erfc}(x)=1-\operatorname{erf}(x) \tag{9.32}$$

The error function is a standard mathematical function given by

$$\operatorname{erf}(x)=\frac{2}{\pi^{0.5}}\int_0^x \exp(-\tau^2)\,\mathrm{d}\tau \tag{9.33}$$

It can be approximated by a series expansion.

If z is set equal to the column height, h, the resulting solution of eq. (9.31) gives the BTC, c (or c/c_0) = f(t).

A deeper inspection of eq. (9.31) shows that there are two basic parameters included that have to be estimated by curve fitting. The first parameter is the retardation factor, R_d, which is included in v_c and D^*_{ax} (eqs. (9.28) and (9.29)). R_d characterizes the sorption equilibrium and is related to the sorption coefficient, K_d, by eq. (9.11). The second fitting parameter is the axial (longitudinal) dispersion coefficient, D_{ax}, included in D^*_{ax}. D_{ax} describes the hydrodynamic dispersion, which summarizes the effects of hydromechanical dispersion and molecular diffusion and can be expressed by

$$D_{ax} = v_W\, \alpha + \frac{D_L}{\varepsilon_B} \tag{9.34}$$

where α is the dispersivity (dispersion length) and D_L is the liquid-phase diffusion coefficient of the sorbate, which can be found, for instance, from eq. (7.131). Since the liquid-phase diffusion coefficient, D_L, is relatively low ($10^{-9} \ldots 10^{-10}$ m^2/s), the second term of the right-hand side of eq. (9.34) can often be neglected

$$D_{ax} \approx v_W\, \alpha \tag{9.35}$$

In practice, the dispersivity, α, is frequently used instead of D_{ax} to describe the dispersion effects because it has the dimension of a length and therefore better indicates the spreading of the concentration front during the transport through the porous sorbent layer. By using eq. (9.34) or (9.35) together with eq. (9.31), curve fitting can also be carried out under variation of α. Here and in the following sections, the dispersivity, α, will be used to describe the dispersion effect.

Both parameters, R_d and α, can be determined from the same experimental BTC because the curve is affected by these parameters in a different manner. Changing R_d causes a shift of the center of the BTC on the time (or relative time) axis. In contrast, the dispersivity, α, influences the steepness of the BTC. The weaker the dispersion is, the smaller is α and the steeper is the BTC. Figure 9.3 shows schematically the influence of R_d and α on the shape and the location of the BTC.

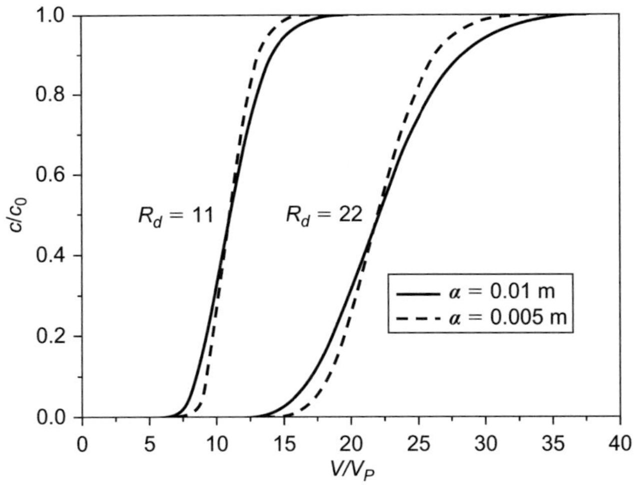

Figure 9.3: Influence of R_d and α on the shape and location of the breakthrough curve.

Typical laboratory-scale simulations of geosorption processes comprise BTC measurements with a conservative tracer as well as with the solute of interest. Equation (9.31) is applicable to both types of BTCs. For the conservative tracer, R_d is 1 and v_c equals v_w. Accordingly, the application of eq. (9.31) to a tracer BTC allows estimating v_w and α by curve fitting. These parameters can then be used for fitting the BTC of the retarded solute.

The pore water velocity, v_w, is connected to other characteristic parameters. Thus, it is needed to relate the velocity v_c to R_d (eq. (9.28)) and can also be used in eq. (9.24) to find the residence time, which is necessary to convert the absolute time axis into a relative time axis. Furthermore, the effective porosity can be estimated from v_w by means of eq. (9.23).

According to the assumptions of the LEM, the dispersivity found from the tracer experiment should be the same as for the retarded substance. If the dispersivity of the sorbed solute, estimated from its BTC, is higher than that for the tracer, then this is an indicator of the additional influence of sorption kinetics. In this case, a model that includes sorption kinetics has to be used instead of the LEM.

9.5.3 Linear driving force model

The linear driving force (LDF) model developed for engineered adsorption (Chapter 7, Section 7.4.4) covers linear and nonlinear isotherms as well as kinetic effects such as film and surface diffusion. This model can be easily extended to geosorption processes by additional consideration of dispersion. Since this BTC model was already described in detail in Chapter 7, only the principle of the supplementary integration of dispersion into the LDF model will be discussed here.

The original LDF model is a plug-flow model in which dispersion is neglected. The additional consideration of dispersion within the plug-flow model can be achieved by a modification of the volumetric mass transfer coefficient for film diffusion, $k_F\, a_{VR}$ (Vermeulen et al. 1973; Raghavan and Ruthven 1983). This simple approach allows maintaining all model equations and also the solution method. Given that external mass transfer (film diffusion) and axial (longitudinal) dispersion act in series with respect to the spreading of the BTC, an effective external mass transfer resistance can be defined that summarizes the effects of film diffusion and dispersion. Since the mass transfer resistances are given by the reciprocal values of the mass transfer coefficients and the resistances have to be added in the case of series connection, the following equation can be derived:

$$\frac{1}{k_{F,eff}\, a_{VR}} = \frac{1}{k_F\, a_{VR}} + \frac{1}{k_D\, a_{VR}} \tag{9.36}$$

where $k_{F,eff}\, a_{VR}$ is the effective volumetric film mass transfer coefficient. The effect caused by dispersion is quantified by an equivalent volumetric mass transfer coefficient, $k_D\, a_{VR}$,

$$k_D\, a_{VR} = \frac{v_F}{\alpha} \tag{9.37}$$

Introducing dimensionless parameters gives

$$\frac{1}{N_{F,eff}} = \frac{1}{N_F} + \frac{1}{N_D} \tag{9.38}$$

with

$$N_D = k_D\, a_{VR} \frac{h}{v_F} = \frac{h}{\alpha} \tag{9.39}$$

where N_D is equivalent to the column Peclet number, *Pe*. As shown in Chapter 7 (Section 7.4.4), N_F is defined as

$$N_F = \frac{k_F\, a_{VR}\, c_0\, t_{b,c}^{id}}{\rho_B\, q_0} \tag{9.40}$$

where $t_{b,c}^{id}$ is the ideal breakthrough time of the retarded compound. The only difference from the original LDF model is that the effective film mass transfer coefficient, $N_{F,eff}$, which combines the effects of dispersion and film diffusion, has to be used in the dimensionless kinetic equation for film diffusion (eq. (7.89)) instead of N_F. To consider the influence of dispersion, the dispersivity, α, must be known. It can be estimated from the BTC of a conservative tracer in the same manner as described in Section 9.5.2. Once α has been estimated, the volumetric dispersion mass transfer coefficient, $k_D\, a_{VR}$, can be calculated by eq. (9.37). The mass transfer coefficient for film diffusion is available from one of the empirical correlations given in Chapter 7, Table 7.7. Consequently, the intraparticle mass transfer coefficient, k_S^*, remains the only mass transfer parameter that has to be found from the sorbate BTC by curve fitting.

Figure 9.4 shows exemplarily the application of the LDF model to an experimentally studied sorbate/sorbent system (dibenzothiophene/sandy aquifer material). In this system, the BTC calculated by the LEM on the basis of the tracer dispersivity is too steep, which indicates that an additional influence of sorption kinetics exists. In contrast, the LDF model, which considers sorption kinetics, can describe the breakthrough behavior satisfactorily.

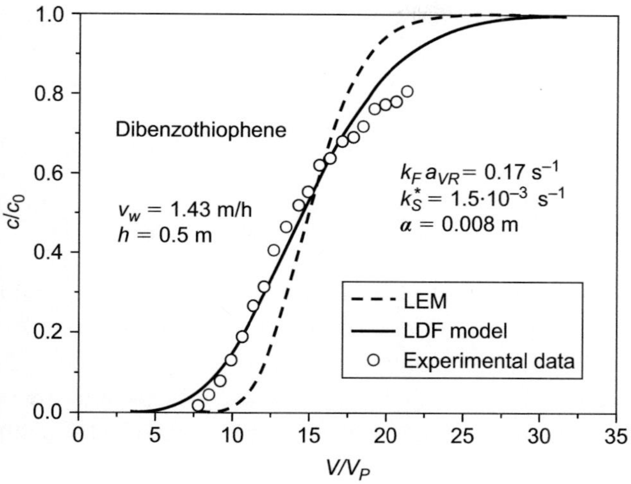

Figure 9.4: Description of an experimental dibenzothiophene breakthrough curve by the LDF model under consideration of sorption kinetics in comparison with the prediction by the LEM based on the tracer dispersivity. Sorbent: sandy aquifer material. Experimental data from Schreiber (2002).

9.5.4 Extension of the local equilibrium model

Given that in the case of a linear isotherm the mass transfer resistances of film and intraparticle diffusion can be added and the resulting overall mass transfer resistance acts in series with the resistance caused by dispersion as discussed in Section 9.5.3, the following equation can be written:

$$\frac{1}{N_{D,eff}} = \frac{1}{N_S} + \frac{1}{N_F} + \frac{1}{N_D} = \frac{1}{N_{S,eff}} + \frac{1}{N_D} \tag{9.41}$$

where $N_{S,eff}$ is the dimensionless effective intraparticle mass transfer coefficient, which is defined analogously to the dimensionless intraparticle mass transfer coefficient, N_S (Chapter 7, Section 7.4.4)

$$N_{S,eff} = k^*_{S,eff}\, t^{id}_{b,c} = k_{kin}\, t^{id}_{b,c} \tag{9.42}$$

Here, the overall mass transfer coefficient, k_{kin}, summarizes the kinetic effects caused by both film and intraparticle diffusion.

Returning to the nondimensionless parameters by using eqs. (9.39), (9.41), and (9.42) gives

$$\frac{\alpha_{eff}}{h} = \frac{1}{k_{kin}\, t^{id}_{b,c}} + \frac{\alpha}{h} = \frac{v_c}{k_{kin}\, h} + \frac{\alpha}{h} \tag{9.43}$$

Multiplying eq. (9.43) by the column height (or transport distance), h, leads to

$$\alpha_{eff} = \frac{v_c}{k_{kin}} + \alpha \tag{9.44}$$

or, with eq. (9.28), to

$$\alpha_{eff} = \frac{v_w}{k_{kin}\, R_d} + \alpha \tag{9.45}$$

The first term on the right-hand side of eq. (9.45) describes the effect of sorption kinetics and the second the effect of dispersion. Both effects together determine the spreading of the BTC. The overall effect is expressed as effective dispersivity, which can be found from the sorbate BTC by using the LEM with α_{eff} instead of α. Given that α is independently available from the tracer BTC, the kinetic coefficient k_{kin} can be found from α_{eff} using eq. (9.45).

This simple approach allows applying the LEM to BTCs that are influenced by sorption kinetics. The only difference from the original LEM is that the effective dispersivity replaces the tracer dispersivity.

In Figure 9.5, the application of the extended LEM is demonstrated for the system dibenzothiophene/sandy aquifer material, which was already shown in Figure 9.4. As can be seen, the extended LEM with α_{eff} as a fitting parameter is able to match the experimental breakthrough curve with the same accuracy as the LDF model, whereas the use of the tracer dispersivity, α, in the LEM yields too steep a BTC. The rate constant, k_{kin}, can be calculated by eq. (9.45) by using the parameters found by fitting the tracer and dibenzothiophene BTCs: effective dispersivity (0.028 m), tracer dispersivity

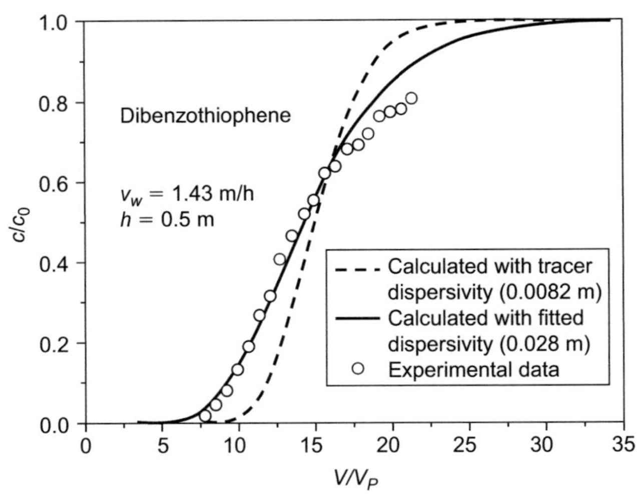

Figure 9.5: Comparison of the conventional and the extended LEM (dibenzothiophene/sandy aquifer material). Experimental data from Schreiber (2002).

(0.0082 m), pore water velocity (1.43 m/h), and retardation factor (15.1). The value of $k_{kin} = 1.33 \cdot 10^{-3}\ \mathrm{s}^{-1}$ is in good agreement with the mass transfer coefficient k_S^* of the LDF model ($1.5 \cdot 10^{-3}\ \mathrm{s}^{-1}$), indicating that in this case the overall sorption rate is mainly determined by intraparticle diffusion.

9.6 Combined sorption and biodegradation

9.6.1 General model approach

During subsurface transport, it is a frequently occurring case that the transported substance is subject to both sorption and biodegradation. To consider this case, an additional degradation term has to be introduced into the transport equation (see also Chapter 7, Section 7.6.2). Frequently, a first-order rate law is used to describe the degradation kinetics. In the most general case, it can be assumed that degradation takes place in the liquid phase as well as in the sorbed phase. The respective kinetic equations are

$$\frac{\partial c}{\partial t} = -\lambda_l\, c \tag{9.46}$$

for the liquid phase and

$$\frac{\partial \bar{q}}{\partial t} = -\lambda_s\, \bar{q} \tag{9.47}$$

for the sorbed phase. Herein, λ_l and λ_s are the first-order rate constants for the degradation in the liquid phase and in the sorbed phase, respectively.

Introducing the rate laws into eq. (9.2) and setting $q = \bar{q}$ (local equilibrium approach) gives

$$v_F \frac{\partial c}{\partial z} + \varepsilon_B \frac{\partial c}{\partial t} + \rho_B \frac{\partial q}{\partial t} + \varepsilon_B\, \lambda_l\, c + \rho_B\, \lambda_s\, q = D_{ax}\, \varepsilon_B \frac{\partial^2 c}{\partial z^2} \tag{9.48}$$

After rearranging and dividing by ε_B we get

$$v_w \frac{\partial c}{\partial z} + \left(1 + \frac{\rho_B}{\varepsilon_B}\frac{\partial q}{\partial c}\right)\frac{\partial c}{\partial t} + \left(\lambda_l + \frac{\rho_B}{\varepsilon_B}\lambda_s \frac{q}{c}\right) c = D_{ax} \frac{\partial^2 c}{\partial z^2} \tag{9.49}$$

Introducing the retardation factor, R_d, and an overall degradation rate constant, λ, gives

$$v_w \frac{\partial c}{\partial z} + R_d \frac{\partial c}{\partial t} + \lambda\, c = D_{ax} \frac{\partial^2 c}{\partial z^2} \tag{9.50}$$

with

$$R_d = 1 + \frac{\rho_B}{\varepsilon_B}\frac{\partial q}{\partial c} = 1 + \frac{\rho_B}{\varepsilon_B} K_d \tag{9.51}$$

and

$$\lambda = \lambda_l + \frac{\rho_B}{\varepsilon_B} \lambda_s \frac{q}{c} = \lambda_l + \frac{\rho_B}{\varepsilon_B} \lambda_s K_d \tag{9.52}$$

It has to be noted that the right-hand sides of eqs. (9.51) and (9.52) are valid for systems with linear isotherms.

Dividing eq. (9.50) by R_d, introducing v_c and D^*_{ax} according to eqs. (9.28) and (9.29), and rearranging leads to

$$\frac{\partial c}{\partial t} = \frac{D_{ax}}{R_d} \frac{\partial^2 c}{\partial z^2} - \frac{v_W}{R_d} \frac{\partial c}{\partial z} - \frac{\lambda}{R_d} c \tag{9.53}$$

$$\frac{\partial c}{\partial t} = D^*_{ax} \frac{\partial^2 c}{\partial z^2} - v_c \frac{\partial c}{\partial z} - \lambda^* c \tag{9.54}$$

where λ^* is the retarded degradation rate constant, λ/R_d.

From eq. (9.52), two special cases can be derived. If the degradation is assumed to occur only in the liquid phase, eq. (9.52) simplifies to

$$\lambda = \lambda_l \tag{9.55}$$

and the transport equation to be solved reads

$$\frac{\partial c}{\partial t} = D^*_{ax} \frac{\partial^2 c}{\partial z^2} - v_c \frac{\partial c}{\partial z} - \lambda^*_l c \tag{9.56}$$

where λ^*_l is λ_l/R_d.

If the degradation rate constant is assumed to be identical in both phases ($\lambda_l = \lambda_s$), eq. (9.52) becomes

$$\lambda = \lambda_l + \frac{\rho_B}{\varepsilon_B} \lambda_l \frac{q}{c} = \lambda_l \left(1 + \frac{\rho_B}{\varepsilon_B} K_d \right) = \lambda_l R_d \tag{9.57}$$

and the resulting transport equation reads

$$\frac{\partial c}{\partial t} = D^*_{ax} \frac{\partial^2 c}{\partial z^2} - v_c \frac{\partial c}{\partial z} - \lambda_l c \tag{9.58}$$

The eqs. (9.54), (9.56), and (9.58) all have the same mathematical form. The only difference lies in the meaning of the respective kinetic parameter. If the degradation rate constant is considered simply as a fitting parameter, it makes no difference which of the equations is used.

For systems with linear isotherms, the analytical solution to eq. (9.54) is (Sun 1996)

$$c(z,t) = \frac{c_0}{2} \exp\left(\frac{v_c z}{2 D^*_{ax}} \right) \left\{ \exp\left[-\frac{z F}{2 D^*_{ax}} \right] \operatorname{erfc}\left[\frac{z - F t}{2\sqrt{D^*_{ax} t}} \right] + \exp\left[\frac{z F}{2 D^*_{ax}} \right] \operatorname{erfc}\left[\frac{z + F t}{2\sqrt{D^*_{ax} t}} \right] \right\} \tag{9.59}$$

with

$$F = \sqrt{v_c^2 + 4\,\lambda^*\,D_{ax}^*} \tag{9.60}$$

According to eq. (9.59), the breakthrough behavior of a solute is determined here by three basic parameters: R_d (included in v_c, D_{ax}^*, and λ^*), α (included in D_{ax}, according to eq. (9.34) or (9.35)), and λ (included in λ^*). R_d, which is related to K_d, quantifies the equilibrium; α, the effect of dispersion; and λ, the degradation rate. If no degradation takes place ($\lambda^* = 0$), eq. (9.59) reduces to eq. (9.31). As long as only dispersion determines the spreading of the BTC (fast sorption kinetics), the dispersivity, α, can be estimated independently from a tracer experiment. In the more general case, if the influence of sorption kinetics cannot be excluded, α has to be replaced by the effective parameter, α_{eff}, which can be estimated by fitting the sorbate BTC (see Section 9.5.4).

In principle, all three parameters can be found from only one experimental BTC. This is possible because the three parameters influence the location and the shape of the BTC in a different manner as illustrated in Figure 9.6. R_d, as an equilibrium parameter, determines the location of the center of the BTC on the time axis. For a given R_d, the steepness of the BTC is influenced by the dispersivity (α) or by the combined effect of dispersivity and sorption rate (α_{eff}). Biodegradation leads to substance loss. Therefore, if biodegradation takes place, the BTC does not reach the initial concentration but ends in a plateau concentration lower than the initial concentration. The height of the concentration plateau is a measure of substance loss. For a given transport length and water velocity (i.e. for a given residence time), the level of the end concentration is determined by the degradation rate constant, λ^*.

It has to be noted that for a degradable solute, a BTC can only be measured if the residence time in the column is shorter than the time needed for a complete degradation. Otherwise, no breakthrough occurs and the outlet concentration is always zero. Consequently, none of the characteristic parameters can be determined. If, nevertheless, the parameters are of interest, then the residence time in the experiment has to be reduced.

As an example, Figure 9.7 shows BTCs calculated for a given degradation rate and different residence times. As can be seen, the longer the residence time is, the more substance can be degraded within the column with the consequence that the level of the concentration plateau decreases.

Furthermore, it has to be noted that the application of the first-order approach to describe biodegradation is a simplification that is appropriate particularly if the following conditions hold:

- Biodegradation is primarily a function of the contaminant concentration.
- The number of degrading microorganisms is constant over time.
- All other nutrients critical to biodegradation processes are in abundance.
- The electron acceptor (in case of oxidative degradation) does not limit the degradation rate.

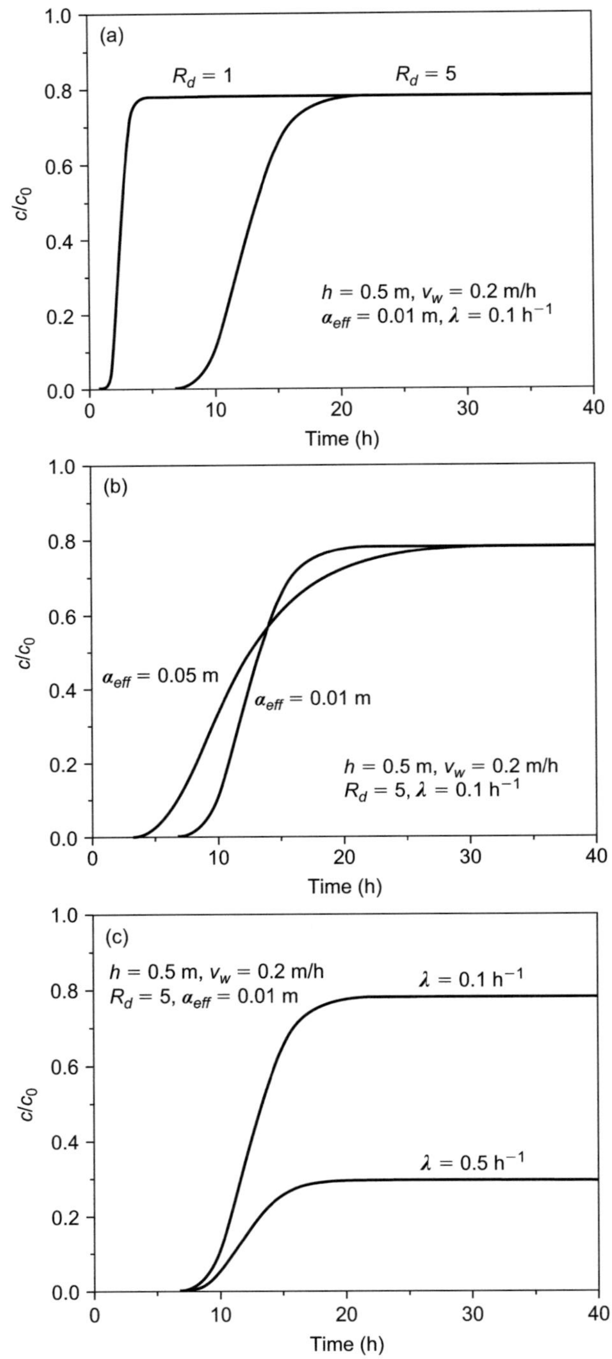

Figure 9.6: Combined sorption and biodegradation: influence of R_d (a), α_{eff} (b), and λ (c) on the shape of the BTC.

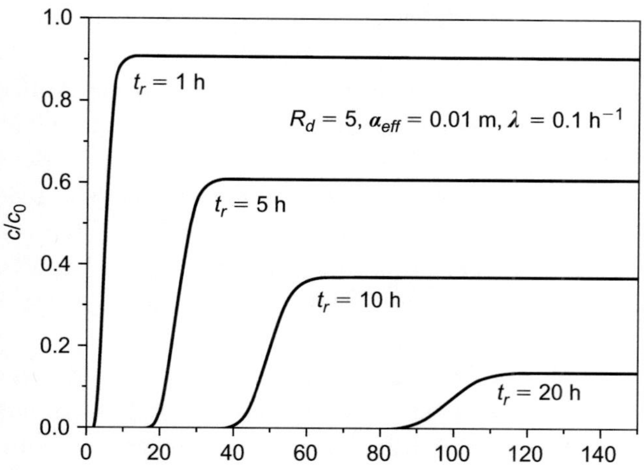

Figure 9.7: Combined sorption and biodegradation: influence of the residence time on the concentration plateau of the BTC.

Despite the simplifying assumptions, the sorption/biodegradation model, based on first-order degradation, often gives good results as shown exemplarily in Figure 9.8.

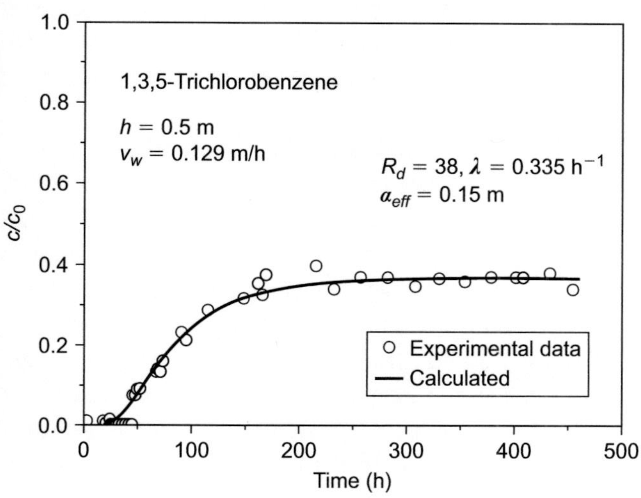

Figure 9.8: Combined sorption and biodegradation: experimental and calculated breakthrough curves of 1,3,5-trichlorobenzene.

In more sophisticated models, other kinetic approaches – for instance, the Monod equation – are used to describe the biodegradation. As already shown in Section 7.6.2 (Chapter 7), the Monod equation reduces to a first-order rate law under certain conditions.

9.6.2 Special case: Sorption and biodegradation of dissolved organic matter

As already discussed for engineered adsorption processes, dissolved organic matter (DOM), typically measured as dissolved organic carbon (DOC), is a multicomponent mixture of unknown composition where the different components may have different adsorption properties. For engineered adsorption, a fictive component approach (adsorption analysis) was developed to overcome the difficulties in modeling DOM adsorption (Section 4.7.2 in Chapter 4). In principle, a comparable approach is possible for DOM sorption onto geosorbents. However, some differences in comparison to the engineered adsorption have to be taken into account.

At first, the modeling of DOM sorption is simpler because linear sorption isotherms for the fictive components can be assumed, which allows neglecting the sorbate competition (Schreiber and Worch 2000). This assumption can be proven by the ideal adsorbed solution theory (see Chapter 4, Section 4.5.4). Under this condition, the sorption zones of the different components travel without any interaction. On the other hand, some fractions of the DOM are typically degraded during the subsurface transport. Thus, the fictive components have to be characterized with respect to not only their sorption behavior but also their degradability. That increases the number of parameters that have to be considered in the model.

As an example, a fictive component approach will be shown here that allows describing sorption and biodegradation of the complex DOM system on the basis of an experimental DOC BTC. This DOC BTC can be considered the sum of the individual BTCs of the fictive components. The general principle of the fictive component approach is described below.

At first, a limited number of fictive components are defined, each of them characterized by a set of the following parameters: retardation factor, R_d; effective dispersivity, α_{eff}; biodegradation rate, λ; and initial concentration, c_0. After that, the values of these parameters have to be found by BTC fitting. Due to the high number of fitting parameters, the number of fictive components should be not higher than three. For a fictive three-component system, the number of fitting parameters would be 11 (9 process parameters and 2 concentrations, the third concentration is given as difference from the total DOC). The number of parameters can be further reduced if the components are characterized in such a manner that they represent limiting cases. For instance, one fictive component can be considered the conservative component ($R_d = 1$, $\lambda = 0$), the second can be assumed to show sorption but no degradation ($\lambda = 0$), and the third can be treated as subject to both sorption and biodegradation. For α_{eff}, as a first guess, the tracer dispersivity can be used. Despite the large number of remaining parameters, curve fitting is, in principle, possible because the parameters R_d, λ, and α_{eff} affect shape and location of the BTC in a different manner, as shown in Section 9.6.1.

Due to the independent transport of the different components, the individual BTCs

$$\frac{c_i}{c_{0,i}} = f(t) \tag{9.61}$$

can be calculated by using eq. (9.59). The concentrations as well as the initial concentrations then have to be added according to

$$c_T = c_1 + c_2 + c_3 \tag{9.62}$$

$$c_{0,T} = c_{0,1} + c_{0,2} + c_{0,3} \tag{9.63}$$

in order to receive the total BTC

$$\frac{c_T}{c_{0,T}} = f(t) \tag{9.64}$$

The BTC calculation has to be repeated under variation of the fitting parameters as long as the best agreement with the experimental data is found.

Despite the strong simplifications, this fictive component approach is appropriate to describe the DOM breakthrough behavior at an acceptable quality as shown exemplarily in Figure 9.9.

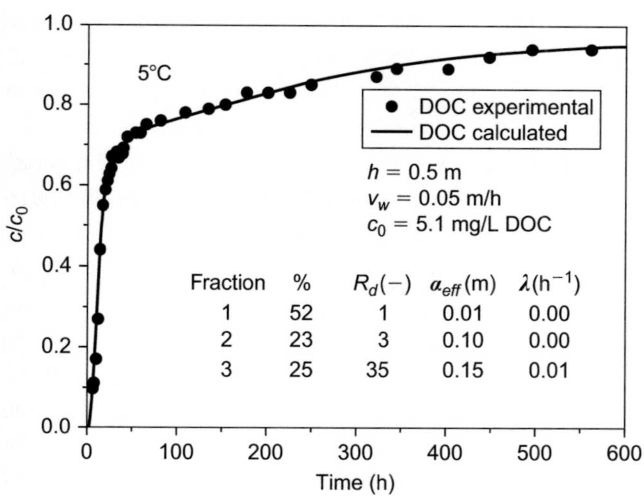

Figure 9.9: Modeling a DOM breakthrough curve by using the fictive component approach. Experimental data from Schoenheinz, 2004.

9.7 The influence of pH and DOM on geosorption processes

9.7.1 pH-dependent sorption

If the sorbate is a weak acid or a weak base, it can occur in neutral or ionized form, depending on the pH value of the solution. The distribution of ionic and neutral species is determined by the acidity constant of the solute, typically written as acidity exponent pK_a, and by the aqueous-phase pH value. Ionized species are much more soluble and thus less hydrophobic than their neutral counterparts. Accordingly, sorption of the ionic and nonionic species will differ, and neutral species are expected to sorb more strongly than ionized species. If the pH of the aqueous phase is within a range of p$K_a \pm 2$, both ionized and neutral species of the acidic or basic compound are present in substantial amounts in solution, and therefore, sorption of both species has to be considered in the sorption model.

The situation is comparable to that described in Chapter 4 (Section 4.6) for engineered systems. However, because the isotherms in geosorption are typically linear, the modeling of the pH-dependent sorption is much easier than in the case of engineered adsorption where the isotherms are typically nonlinear and a competitive adsorption model, such as the thermodynamic IAST, has to be applied.

Hereinafter, the modeling of the pH-dependent sorption will be considered exemplarily for an acid. The dissociation of an acid (HA), consisting of the proton and the anion, can be described by the general reaction equation

$$\mathrm{HA} \rightleftharpoons \mathrm{H}^+ + \mathrm{A}^-$$

Since the concentrations of neutral (HA) and ionic species (A^-) cannot be determined separately, sorption equilibria can only be measured as overall isotherms as given by eqs. (9.65) and (9.66) for the case of a linear isotherm

$$q(\mathrm{HA} + \mathrm{A}^-) = K_{d,app}\, c(\mathrm{HA} + \mathrm{A}^-) \tag{9.65}$$

or

$$q(\mathrm{HA}) + q(\mathrm{A}^-) = K_{d,app}[c(\mathrm{HA}) + c(\mathrm{A}^-)] \tag{9.66}$$

where $K_{d,app}$ is the apparent (observed) sorption coefficient.

The portion of neutral species HA contributing to the total concentration can be expressed by the degree of protonation, α_P, according to

$$\alpha_P = \frac{c(\mathrm{HA})}{c(\mathrm{HA}) + c(\mathrm{A}^-)} = \frac{1}{1 + 10^{\mathrm{pH} - \mathrm{p}K_a}} \tag{9.67}$$

Note that the degree of protonation of an acid, α_P, is different from the degree of protolysis, α, defined in Section 4.6. Both parameters are related by $\alpha_P = 1 - \alpha$.

Introduction of sorption coefficients for neutral ($K_{d,n}$) and ionic species ($K_{d,i}$) allows formulating the total isotherm as

$$q(\text{HA}) + q(\text{A}^-) = K_{d,app}[c(\text{HA}) + c(\text{A}^-)] = K_{d,n}\, c(\text{HA}) + K_{d,i}\, c(\text{A}^-) \qquad (9.68)$$

with $K_{d,app} = K_{d,n}$ for the limiting case $\alpha_P = 1$ and $K_{d,app} = K_{d,i}$ for $\alpha_P = 0$. Combining eqs. (9.67) and (9.68) gives the following relationship between the apparent sorption coefficient, $K_{d,app}$, and the sorption coefficients of the neutral and ionic species:

$$K_{d,app} = (K_{d,n} - K_{d,i})\ \alpha_P + K_{d,i} \qquad (9.69)$$

With this equation, the apparent sorption coefficient can be resolved into the sorption coefficients of ionic and neutral species provided that the sorption experiments with the compound of interest were carried out at different pH values. Plotting $K_{d,app}$ versus α_P will enable the calculation of $K_{d,n}$ and $K_{d,i}$. Once $K_{d,n}$ and $K_{d,i}$ are known, the apparent sorption coefficient $K_{d,app}$ can be predicted for any other pH value.

If the ionized species does not significantly contribute to the overall sorption process (i.e. $K_{d,n} \gg K_{d,i}$), eq. (9.69) reduces to

$$K_{d,app} = K_{d,n}\ \alpha_P \qquad (9.70)$$

Figure 9.10 shows exemplarily BTCs of pentachlorophenol ($pK_a = 4.75$) sorbed at different pH values onto sandy aquifer material. With decreasing pH, the BTCs are shifted to longer times, indicating an increasing overall sorption with increasing portion of

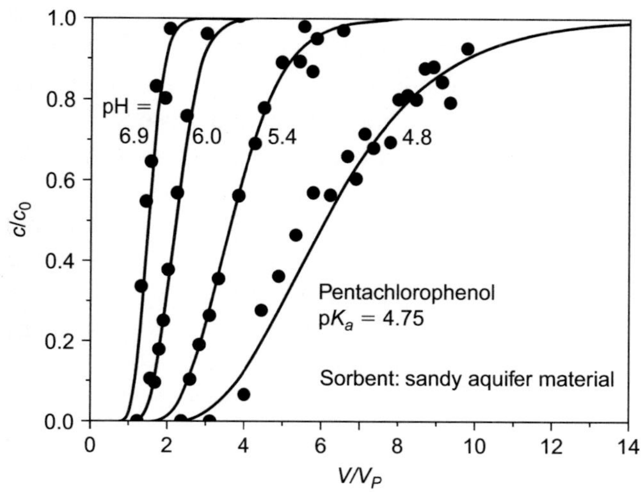

Figure 9.10: Breakthrough curves of pentachlorophenol ($pK_a = 4.75$) at different pH values (Amiri et al. 2004).

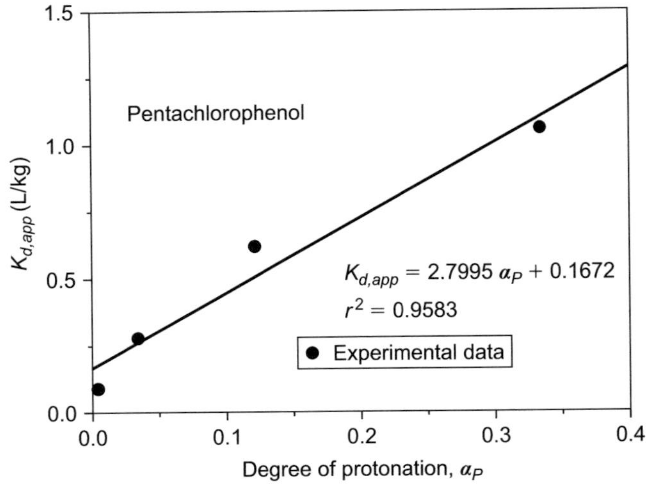

Figure 9.11: Determination of the sorption coefficients for neutral and ionized species according to eq. (9.69).

neutral species. From the BTCs, $K_{d,app}$ can be determined by means of a conventional BTC model (e.g. LEM or extended LEM). In Figure 9.11, the related $K_{d,app} - \alpha_P$ relationship (according to eq. (9.69)) is depicted.

9.7.2 Influence of DOM on micropollutant sorption

In natural systems, DOM (measured as DOC) is always present. Therefore, it is necessary to inspect the possible influence of DOM on micropollutant (MP) sorption. In the complex system consisting of DOM, MP, and geosorbent, a number of different interactions are possible (Figure 9.12). Besides different sorption processes, in particular complex formation between micropollutants and humic substances, which are the main components of DOM, has to be taken into account.

Hereafter, it will be assumed that the geosorbent is already in equilibrium with the DOM and no more DOM can be sorbed. This is a reasonable assumption for sorbents that are in continuous contact with the DOM-containing water over a long period of time as it is the case, for instance, at bank filtration or infiltration sites. Under this condition, the interactions to be considered can be restricted to sorption and complex formation of the micropollutant, which is assumed to be a newly occurring component in the considered water. Accordingly, sorption of the micropollutant onto the geosorbent and complex formation with the DOM in the liquid phase are competitive processes.

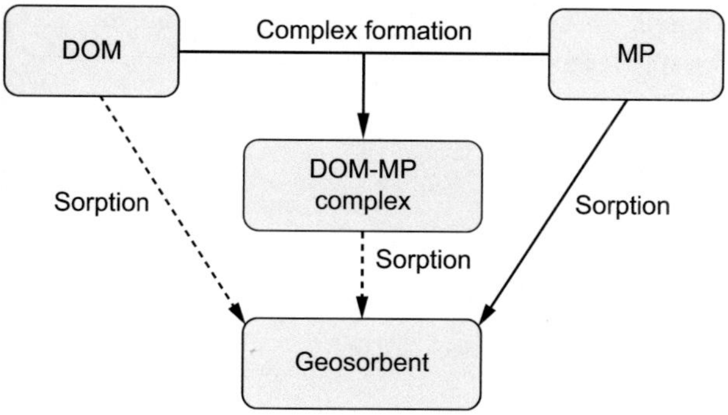

Figure 9.12: Possible interactions in the system MP/DOM/geosorbent. Dashed lines indicate interactions that are of minor relevance if the dissolved organic matter is in equilibrium with the organic fraction of the solid material.

An appropriate model that describes the influence of complex formation of a neutral solute with dissolved humic substances (DHS) on its sorption was developed by Rebhun et al. (1996). The law of mass action for the complex formation reads

$$K_c = \frac{c_{bound}}{c_{free}\,[\text{DHS}]} \tag{9.71}$$

where K_c is the equilibrium constant for complex formation, c_{bound} is the concentration of the bound solute, c_{free} is the concentration of the free solute, and [DHS] is the concentration of the DHS.

By using the material balance with c as total concentration

$$c = c_{bound} + c_{free} = (1 + K_c\,[\text{DHS}])\,c_{free} \tag{9.72}$$

the concentration of the free solute, c_{free}, can be written as

$$c_{free} = \frac{c}{1 + K_c\,[\text{DHS}]} \tag{9.73}$$

Neglecting additional sorption of DHS to solid material, the apparent sorption coefficient, $K_{d,app}$, related to the total solute concentration, c, can be expressed as

$$K_{d,app} = \frac{q}{c} = \frac{q}{c_{free}(1 + K_c[\text{DHS}])} = \frac{K_d}{1 + K_c[\text{DHS}]} \tag{9.74}$$

where K_d (= q/c_{free}) is the intrinsic sorption coefficient related to the micropollutant sorption from DOM-free water.

According to eq. (9.74), the sorption coefficient is reduced in the presence of DHS, and the extent of reduction depends on the equilibrium constant for the solute binding to DHS and on the concentration of DHS. Assuming a constant ratio of DOM and DHS and using dissolved organic carbon concentration, c(DOC), as a measure of DOM, eq. (9.74) can also be written in a modified form as

$$K_{d,app} = \frac{K_d}{1 + K_c\, c(\mathrm{DOC})} \tag{9.75}$$

Rearranging gives

$$K_c\, c(\mathrm{DOC}) = \frac{(K_d - K_{d,app})}{K_{d,app}} \tag{9.76}$$

According to eq. (9.76), the complex formation constant of the micropollutant, K_c, can be calculated from the relative reduction of the sorption coefficient.

Equations. (9.75) and (9.76) can be used to gain an impression of the relevance of complex formation with DOM for the strength of sorption in natural systems. Given that the concentration of DOC in natural waters is low (0.1–1.5 mg/L in groundwaters and 1–10 mg/L in rivers and lakes), it can be expected that the sorption is only affected by complex formation if the binding constant is very high. For example, it can be derived from eq. (9.75) or (9.76) that in water with c(DOC) = 5 mg/L, the complex formation constant of the sorbate must be at least $2 \cdot 10^4$ L/kg to reduce the sorption coefficient by approximately 10%.

Summarizing, only very strong complex formation can substantially affect the sorption. Whether such a strong complex formation is possible depends on the chemical nature of the micropollutant. For example, it was found that the sorption of several nitrophenols was affected by complex formation, whereas the sorption of chlorophenols under the same conditions was unaffected (Amiri et al. 2005). Here, the ability of the nitro group to form strong charge-transfer complexes was assumed as a possible reason for the different behavior.

9.8 Practical aspects: Prediction of subsurface solute transport

9.8.1 General considerations

During bank filtration or infiltration, the subsurface transport proceeds in a preferred direction. Therefore, the assumption of one-dimensional transport is a feasible approximation, even if the transport under field conditions is not strictly one-dimensional. Assuming one-dimensional transport, the same model approaches as used for modeling column BTCs can be applied to predict the solute transport during bank filtration or infiltration. For the application of the models, the characteristic parameters that describe sorption, dispersion, and biodegradation, are necessary. Here and in the next sections, the possibilities of parameter estimation will be discussed.

If only sorption plays a role and, furthermore, the influence of sorption kinetics can be neglected, the required input data are only the sorption coefficient and the dispersivity. Otherwise, the rate constant for sorption and, if relevant, for biodegradation is also required. As shown before, all these parameters can be determined by laboratory-scale column experiments. However, it has to be noted that the dispersivity is scale-dependent and is therefore not transferable to field conditions. This aspect will be discussed in more detail in Section 9.8.3.

In view of the high effort needed for sorption experiments, it is reasonable to look for methods that allow predicting the parameters required for transport modeling. However, the possibilities of predicting transport parameters are limited. Prediction tools are only available for the sorption coefficient and the dispersivity. Neither the biodegradation rate constant nor the sorption rate constant can be predicted. Thus, the possibility of a priori prediction of the transport behavior is restricted to systems where only sorption but no degradation is relevant and where the influence of sorption kinetics on the breakthrough behavior can be neglected.

Further restrictions exist regarding the sorption coefficient. Whether the sorption coefficient can be predicted or not depends on the dominating sorption mechanism. Natural sorbents, which are relevant for bank filtration and infiltration, in particular soil and aquifer material, typically have a heterogeneous composition and consist of organic and inorganic components. Therefore, different sorption mechanisms are possible. In principle, the following interactions can be distinguished: interactions of inorganic ions with mineral surfaces (electrostatic interactions, ion exchange) or with solid organic material (complex formation) and interactions of organic solutes with solid organic matter or with mineral surfaces (hydrophobic interactions, van der Waals forces, hydrogen bond formation). Ionized organic species take an intermediate position because the binding forces can include electrostatic interactions as well as weak intermolecular forces.

Up until now, prediction tools were developed only for the interactions between organic solutes with organic solid matter. However, this covers a broad range of sorption processes relevant for bank filtration and infiltration. For organic solutes, the accumulation on and within the organic fraction of the sorbent is the dominating binding mechanism. Sorption of organic solutes onto mineral surfaces becomes relevant only if the content of organic material in the sorbent is very low. According to Schwarzenbach et al. (1993), the sorption to mineral surfaces becomes dominant if the organic carbon fraction of the sorbent, f_{oc}, is less than approximately 0.001, where f_{oc} is defined as

$$f_{oc} = \frac{m_{oc}}{m_{solid}} \tag{9.77}$$

Herein, m_{oc} is the mass of organic carbon in the solid material and m_{solid} is the total mass of the solid material. It has to be noted that the given limit is only true for neutral species. In the case of ionized species, the contribution of ionic interactions between charged species and mineral surfaces may be higher, and these interactions may dominate even if the organic carbon fraction is higher than 0.001. Prediction methods for sorption coefficients of organic solutes that are sorbed to organic solid material are presented in Section 9.8.2.

Besides the sorption coefficient, the dispersivity is also needed to predict the sorption-affected solute transport because the dispersivity is responsible for the spreading of the concentration front. In numerous studies, dispersivity was found to be scale-dependent. This is a result of the different heterogeneities in the different scales. Simple equations that can be used to predict dispersivities for the field scale will be given in Section 9.8.3.

9.8.2 Prediction of sorption coefficients

Under the assumption that interaction between the organic solute and the organic fraction of the solid is the dominating sorption mechanism, it is reasonable to normalize the sorption coefficient, K_d, to the organic carbon content, f_{oc}, of the sorbent

$$K_{oc} = \frac{K_d}{f_{oc}} \tag{9.78}$$

This normalization makes the sorption coefficient independent of the sorbent type if the following conditions are fulfilled: (a) the sorption onto the solid organic matter is the only sorption mechanism, and (b) the organic material of different sorbents always has the same sorption properties. Under these idealized conditions, the normalized sorption coefficient, K_{oc}, depends only on the sorbate properties.

Given that the sorption of organic solutes is dominated by hydrophobic interactions, it can be expected that the sorption increases with increasing hydrophobicity of the sorbate. The hydrophobicity can be characterized by the n-octanol-water partition coefficient, K_{ow}. Consequently, K_{oc} should be strongly correlated with K_{ow}. Indeed, such correlations were found in numerous studies. The general form of all these correlations is

$$\log K_{oc} = a \log K_{ow} + b \tag{9.79}$$

where a and b are empirical parameters.

Depending on the substances included in the studies, two groups of correlations can be distinguished: class-specific correlations and nonspecific correlations. Table 9.1 gives a selection of $\log K_{oc} - \log K_{ow}$ correlations. More correlations can be found in the review paper of Gawlik et al. (1997). The values of the parameters in the correlations are slightly different, which indicates that the preconditions mentioned previously are obviously not strictly fulfilled. On the other hand, the deviations are mostly smaller than one order of magnitude. Taking $\log K_{ow} = 3$ as an example, the $\log K_{oc}$ values calculated by the nonspecific correlations given in Table 9.1 are in the range of 2.70 to 3.01. Slightly lower values are found if the class-specific correlations are used ($\log K_{oc} = 2.23 \dots 2.79$). In general, it has to be stated that the application of such empirical correlations can give only a rough estimate of K_{oc}. However, it can be expected that at least the right order of magnitude for K_{oc} can be found from the correlations. The n-octanol-water partition coefficient is available for most solutes from databases or can be estimated by special prediction methods.

Table 9.1: Selection of $\log K_{oc} - \log K_{ow}$ correlations.

Correlation	Valid for substance class	Authors
$\log K_{oc} = 0.544 \log K_{ow} + 1.377$	Not specified	Kenaga and Goring (1980)
$\log K_{oc} = 0.909 \log K_{ow} + 0.088$	Not specified	Hassett et al. (1983)
$\log K_{oc} = 0.679 \log K_{ow} + 0.663$	Not specified	Gerstl (1990)
$\log K_{oc} = 0.903 \log K_{ow} + 0.094$	Not specified	Baker et al. (1997)
$\log K_{oc} = 1.00 \log K_{ow} - 0.21$	Benzenes, polycyclic aromatic hydrocarbons	Karickhoff et al. (1979)
$\log K_{oc} = 0.72 \log K_{ow} + 0.49$	Chlorobenzenes, methylbenzenes	Schwarzenbach and Westall (1981)
$\log K_{oc} = 0.89 \log K_{ow} - 0.32$	Chlorophenols, chlorobenzenes	van Gestel et al. (1991)
$\log K_{oc} = 0.63 \log K_{ow} + 0.90$	Substituted phenols, anilines, nitrobenzenes, chlorinated benzonitriles	Sabljiç et al. (1995)
$\log K_{oc} = 1.07 \log K_{ow} - 0.98$	Polychlorinated biphenyls	Girvin and Scott (1997)
$\log K_{oc} = 0.42 \log K_{ow} + 1.49$	Aromatic amines	Worch et al. (2002)

If the fraction f_{oc} for the considered sorbent is known, K_d can be calculated from K_{oc}, and, knowing the bulk density and the porosity, R_d is also available (eq. (9.11)). The porosity, ε_B, is often in the range of 0.3 to 0.4. Instead of the bulk density (ρ_B), the particle density (ρ_P), which is for nonporous material equal to the material density, can also be used to estimate R_d because both densities are related according to

$$\varepsilon_B = 1 - \frac{\rho_B}{\rho_P} \tag{9.80}$$

(see Section 6.3 in Chapter 6). R_d is then given by

$$R_d = 1 + \frac{\rho_B}{\varepsilon_B} K_d = 1 + \frac{(1-\varepsilon_B)\,\rho_P\, K_d}{\varepsilon_B} \tag{9.81}$$

From R_d, the ideal breakthrough time (the time at the center of mass of the BTC) can be predicted by using eq. (9.21) in the form

$$t_{b,c}^{id} = R_d\, t_r \tag{9.82}$$

with

$$t_r = \frac{L}{v_w} \tag{9.83}$$

where L is the transport distance and v_w is the pore water velocity.

9.8.3 Prediction of the dispersivity

As shown in Section 9.8.2, R_d alone allows only calculating the ideal breakthrough time. For a more realistic prediction of the transport behavior, for instance, by using the LEM, at least the dispersivity has to be known. As already mentioned in Section 9.8.1, the dispersivity was found to be scale-dependent. According to the variety of factors that may affect the dispersion, the dispersivities found from experiments show a broad variation even within the same scale. Therefore, the available prediction methods can give only a rough estimate of dispersivity as a function of the transport length.

The simplest approach is to assume that the dispersivity is one-tenth of the transport length, L (Pickens and Grisak 1981),

$$\alpha = 0.1\, L \tag{9.84}$$

Another equation was proposed by Xu and Eckstein (1995)

$$\alpha = 0.83\, (\log\, L)^{2.414} \tag{9.85}$$

The results of the different equations diverge increasingly with increasing transport distance. While for a transport length of 10 m the results are still comparable (1 m vs. 0.83 m), the dispersivities calculated for 100 m already differ considerably (10 m vs. 4.4 m).

10 Appendix

10.1 Conversion of Freundlich coefficients

The Freundlich isotherm

$$q = K\,c^n \tag{10.1}$$

is the isotherm equation that is most frequently used to describe adsorption from aqueous solutions. Freundlich coefficients, K, can be expressed in different units depending on the units used for the liquid-phase (c) and solid-phase concentrations (q). The following tables (Tables 10.1–10.3) present the most important conversion equations for Freundlich coefficients.

Table 10.1: Conversion of Freundlich coefficients given in different units (liquid-phase and solid-phase concentrations expressed as mass concentrations, molar concentrations, and organic carbon concentrations).

K_1↓ K_2→	(mg/g)/(mg/L)^n	(mg C/g)/(mg C/L)^n	(mmol/g)/(mmol/L)^n
(mg/g)/(mg/L)^n	–	$K_1 = K_2\,w_C^{n-1}$	$K_1 = K_2\,M^{1-n}$
(mg C/g)/(mg C/L)^n	$K_1 = K_2\,w_C^{1-n}$	–	$K_1 = K_2\,w_C^{1-n}\,M^{1-n}$
(mmol/g)/(mmol/L)^n	$K_1 = K_2\,M^{n-1}$	$K_1 = K_2\,w_C^{n-1}\,M^{n-1}$	–

M, molecular weight of the adsorbate; w_C, carbon fraction of the adsorbate ($w_C = M_C/M$; M_C = number of carbon atoms in the adsorbate molecule × 12 g/mol); n, Freundlich exponent.

Table 10.2: Conversion of Freundlich coefficients given in different mass concentrations.

K_1↓ K_2→	(mg/g)/(mg/L)^n	(µg/g)/(ng/L)^n
(mg/g)/(mg/L)^n	–	$K_1 = 10^{6n-3}\,K_2$
(µg/g)/(ng/L)^n	$K_1 = 10^{3-6n}\,K_2$	–

Table 10.3: Conversion of Freundlich coefficients given in different molar concentrations.

K_1↓ K_2→	(mmol/g)/(mmol/L)^n	(mmol/g)/(µmol/L)^n
(mmol/g)/(mmol/L)^n	–	$K_1 = 10^{3n}\,K_2$
(mmol/g)/(µmol/L)^n	$K_1 = 10^{-3n}\,K_2$	–

https://doi.org/10.1515/9783110715507-010

10.2 Alternative solution equations of the ideal adsorbed solution theory

As described in Chapter 4, the ideal adsorbed solution theory (IAST) can be considered the standard method for predicting multisolute equilibrium data. Furthermore, the IAST provides the basis for characterizing the adsorption behavior of unknown multicomponent systems by the fictive component approach (adsorption analysis) and for modeling the competitive adsorption of micropollutants and background organic matter. To solve the set of the basic IAST equations, different solution methods are possible. In addition to the methods demonstrated in Chapter 4 (Sections 4.5.2 and 4.5.3), an alternative approach, especially for the Freundlich isotherm, is given here. The solution equations that are used in this approach are based on the same basic IAST equations and are, therefore, equivalent to the solution equations given in Chapter 4.

We start with eq. (4.23) (Chapter 4) in the form

$$\frac{1}{q_T} = \sum_{i=1}^{N} \frac{z_i}{q_i^0} = \sum_{i=1}^{N} \frac{q_i}{q_T\, q_i^0} \tag{10.2}$$

After multiplying eq. (10.2) with q_T and considering the relationship between q_i^0 and the spreading pressure term, φ_i (Table 4.1)

$$q_i^0 = \varphi_i\, n_i \tag{10.3}$$

we get

$$\sum_{i=1}^{N} \frac{q_i}{q_i^0} = \sum_{i=1}^{N} \frac{q_i}{\varphi\, n_i} = 1 \tag{10.4}$$

Note that we can replace the individual φ_i by the constant φ because the IAST equations are valid for φ = constant (Chapter 4).

Rearranging eq. (10.4) gives

$$\sum_{i=1}^{N} \frac{q_i}{n_i} = \varphi \tag{10.5}$$

Introducing the definition of the solid-phase mol fraction, z_i,

$$z_i = \frac{q_i}{q_T} = \frac{q_i}{\sum_{j=1}^{N} q_j} \tag{10.6}$$

and the relationship between c_i^0 and φ (Table 4.1 in Chapter 4)

$$c_i^0 = \left(\frac{\varphi\, n_i}{K_i}\right)^{1/n_i} \tag{10.7}$$

into the basic IAST equation (eq. (4.22) in Chapter 4)

$$c_i = z_i\, c_i^0 \tag{10.8}$$

gives the following relationship for the component *i* in an N-component system:

$$c_i = z_i\, c_i^0 = \frac{q_i}{\sum\limits_{j=1}^{N} q_j} \left(\frac{\varphi\, n_i}{K_i}\right)^{1/n_i} \tag{10.9}$$

Finally, we can replace φ by eq. (10.5)

$$c_i = \frac{q_i}{\sum\limits_{j=1}^{N} q_j} \left(\frac{\sum\limits_{j=1}^{N} \frac{q_j}{n_j}}{\frac{K_i}{n_i}}\right)^{1/n_i} \tag{10.10}$$

Equation (10.10) has to be formulated for all components ($i = 1 \ldots N$), and the resulting set of equations has to be solved numerically to find the equilibrium adsorbent loadings, q_i, for given equilibrium concentrations, c_i.

For given initial concentrations, $c_{0,i}$, in batch systems, the equilibrium concentration, c_i, on the left-hand side of eq. (10.10) has to be replaced by a term derived from the material balance equation for batch systems (eq. (4.3) in Chapter 4)

$$q_i = \frac{V_L}{m_A}(c_{0,i} - c_i) \tag{10.11}$$

$$c_i = c_{0,i} - \frac{m_A}{V_L} q_i \tag{10.12}$$

10.3 Theoretical basics of the simplified equivalent background compound model

The simplified equivalent background compound model (SEBCM) can be used to describe the micropollutant adsorption in the presence of background organic matter in a batch system (Knappe et al. 1998, Qi et al. 2007). The SEBCM was presented in Chapter 4, Section 4.8.3 with the focus on the equations that are necessary for the practical application. As a supplement, a brief overview of the theoretical basics is given here.

The SEBCM is based on the ideal adsorbed solution theory (IAST), in particular on the equivalent background model (EBCM) that describes the micropollutant adsorption in the presence of background organic matter by a two-component IAST approach (Chapter 4, Section 4.7.3). Component 1 is the micropollutant and component 2 is the equivalent background compound (EBC) that stands for the background organic matter (NOM in drinking water treatment or EfOM in wastewater treatment).

In the SEBCM, it is assumed that the EBC is much better adsorbed than the micropollutant ($q_2 \gg q_1$) and the Freundlich exponents of micropollutant and EBC have comparable values. Under this condition, eq. (10.10), given in Section 10.2, can be written for the micropollutant and the EBC in a simplified manner. For the micropollutant, we find

$$c_1 = \frac{q_1}{q_1 + q_2}\left(\frac{\dfrac{q_1}{n_1} + \dfrac{q_2}{n_2}}{\dfrac{K_1}{n_1}}\right)^{1/n_1} \approx \frac{q_1}{q_2}\left(\frac{\dfrac{q_2}{n_2}}{\dfrac{K_1}{n_1}}\right)^{1/n_1} = q_1\, q_2^{(1/n_1)-1}\left(\frac{n_1}{K_1\, n_2}\right)^{1/n_1} \tag{10.13}$$

where c_1 and c_2 are the equilibrium concentrations, q_1 and q_2 are the adsorbent loadings, n_1 and n_2 are the Freundlich exponents, and K_1 is the Freundlich coefficient of the micropollutant. Rearranging eq. (10.13) gives

$$q_1 = c_1\, q_2^{1-(1/n_1)}\left(\frac{K_1\, n_2}{n_1}\right)^{1/n_1} \tag{10.14}$$

The simplified eq. (10.10) for the EBC reads

$$c_2 = \frac{q_2}{q_1 + q_2}\left(\frac{\dfrac{q_1}{n_1} + \dfrac{q_2}{n_2}}{\dfrac{K_2}{n_2}}\right)^{1/n_2} \approx \left(\frac{q_2}{K_2}\right)^{1/n_2} \tag{10.15}$$

Rearranging eq. (10.15) gives

$$q_2 = K_2\, c_2^{n_2} \tag{10.16}$$

It can be derived from eqs. (10.14) and (10.15) that under the given conditions, the micropollutant shows a linear adsorption behavior in the presence of a strongly competing EBC and that the adsorbed amount of the EBC is not influenced by the micropollutant.

The material balance equation for the micropollutant reads

$$q_1 = \frac{V_L}{m_A}(c_{0,1} - c_1) \tag{10.17}$$

where m_A is the adsorbent mass, and V_L is the volume of the solution. Since for the EBC a strong adsorption is assumed ($c_2 \ll c_{0,2}$), the respective material balance equation can be written as

$$q_2 \approx \frac{V_L}{m_A} c_{0,2} \tag{10.18}$$

Combining eqs. (10.17) and (10.18) gives an expression that can be used to substitute q_2 in eq. (10.14)

$$q_2 = \frac{q_1 c_{0,2}}{c_{0,1} - c_1} \tag{10.19}$$

Introducing eqs. (10.19) into (10.14) gives

$$q_1 = c_1 \left(\frac{q_1 c_{0,2}}{c_{0,1} - c_1} \right)^{1-(1/n_1)} \left(\frac{K_1 n_2}{n_1} \right)^{1/n_1} \tag{10.20}$$

or

$$q_1^{1/n_1} = c_1 c_{0,2}^{1-(1/n_1)} (c_{0,1} - c_1)^{(1/n_1)-1} \left(\frac{K_1 n_2}{n_1} \right)^{1/n_1} \tag{10.21}$$

Raising the equation to the power of n_1 gives a pseudo-single isotherm of the micropollutant

$$q_1 = c_1^{n_1} c_{0,2}^{n_1-1} (c_{0,1} - c_1)^{1-n_1} \left(\frac{K_1 n_2}{n_1} \right) \tag{10.22}$$

Finally, the following expression for the removal of the micropollutant as a function of the adsorbent dose can be derived by combining the pseudo-single isotherm equation with the related material balance equation (eq. (10.17)):

$$\frac{c_1}{c_{0,1}} = \frac{c_{0,2}^{(1/n_1)-1} \left(\dfrac{n_1}{n_2 K_1} \right)^{1/n_1}}{\left(\dfrac{m_A}{V} \right)^{1/n_1} + c_{0,2}^{(1/n_1)-1} \left(\dfrac{n_1}{n_2 K_1} \right)^{1/n_1}} = \frac{A}{\left(\dfrac{m_A}{V_L} \right)^{1/n_1} + A} \tag{10.23}$$

with

$$A = c_{0,2}^{(1/n_1)-1} \left(\frac{n_1}{n_2 K_1} \right)^{1/n_1} \tag{10.24}$$

The eqs. (10.23) and (10.24) correspond to eqs. (4.66) and (4.67) in Chapter 4 (Section 4.8.3) where their relevance and practical application are discussed.

10.4 Evaluation of surface diffusion coefficients from experimental data

As shown in Chapter 5 (Section 5.4.3), Zhang et al. (2009) have approximated the solutions to the homogeneous surface diffusion model (HSDM) by empirical polynomials. These polynomials can be used to evaluate the surface diffusion coefficients from experimentally determined kinetic curves. The general polynomial equation reads

$$\bar{C} = A_0 + A_1 \ln T_B + A_2 (\ln T_B)^2 + A_3 (\ln T_B)^3 \tag{10.25}$$

where $\bar{C}$ is a dimensionless concentration, defined as

$$\bar{C} = \frac{c - c_{eq}}{c_0 - c_{eq}} \tag{10.26}$$

T_B is the dimensionless time, given by

$$T_B = \frac{D_S\, t}{r_P^2} \tag{10.27}$$

where D_S is the surface diffusion coefficient, t is the time, and r_P is the particle radius. The coefficients A_i are listed in Table 10.4 for different relative equilibrium concentrations, c_{eq}/c_0, and Freundlich exponents, n.

The dimensionless concentrations, $\bar{C}$, can be converted into the relative concentrations, c/c_0, by

$$\frac{c}{c_0} = \bar{C}\left(1 - \frac{c_{eq}}{c_0}\right) + \frac{c_{eq}}{c_0} \tag{10.28}$$

The diffusion coefficient, D_S, which best describes the experimental kinetic data, can be found from a fitting procedure under variation of D_S by using the equations given previously and the parameters listed in the table. The procedure is as follows: find the parameters A_i for the given n and c_{eq}/c_0 from Table 10.4; assume a value for D_S; calculate the dimensionless times, T_B, for different times within the range of the experimental kinetic curve; and take the dimensionless times, T_B, to find $\bar{C}$ and c/c_0 from eqs. (10.25) and (10.28). Only T_B values in the given validity range should be used. As a result, a kinetic curve $c/c_0 = \mathrm{f}(t)$ is calculated, which can be compared with the experimental data. If necessary, the calculation has to be repeated with other values of D_s to minimize the mean deviations between predicted and experimental data.

Table 10.4: Parameters of eq. (10.25).

n	c_{eq}/c_0	A_0	A_1	A_2	A_3	Lower limit T_B	Upper limit T_B
0.1	0.001	7.76201 E-1	7.56773 E-1	2.16870 E-1	1.53227 E-2	6.85 E-4	7.40 E-2
	0.005	4.99720 E-1	5.65235 E-1	1.87201 E-1	1.40911 E-2	6.85 E-4	1.35 E-1
	0.01	3.64919 E-1	4.59934 E-1	1.68512 E-1	1.32161 E-2	7.80 E-4	1.60 E-1
	0.05	1.41974 E-1	2.48942 E-1	1.24660 E-1	1.05571 E-2	7.80 E-4	2.70 E-1
	0.1	8.53623 E-2	1.82932 E-1	1.10190 E-1	9.65166 E-3	8.50 E-4	3.00 E-1
	0.2	2.65210 E-2	1.09731 E-1	9.26188 E-2	8.46678 E-3	9.90 E-4	3.00 E-1
	0.3	−5.47219 E-3	6.86045 E-2	8.22572 E-2	7.71999 E-3	9.90 E-4	3.00 E-1
	0.4	−2.53237 E-2	4.28910 E-2	7.58529 E-2	7.26355 E-3	9.90 E-4	3.00 E-1
	0.5	−3.91816 E-2	2.48838 E-2	7.14332 E-2	6.95285 E-3	9.90 E-4	3.00 E-1
	0.6	−4.95959 E-2	1.13457 E-2	6.81093 E-2	6.71882 E-3	9.90 E-4	3.00 E-1
	0.7	−5.77042 E-2	8.50961 E-4	6.55653 E-2	6.54166 E-3	9.90 E-4	3.00 E-1
	0.8	−6.42211 E-2	−7.68734 E-3	6.34888 E-2	6.39649 E-3	9.90 E-4	3.00 E-1
	0.9	−6.94874 E-2	−1.47172 E-2	6.17973 E-2	6.27959 E-3	9.90 E-4	3.00 E-1
0.2	0.001	1.17237 E-0	7.85118 E-1	1.58099 E-1	8.23844 E-3	5.05 E-4	3.50 E-2
	0.005	6.45253 E-1	5.50680 E-1	1.39005 E-1	8.36855 E-3	2.05 E-4	8.50 E-2
	0.01	5.13173 E-1	4.92388 E-1	1.39344 E-1	9.08314 E-3	3.65 E-4	1.10 E-1
	0.05	2.24322 E-1	3.05970 E-1	1.21216 E-1	9.26458 E-3	7.05 E-4	2.00 E-1
	0.1	1.22475 E-1	2.12696 E-1	1.05034 E-1	8.51555 E-3	8.50 E-4	3.00 E-1
	0.2	8.97853 E-2	1.84347 E-1	1.08293 E-1	9.48046 E-3	9.90 E-4	3.00 E-1
	0.3	5.48862 E-2	1.44294 E-1	1.01682 E-1	9.26875 E-3	9.90 E-4	3.00 E-1
	0.4	2.47976 E-2	1.07206 E-1	9.33323 E-2	8.72285 E-3	9.90 E-4	3.00 E-1
	0.5	1.84338 E-3	7.83684 E-2	8.66479 E-2	8.27250 E-3	9.90 E-4	3.00 E-1
	0.6	−1.62317 E-2	5.53300 E-2	8.12304 E-2	7.90238 E-3	9.90 E-4	3.00 E-1
	0.7	−3.07867 E-2	3.68066 E-2	7.68789 E-2	7.60532 E-3	9.90 E-4	3.00 E-1
	0.8	−4.28149 E-2	2.13154 E-2	7.32217 E-2	7.35463 E-3	9.90 E-4	3.00 E-1
	0.9	−5.29236 E-2	8.23859 E-3	7.01365 E-2	7.14335 E-3	9.90 E-4	3.00 E-1

Table 10.4 (continued)

n	c_{eq}/c_0	A_0	A_1	A_2	A_3	Lower limit T_B	Upper limit T_B
0.3	0.001	8.48276 E-1	4.69793 E-1	7.59594 E-2	2.81270 E-3	1.10 E-4	2.20 E-2
	0.005	3.73741 E-1	2.78799 E-1	5.91368 E-2	2.42831 E-3	2.00 E-4	5.65 E-2
	0.01	3.91829 E-1	3.26121 E-1	7.86862 E-2	4.00352 E-3	2.80 E-4	9.20 E-2
	0.05	2.01974 E-1	2.48109 E-1	8.80228 E-2	5.81922 E-3	2.25 E-4	1.80 E-1
	0.1	1.35657 E-1	2.04684 E-1	8.84873 E-2	6.38200 E-3	2.95 E-4	2.35 E-1
	0.2	9.58145 E-2	1.78899 E-1	9.53563 E-2	7.66911 E-3	5.40 E-4	3.00 E-1
	0.3	6.80770 E-2	1.52203 E-1	9.51831 E-2	8.08638 E-3	7.80 E-4	3.00 E-1
	0.4	4.94239 E-2	1.34897 E-1	9.69648 E-2	8.73526 E-3	9.90 E-4	3.00 E-1
	0.5	2.19306 E-2	1.01747 E-1	9.01030 E-2	8.32783 E-3	9.90 E-4	3.00 E-1
	0.6	−1.62994 E-3	7.25973 E-2	8.36763 E-2	7.91697 E-3	9.90 E-4	3.00 E-1
	0.7	−2.17635 E-2	4.73349 E-2	7.79470 E-2	7.53985 E-3	9.90 E-4	3.00 E-1
	0.8	−3.88441 E-2	2.56893 E-2	7.29750 E-2	7.20863 E-3	9.90 E-4	3.00 E-1
	0.9	−5.34998 E-2	6.64081 E-3	6.85331 E-2	6.90873 E-3	9.90 E-4	3.00 E-1
0.4	0.001	2.75698 E-1	9.91220 E-2	4.14830 E-3	−1.00802 E-3	5.30 E-5	1.34 E-2
	0.005	2.57288 E-1	1.58098 E-1	2.55731 E-2	2.21928 E-4	1.30 E-4	4.95 E-2
	0.01	2.94919 E-1	2.17595 E-1	4.53972 E-2	1.63889 E-3	1.65 E-4	7.45 E-2
	0.05	1.21340 E-1	1.49889 E-1	5.20309 E-2	2.80540 E-3	3.30 E-4	2.00 E-1
	0.1	1.19490 E-1	1.71903 E-1	7.02191 E-2	4.60326 E-3	1.55 E-4	2.30 E-1
	0.2	9.25248 E-2	1.64374 E-1	8.27308 E-2	6.22468 E-3	3.30 E-4	3.00 E-1
	0.3	8.23744 E-2	1.63828 E-1	9.29291 E-2	7.60006 E-3	5.75 E-4	3.00 E-1
	0.4	5.71644 E-2	1.38423 E-1	9.16432 E-2	7.83310 E-3	6.80 E-4	3.00 E-1
	0.5	4.16952 E-2	1.24865 E-1	9.42554 E-2	8.52955 E-3	9.90 E-4	3.00 E-1
	0.6	1.54332 E-2	9.33253 E-2	8.79684 E-2	8.17369 E-3	9.90 E-4	3.00 E-1
	0.7	−8.92361 E-3	6.33630 E-2	8.14557 E-2	7.76267 E-3	9.90 E-4	3.00 E-1
	0.8	−3.05202 E-2	3.63197 E-2	7.53774 E-2	7.36578 E-3	9.90 E-4	3.00 E-1
	0.9	−4.94879 E-2	1.24899 E-2	6.99681 E-2	7.00956 E-3	9.90 E-4	3.00 E-1

Table 10.4 (continued)

n	c_{eq}/c_0	A_0	A_1	A_2	A_3	Lower limit T_B	Upper limit T_B
0.5	0.001	−2.31520 E-1	−1.50063 E-1	−3.25959 E-2	−2.41622 E-3	3.18 E-5	1.49 E-2
	0.005	−7.52194 E-2	−7.29454 E-2	−2.35987 E-2	−2.68907 E-3	9.55 E-5	4.24 E-2
	0.01	9.57667 E-2	5.99179 E-2	7.87512 E-3	−6.91859 E-4	1.17 E-4	6.36 E-2
	0.05	1.39683 E-1	1.50472 E-1	4.61555 E-2	2.23837 E-3	1.59 E-4	1.56 E-1
	0.1	1.13944 E-1	1.53602 E-1	5.88242 E-2	3.53550 E-3	5.30 E-5	2.12 E-1
	0.2	9.41077 E-2	1.57242 E-1	7.45104 E-2	5.30910 E-3	1.87 E-4	2.76 E-1
	0.3	8.27436 E-2	1.56866 E-1	8.46437 E-2	6.58597 E-3	4.14 E-4	3.00 E-1
	0.4	7.18693 E-2	1.53217 E-1	9.24723 E-2	7.73377 E-3	5.97 E-4	3.00 E-1
	0.5	5.69988 E-2	1.41829 E-1	9.65499 E-2	8.57838 E-3	9.93 E-4	3.00 E-1
	0.6	2.97216 E-2	1.10258 E-1	9.10755 E-2	8.32978 E-3	9.93 E-4	3.00 E-1
	0.7	2.79026 E-3	7.75762 E-2	8.43987 E-2	7.93809 E-3	9.93 E-4	3.00 E-1
	0.8	−2.24102 E-2	4.63400 E-2	7.75848 E-2	7.50667 E-3	9.93 E-4	3.00 E-1
	0.9	−4.53707 E-2	1.69423 E-2	7.09081 E-2	7.06728 E-3	9.93 E-4	3.00 E-1
0.6	0.001	−2.27013 E-1	−1.31908 E-1	−2.55075 E-2	−1.67910 E-3	2.09 E-4	1.56 E-2
	0.005	−1.02651 E-1	−8.25863 E-2	−2.23388 E-2	−2.15290 E-3	6.01 E-5	4.95 E-2
	0.01	−2.32271 E-2	−2.77669 E-2	−1.12094 E-2	−1.70026 E-3	1.59 E-4	7.42 E-2
	0.05	4.46682 E-2	5.35561 E-2	1.67205 E-2	2.68478 E-5	3.29 E-4	1.98 E-1
	0.1	7.69429 E-2	1.06743 E-1	4.11098 E-2	2.08459 E-3	1.52 E-4	2.33 E-1
	0.2	7.77036 E-2	1.31279 E-1	6.19315 E-2	4.15330 E-3	1.94 E-4	3.00 E-1
	0.3	8.95803 E-2	1.61488 E-1	8.28642 E-2	6.37445 E-3	5.41 E-4	3.00 E-1
	0.4	8.02144 E-2	1.59763 E-1	9.09102 E-2	7.44596 E-3	6.12 E-4	3.00 E-1
	0.5	6.80455 E-2	1.53078 E-1	9.71191 E-2	8.48442 E-3	9.93 E-4	3.00 E-1
	0.6	4.17728 E-2	1.23877 E-1	9.31918 E-2	8.40527 E-3	9.93 E-4	3.00 E-1
	0.7	1.33100 E-2	9.01344 E-2	8.68366 E-2	8.07126 E-3	9.93 E-4	3.00 E-1
	0.8	−1.49279 E-2	5.53146 E-2	7.94822 E-2	7.62253 E-3	9.93 E-4	3.00 E-1
	0.9	−4.18125 E-2	2.16511 E-2	7.19665 E-2	7.13504 E-3	9.93 E-4	3.00 E-1

Table 10.4 (continued)

n	c_{eq}/c_0	A_0	A_1	A_2	A_3	Lower limit T_B	Upper limit T_B
0.7	0.001	−1.34083 E-1	−7.58932 E-2	−1.42521 E-2	−9.15977 E-4	1.59 E-4	1.56 E-2
	0.005	−8.02396 E-2	−6.17352 E-2	−1.58428 E-2	−1.45553 E-3	1.31 E-4	4.95 E-2
	0.01	−4.94192 E-2	−4.43828 E-2	−1.35692 E-2	−1.59067 E-3	1.59 E-4	7.42 E-2
	0.05	2.50300 E-2	2.96632 E-2	8.40224 E-3	−5.36562 E-4	3.29 E-4	1.98 E-1
	0.1	5.75694 E-2	7.94981 E-2	2.99890 E-2	1.17739 E-3	1.52 E-4	2.33 E-1
	0.2	6.66125 E-2	1.11531 E-1	5.17050 E-2	3.19101 E-3	4.60 E-5	3.00 E-1
	0.3	8.60269 E-2	1.51660 E-1	7.57571 E-2	5.62554 E-3	3.99 E-4	3.00 E-1
	0.4	8.44706 E-2	1.61100 E-1	8.79160 E-2	7.05039 E-3	4.91 E-4	3.00 E-1
	0.5	7.62631 E-2	1.60491 E-1	9.65951 E-2	8.30479 E-3	9.93 E-4	3.00 E-1
	0.6	5.19715 E-2	1.35101 E-1	9.45927 E-2	8.42186 E-3	9.93 E-4	3.00 E-1
	0.7	2.30088 E-2	1.01810 E-1	8.90651 E-2	8.19056 E-3	9.93 E-4	3.00 E-1
	0.8	−7.70658 E-3	6.40726 E-2	8.13074 E-2	7.73129 E-3	9.93 E-4	3.00 E-1
	0.9	−3.79987 E-2	2.66329 E-2	7.31275 E-2	7.21264 E-3	9.93 E-4	3.00 E-1
0.8	0.001	−1.36815 E-1	−7.18166 E-2	−1.24922 E-2	−7.42617 E-4	1.10 E-4	1.13 E-2
	0.005	−5.22713 E-2	−3.93750 E-2	−9.86987 E-3	−9.05817 E-4	1.59 E-4	4.95 E-2
	0.01	−4.82363 E-2	−4.06025 E-2	−1.14523 E-2	−1.24086 E-3	1.59 E-4	7.42 E-2
	0.05	1.33720 E-2	1.57423 E-2	3.75829 E-3	−7.79770 E-4	2.02 E-4	1.98 E-1
	0.1	4.28220 E-2	5.90646 E-2	2.18305 E-2	5.45716 E-4	1.52 E-4	2.33 E-1
	0.2	5.69035 E-2	9.45176 E-2	4.31204 E-2	2.40200 E-3	3.29 E-4	3.00 E-1
	0.3	8.02748 E-2	1.39498 E-1	6.83292 E-2	4.87245 E-3	3.64 E-4	3.00 E-1
	0.4	8.54937 E-2	1.58502 E-1	8.39171 E-2	6.58361 E-3	6.12 E-4	3.00 E-1
	0.5	8.13758 E-2	1.64018 E-1	9.49454 E-2	8.03418 E-3	9.93 E-4	3.00 E-1
	0.6	6.03068 E-2	1.44020 E-1	9.53494 E-2	8.38808 E-3	9.93 E-4	3.00 E-1
	0.7	3.18482 E-2	1.12089 E-1	9.08525 E-2	8.27224 E-3	9.93 E-4	3.00 E-1
	0.8	−8.12011 E-4	7.28461 E-2	8.32013 E-2	7.84895 E-3	9.93 E-4	3.00 E-1
	0.9	−3.44824 E-2	3.00231 E-2	7.37587 E-2	7.24504 E-3	9.93 E-4	3.00 E-1

Table 10.4 (continued)

n	c_{eq}/c_0	A_0	A_1	A_2	A_3	Lower limit T_B	Upper limit T_B
0.9	0.001	−9.91606 E-2	−5.08129 E-2	−8.61272 E-3	−5.00782 E-4	1.10 E-4	1.06 E-2
	0.005	−4.34533 E-2	−3.12043 E-2	−7.41242 E-3	−6.54366 E-4	1.31 E-4	4.95 E-2
	0.01	−3.80824 E-2	−3.09971 E-2	−8.39448 E-3	−8.93772 E-4	1.59 E-4	7.42 E-2
	0.05	6.57636 E-3	7.90077 E-3	1.32920 E-3	−8.37313 E-4	3.29 E-4	1.98 E-1
	0.1	3.23620 E-2	4.46623 E-2	1.61704 E-2	1.42772 E-4	1.52 E-4	2.33 E-1
	0.2	4.81562 E-2	7.96819 E-2	3.58699 E-2	1.75445 E-3	3.29 E-4	3.00 E-1
	0.3	7.32636 E-2	1.26236 E-1	6.09842 E-2	4.14919 E-3	3.29 E-4	3.00 E-1
	0.4	8.45046 E-2	1.53374 E-1	7.93554 E-2	6.08277 E-3	5.41 E-4	3.00 E-1
	0.5	8.43750 E-2	1.64762 E-1	9.25243 E-2	7.70358 E-3	9.93 E-4	3.00 E-1
	0.6	6.69285 E-2	1.50258 E-1	9.53075 E-2	8.28898 E-3	9.93 E-4	3.00 E-1
	0.7	3.93029 E-2	1.20439 E-1	9.20515 E-2	8.30405 E-3	9.93 E-4	3.00 E-1
	0.8	5.77253 E-3	8.15003 E-2	8.50816 E-2	7.96580 E-3	9.93 E-4	3.00 E-1
	0.9	−3.07193 E-2	3.54463 E-2	7.51094 E-2	7.34009 E-3	9.93 E-4	3.00 E-1

10.5 Constant pattern solution to the homogeneous surface diffusion model

As described in Chapter 7 (Section 7.4.3), Hand et al. (1984) have approximated the constant pattern solution to the HSDM (CPHSDM) for the minimum Stanton number, St^*_{min}, by polynomials of the form

$$T(n, Bi, St^*_{min}) = A_0 + A_1\left(\frac{c}{c_0}\right)^{A_2} + \frac{A_3}{1.01 - \left(\frac{c}{c_0}\right)^{A_4}} \tag{10.29}$$

where T is the throughput ratio and c/c_0 is the normalized concentration. The minimum Stanton number defines the condition under which the constant pattern occurs for the first time in the fixed-bed adsorber. It is related to the minimum empty bed contact time, $EBCT_{min}$, by

$$EBCT_{min} = \frac{t_{r,min}}{\varepsilon_B} = \frac{St^*_{min}\, r_P}{k_F(1-\varepsilon_B)} \tag{10.30}$$

where $EBCT_{min}$ is the minimum empty bed contact time required for constant pattern formation (see eq. (6.16) in Chapter 6 and Table 7.3 in Chapter 7). Equation (10.29) allows calculating the breakthrough curve (BTC) for the minimum empty bed contact time. BTCs for longer empty bed contact times can be received by a simple parallel translation of the calculated BTC as described in Chapter 7, Section 7.4.3.

The minimum Stanton number depends on the Biot number, *Bi*. This relationship can also be expressed by an empirical equation

$$St^*_{min} = A_0\, Bi + A_1 \tag{10.31}$$

For the definition of *Bi* see Table 7.3 (Chapter 7).

Table 10.5 contains the parameters A_0 and A_1 necessary to calculate the minimum Stanton number, whereas Table 10.6 lists the parameters A_0–A_4 for the BTC calculation. The parameters in Table 10.6 are valid in the concentration range $0.02 < c/c_0 < 0.98$.

Table 10.5: Parameters for calculating the minimum Stanton number required to achieve the constant pattern conditions (eq. (10.31)).

n	**Minimum Stanton number required for constant pattern**				**Minimum Stanton number required to be within 10% of constant pattern**			
	0.5 ≤ *Bi* ≤ 10		**10 ≤ *Bi* ≤ ∞**		**0.5 ≤ *Bi* ≤ 10**		**10 ≤ *Bi* ≤ ∞**	
	A_0	A_1	A_0	A_1	A_0	A_1	A_0	A_1
0.05	2.10526 E-2	1.98947	0.22	0	1.05263 E-2	1.39474	1.278 E-1	0.22
0.10	2.10526 E-2	2.18947	0.24	0	3.15789 E-2	1.38421	1.367 E-1	0.33
0.20	4.21053 E-2	2.37895	0.28	0	6.31578 E-2	1.36842	1.625 E-1	0.38
0.30	1.05263 E-1	2.54737	0.36	0	9.47368 E-2	1.35263	1.800 E-1	0.50
0.40	2.31579 E-1	2.68421	0.50	0	1.68421 E-1	1.41579	2.475 E-1	0.63
0.50	5.26316 E-1	2.73684	0.80	0	2.78947 E-1	1.46053	3.438 E-1	0.81
0.60	1.15789	3.42105	1.50	0	4.52631 E-1	1.97368	5.056 E-1	1.40
0.70	1.78947	7.10526	2.50	0	6.84211 E-1	3.65789	7.722 E-1	2.80
0.80	3.68421	13.1579	5.00	0	1.21053	5.89474	1.355	4.40
0.90	6.31579	56.8421	12.00	0	2.84211	15.5789	3.122	12.80

Table 10.6: Parameters for calculating breakthrough curves by eq. (10.29).

n	*Bi*	A_0	A_1	A_2	A_3	A_4
0.05	0.5	−5.447214	6.598598	0.026569	0.019384	20.450470
0.05	2.0	−5.465811	6.592484	0.004989	0.004988	0.503250
0.05	4.0	−5.531155	6.584935	0.023580	0.009019	0.273076
0.05	6.0	−5.606508	6.582188	0.022088	0.013126	0.214246
0.05	8.0	−5.606500	6.504701	0.020872	0.017083	0.189537
0.05	10.0	−5.664173	6.456597	0.018157	0.019935	0.149314
0.05	14.0	−0.662780	1.411252	0.060709	0.020229	0.143293
0.05	25.0	−0.662783	1.350940	0.031070	0.020350	0.129998
0.05	≥100.0	0.665879	0.711310	2.987309	0.016783	0.361023
0.10	0.5	−1.919873	3.055368	0.055488	0.024284	15.311766
0.10	2.0	−2.278950	3.393925	0.046838	0.004751	0.384675
0.10	4.0	−2.337178	3.379926	0.043994	0.008650	0.243412
0.10	6.0	−2.407407	3.374131	0.041322	0.012552	0.196565
0.10	8.0	−2.477819	3.370954	0.038993	0.016275	0.176437
0.10	10.0	−2.566414	3.370950	0.035003	0.019386	0.150788
0.10	16.0	−2.567201	3.306341	0.020940	0.019483	0.136813
0.10	30.0	−2.568618	3.241783	0.009595	0.019610	0.121746
0.10	≥100.0	−2.568360	3.191482	0.001555	0.019682	0.110113
0.20	0.5	−1.441000	2.569000	0.060920	0.002333	0.371100
0.20	2.0	−1.474313	2.558300	0.058480	0.005026	0.241265
0.20	4.0	−1.506696	2.519259	0.055525	0.008797	0.187510
0.20	6.0	−1.035395	1.983018	0.069283	0.012302	0.167924
0.20	8.0	−0.169192	1.077521	0.144879	0.015500	0.168083
0.20	10.0	−1.402932	2.188339	0.052191	0.018422	0.133574
0.20	13.0	−1.369220	2.118545	0.039492	0.018453	0.127565
0.20	25.0	−1.514159	2.209450	0.017937	0.018510	0.118517
0.20	≥100.0	0.680346	0.649006	2.570086	0.014947	0.369818
0.30	0.5	−1.758696	2.846576	0.049530	0.003022	0.156816

Table 10.6 (continued)

n	Bi	A_0	A_1	A_2	A_3	A_4
0.30	2.0	−1.657862	2.688895	0.048409	0.005612	0.140937
0.30	4.0	−0.565664	1.537833	0.084451	0.008808	0.139086
0.30	6.0	−0.197077	1.118564	0.117894	0.011527	0.135874
0.30	8.0	−0.197070	1.069216	0.119760	0.013925	0.132691
0.30	10.0	−0.173358	1.000000	0.120311	0.015940	0.133973
0.30	15.0	−0.173350	0.919411	0.071768	0.014156	0.086270
0.30	35.0	0.666471	0.484570	1.719440	0.013444	0.259545
0.30	≥100.0	0.696161	0.516951	2.054587	0.012961	0.303218
0.40	0.5	−0.534251	1.603834	0.094055	0.004141	0.137797
0.40	2.0	−0.166270	1.190897	0.122280	0.006261	0.134278
0.40	4.0	−0.166270	1.131946	0.115513	0.008634	0.126813
0.40	6.0	−0.166270	1.089789	0.112284	0.010463	0.124307
0.40	9.0	0.491912	0.491833	0.487414	0.011371	0.147747
0.40	12.0	0.564119	0.419196	0.639819	0.011543	0.149005
0.40	15.0	0.640669	0.432466	1.048056	0.011616	0.212726
0.40	25.0	0.672353	0.397007	1.153169	0.011280	0.216883
0.40	≥100.0	0.741435	0.448054	1.929879	0.010152	0.306448
0.50	0.5	−0.040800	1.099652	0.158995	0.005467	0.139116
0.50	4.0	−0.040800	0.982757	0.111618	0.008072	0.111404
0.50	10.0	0.094602	0.754878	0.092069	0.009877	0.090763
0.50	14.0	0.023000	0.802068	0.057545	0.009662	0.084532
0.50	25.0	0.023000	0.793673	0.039324	0.009326	0.082751
0.50	≥100.0	0.529213	0.291801	0.082428	0.008317	0.075461
0.60	0.5	0.352536	0.692114	0.263134	0.005482	0.121775
0.60	2.0	0.521979	0.504220	0.327290	0.005612	0.128679
0.60	6.0	0.676253	0.334583	0.482297	0.005898	0.138946
0.60	14.0	0.769531	0.259497	0.774068	0.005600	0.165513
0.60	50.0	0.849057	0.215799	1.343183	0.004725	0.223759

Table 10.6 (continued)

n	*Bi*	A_0	A_1	A_2	A_3	A_4
0.60	≥100.0	0.831231	0.227304	1.174756	0.004961	0.212109
0.70	0.5	0.575024	0.449062	0.278452	0.004122	0.121682
0.70	4.0	0.715269	0.307172	0.442104	0.004371	0.138351
0.70	12.0	0.787940	0.243548	0.661599	0.004403	0.162595
0.70	25.0	0.829492	0.204078	0.784529	0.004050	0.179003
0.70	≥100.0	0.847012	0.190678	0.931686	0.003849	0.183239
0.80	0.5	0.708905	0.314101	0.357499	0.003276	0.119300
0.80	4.0	0.784576	0.239663	0.484422	0.003206	0.134987
0.80	14.0	0.839439	0.188966	0.648124	0.003006	0.157697
0.80	≥100.0	0.882747	0.146229	0.807987	0.002537	0.174543
0.90	0.5	0.865453	0.157618	0.444973	0.001650	0.148084
0.90	4.0	0.854768	0.171434	0.495042	0.001910	0.142251
0.90	16.0	0.866180	0.163992	0.573946	0.001987	0.157594
0.90	≥100.0	0.893192	0.133039	0.624100	0.001740	0.164248

10.6 Viscosity of water

Viscosity data (kinematic and dynamic viscosity) are important input data for the estimation of parameters required for fixed-bed adsorber modeling. The kinematic viscosity, ν, is necessary to calculate the Reynolds number and the Schmidt number, which are constituents of a number of empirical correlations that can be used to calculate film mass transfer coefficients (Chapter 7, Table 7.7). The dynamic viscosity, η, is necessary to estimate the aqueous phase diffusion coefficient, D_L (Chapter 7, Table 7.8). Kinematic and dynamic viscosities are related by the density, ρ,

$$\eta = \nu \rho \tag{10.32}$$

Table 10.7 lists some viscosity data for water in the relevant temperature range.

Table 10.7: Kinematic and dynamic viscosities of water.

Temperature (°C)	Kinematic viscosity, v (m^2/s)	Dynamic viscosity, η (Pa s = kg/(m s))
0	$1.792 \cdot 10^{-6}$	$1.792 \cdot 10^{-3}$
5	$1.520 \cdot 10^{-6}$	$1.520 \cdot 10^{-3}$
10	$1.308 \cdot 10^{-6}$	$1.308 \cdot 10^{-3}$
15	$1.140 \cdot 10^{-6}$	$1.139 \cdot 10^{-3}$
20	$1.004 \cdot 10^{-6}$	$1.003 \cdot 10^{-3}$
25	$8.936 \cdot 10^{-7}$	$8.908 \cdot 10^{-4}$
30	$8.014 \cdot 10^{-7}$	$7.977 \cdot 10^{-4}$

Nomenclature

Preliminary notes

Note 1: In the parameter list, general dimensions are given instead of special units. The dimensions for the basic physical quantities are indicated as follows:

I	electric current
L	length
M	mass
N	amount of substance (mol)
T	time
Θ	temperature

Additionally, the following symbols for derived types of measures are used:

E	energy (L^2 M T^{-2})
F	force (L M T^{-2})
P	pressure (M L^{-1} T^{-2})
U	voltage (L^2 M T^{-3} I^{-1})

Note 2: Empirical parameters, typically named as A, B, a, b, k, α, or β, are not listed here. They are explained in context with the respective equations.

English alphabet

A surface area (L^2)
A_{BET} specific surface area determined by the Brunauer-Emmett-Teller (BET) method (L^2 M^{-1})
A_M surface area occupied by a molecule (L^2)
A_m surface area per mass (L^2 M^{-1})
A_R cross-sectional area of the fixed-bed adsorber (L^2)
A_s external adsorbent surface area (L^2)
subscript:
P particle
a external adsorbent particle surface area related to mass or volume (L^2 M^{-1} or L^2 L^{-3})
subscripts:
m related to the adsorbent mass
VA related to the adsorbent volume
VR related to the reactor volume
a activity (N L^{-3})
B empirical parameter in the Dubinin-type isotherm equation (dimensionless)
Bi Biot number, characterizes the ratio of internal and external mass transfer resistances (dimensionless)
BV bed volumes, measure of throughput in fixed-bed adsorbers (dimensionless)
subscript:
st stoichiometric
b isotherm parameter in several isotherm equations (L^3 M^{-1} or L^3 N^{-1})
b_0 preexponential factor in eq. (3.61) (L^3 M^{-1} or L^3 N^{-1})

https://doi.org/10.1515/9783110715507-011

b_1, b_2 isotherm parameters in several isotherm equations ($L^3\ M^{-1}$ or $L^3\ N^{-1}$)
b^* $= b^n$, isotherm parameter in the Langmuir-Freundlich isotherm ($L^{3n}\ M^{-n}$ or $L^{3n}\ N^{-n}$)
C_B parameter in the BET isotherm equation (dimensionless)
$\overline{C}$ dimensionless concentration, $(c-c_{eq})/(c_0-c_{eq})$
CUR carbon usage rate ($M\ L^{-3}$)
c concentration ($M\ L^{-3}$ or $N\ L^{-3}$)
subscripts:
- 0 initial
- a acid
- BM biomass, Monod equation
- b base
- eq equilibrium
- $Feed$ feed
- in inlet
- out outlet
- p in the pore liquid
- pl plateau
- s at external particle surface
- sat saturation
- $solv$ solvent
- T total

superscript:
- 0 single-solute adsorption

D diffusion coefficient ($L^2\ T^{-1}$)
subscripts:
- 0 initial value
- a apparent
- eff effective
- L in the free liquid
- P pore
- S surface

D pH-dependent n-octanol-water partition coefficient (dimensionless), typically given as log D
D_{ax} axial (longitudinal) dispersion coefficient ($L^2\ T^{-1}$)
superscript:
- $*$ retarded (D_{ax}/R_d)

D_B distribution parameter, batch reactor (dimensionless)
D_g distribution parameter, fixed-bed adsorber (dimensionless)
$D_S(0)$ intrinsic surface diffusion coefficient ($L^2\ T^{-1}$)
d diameter (L)
subscripts:
- P particle
- R reactor (adsorber)

d parameter in the Fritz-Schlünder isotherm (dimensionless)
E parameter used to simplify some ideal adsorbed solution theory (IAST) equations (dimensionless)
E_1, E_2 parameters in the multisolute isotherms given by eqs. (4.6) and (4.7) (dimensionless)
$E_{A,des}$ activation energy for desorption ($E\ N^{-1}$)

EBCT empty bed contact time (T)
subscript:
min minimum for constant pattern formation

E_C characteristic adsorption energy (E N^{-1})
E_d diffusion modulus (dimensionless)
F fractional uptake (dimensionless)
F Faraday constant (96,485 C/mol)
FI fouling index used in the rapid small-scale column test (dimensionless)
F_S symmetry factor of the breakthrough curve (dimensionless)
$\overline{F}$ mean fractional uptake (dimensionless)
f fraction (dimensionless)
subscript:
oc organic carbon

G Gibbs free energy (E) or molar free energy (E N^{-1})
subscript:
ads adsorption

H enthalpy (E) or molar enthalpy (E N^{-1})
subscripts:
ads adsorption
R reaction
sol dissolution
superscript:
iso isosteric (only for H_{ads})

h adsorber (adsorbent bed) height (L)
h_{st} location of the stoichiometric front (L)
h_z height of the mass transfer zone (L)
K isotherm parameter in the Freundlich isotherm [(N M^{-1})/(N L^{-3})^n or (M M^{-1})/(M L^{-3})^n]
K equilibrium constant (N^x L^{-3x}, x depends on the specific law of mass action)
subscripts:
a acidity
b basicity
superscripts:
app apparent
int intrinsic

$K_{1,2}$ competition coefficient in the multisolute Freundlich equation (dimensionless)
K_c complex formation constant (L^3 M^{-1})
K_d distribution coefficient [(M M^{-1})/(M L^{-3}) or L^3 M^{-1}]
subscripts:
app apparent
i ionic species
n neutral species

K_H isotherm parameter, Henry isotherm [(M M^{-1})/(M L^{-3}) or (N M^{-1})/(N L^{-3})]
K_{oc} organic carbon normalized distribution coefficient [(M M^{-1})/(M L^{-3}) or L^3 M^{-1}]
K_{ow} n-octanol-water partition coefficient (dimensionless), typically given as log K_{ow}
K_s half saturation constant, Monod equation (M L^{-3})
K^* isotherm parameter in the solubility-normalized Freundlich model (M M^{-1} or N M^{-1})

k	mass transfer coefficient (L T^{-1}) *subscripts:* 0 initial value D dispersion eff effective F film diffusion S intraparticle (surface) diffusion
k	rate constant (dimension depending on the rate law) *subscripts:* 1 first-order rate law 2 second-order rate law $decay$ biomass decay (first order) des desorption
k_A	preexponential factor (frequency factor) of the Arrhenius equation (same dimension as the related rate constant)
k_{kin}	sorption rate constant in the extended local equilibrium model (T^{-1})
k_s^*	modified mass transfer coefficient (intraparticle diffusion), volumetric mass transfer coefficient (T^{-1}) *subscript:* eff effective
$k_s^*(0)$	intrinsic volumetric intraparticle mass transfer coefficient (T^{-1})
$kx_{F,min}$	minimum film mass transfer coefficient expressed as a fraction of its initial value (dimensionless)
L	transport length (L)
LUB	length of the unused bed (L)
M	molecular weight (M N^{-1}) *subscript:* $solv$ solvent
m	mass (M) *subscripts:* 0 initial A adsorbent eq equilibrium oc organic carbon P particle pyc pycnometer S sample $solid$ solid T total wet wet adsorbent *superscripts:* a adsorbed phase l liquid phase
m	isotherm parameter (exponent) in Dubinin-Astachov, generalized Langmuir, and Fritz-Schlünder isotherm equations (dimensionless)
$\dot{m}$	mass flow (M T^{-1}) *subscript:* A adsorbent

N normalizing factor in eq. (3.49)

N dimensionless mass transfer coefficient

subscripts:

- D dispersion
- eff effective
- F film diffusion
- min minimum
- S intraparticle (surface) diffusion

N_A Avogadro's number ($6.022 \cdot 10^{23}$ mol^{-1})

$\dot{N}$ mass transfer rate (M T^{-1} or N T^{-1})

subscripts:

- acc accumulation
- ads adsorption
- adv advection
- $disp$ dispersion
- F film diffusion
- P pore diffusion
- S surface diffusion

n amount of substance (N)

subscripts:

- a adsorbed
- $solv$ solvent
- w water

superscripts:

- a adsorbed phase
- l liquid phase

n isotherm parameter (exponent) in several isotherm equations (dimensionless)

$\dot{n}$ flux (M L^{-2} T^{-1} or N L^{-2} T^{-1})

subscripts:

- F film diffusion
- P pore diffusion
- S surface diffusion
- T total

n^* isotherm parameter in the solubility-normalized Freundlich model (dimensionless)

Pe Peclet number (dimensionless)

p pressure or partial pressure (P)

subscript:

- 0 saturation vapor pressure

pH negative decadic logarithm of the proton activity (dimensionless)

pK_a negative decadic logarithm of the acidity constant (dimensionless)

pK_b negative decadic logarithm of the basicity constant (dimensionless)

pK_w negative decadic logarithm of the water dissociation constant (dimensionless)

pOH negative decadic logarithm of the hydroxide ion activity (dimensionless)

Q_{ads} heat of adsorption (E) or molar heat of adsorption (E N^{-1})

subscripts:

- 0 zero loading
- net net
- w water

superscripts:
diff differential
iso isosteric

Q_s surface charge (N M^{-1})

q adsorbed amount (adsorbent loading) (M M^{-1} or N M^{-1})
subscripts:
0 initial or related to c_0
b breakthrough
cr critical
des desorbed
eq equilibrium
exp experimental
mono monolayer
pl plateau
pred predicted
s at external particle surface
T total
superscript:
0 single-solute adsorption

q_m isotherm parameter in several isotherm equations, maximum adsorbent loading (M M^{-1} or N M^{-1})

$\overline{q}$ mean adsorbent loading (M M^{-1} or N M^{-1})

R dimensionless radial coordinate

R universal gas constant (8.314 J/(mol K), 8.314 Pa m^3/(mol K), 0.08314 bar L/(mol K))

R_d retardation factor (dimensionless)

Re Reynolds number (dimensionless)
subscript:
min minimum

R_T reaction term in the differential mass balance equation for fixed adsorbers (M L^{-3} T^{-1} or N $L^{-3}T^{-1}$)

R^* separation factor (dimensionless)

r radial coordinate (L)

r radius (L)
subscripts:
K related to the Kelvin equation
P particle
pore pore

r correlation coefficient (dimensionless)

S entropy (E Θ^{-1}) or molar entropy (E Θ^{-1} N^{-1})
subscript:
ads adsorption

S dimensionless distance

Sc Schmidt number (dimensionless)

Sh Sherwood number (dimensionless)
subscripts:
L laminar
T turbulent

SPDFR surface-to-pore diffusion flux ratio (dimensionless)

St^* Stanton number (dimensionless)
subscript:
min minimum for constant pattern formation

T absolute temperature (Θ)

T throughput ratio (dimensionless)

T_B dimensionless time, batch reactor
subscript:
min minimum

T^* transformed throughput ratio (dimensionless)

t statistical monolayer thickness (L)

t time (T)
subscripts:
b breakthrough
c (ad)sorbed compound
eq equilibrium, equilibration
f formation of mass transfer zone
min minimum
r retention
s saturation
st stoichiometric
z mass transfer zone
superscript:
id ideal

$\bar{t}_r$ mean hydraulic residence time (T)

t^* transformed time (T)

u_F effective (interstitial) filter velocity ($L\ T^{-1}$)

V volume (L^3)
subscripts:
A adsorbent
ads adsorbed
$Feed$ feed
L liquid
m molar
mat material
$mono$ monolayer
P (external) pores (in geosorption)
$pore$ (internal) pores
pyc pycnometer
R reactor (adsorber)
wet wet adsorbent

V_0 volume of the micropores (DR equation) ($L^3\ M^{-1}$)

V_b molar volume at boiling point ($L^3\ N^{-1}$)

V_i intrinsic (van der Waals) molar volume ($L^3\ N^{-1}$)

V_m molar volume ($L^3\ N^{-1}$)

V_{mono} adsorbed volume in the monolayer ($L^3\ M^{-1}$)

V_{sp} specific throughput ($L^3\ M^{-1}$)
subscript:
b until breakthrough

$\dot{V}$ volumetric flow rate ($L^3\ T^{-1}$)
v velocity ($L\ T^{-1}$)
subscripts:
- c concentration point
- F filter (superficial velocity)
- w pore water (interstitial velocity)
- z mass transfer zone

v_H heating rate ($\Theta\ T^{-1}$)
w_C weight fraction of carbon in the adsorbate (dimensionless)
X dimensionless concentration
subscripts:
- s at external particle surface
- p in the pore fluid

Y dimensionless adsorbent loading
subscript:
- s at external particle surface

Y fouling factor used in the rapid small-scale column test (dimensionless)
Y_C yield coefficient ($M\ M^{-1}$)
Y^* fictive dimensionless adsorbent loading, defined in eq. (5.110)
$\overline{Y}$ mean dimensionless adsorbent loading
Z number of particles (dimensionless)
subscripts:
- S in the sample
- T total

Z_{sp} number of particles per unit adsorbent mass (M^{-1})
z charge
subscripts:
- s surface layer
- β beta layer

z distance (L)
z adsorbed-phase mole fraction (dimensionless)
z^* transformed spatial coordinate, defined in eq. (6.48) (L)

Greek alphabet

α concentration term, defined in eq. (5.102) (dimensionless)
α dispersivity (L)
subscript:
- eff effective

α degree of protolysis (dimensionless)
α hydrogen-bonding donor parameter, LSER parameter (dimensionless)
α_P degree of protonation (dimensionless)
β concentration term, defined in eq. (5.103) (dimensionless)
β $= 1/b^n$, isotherm parameter in the Tóth isotherm ($M^n\ L^{-3n}$ or $N^n\ L^{-3n}$)
β hydrogen-bonding acceptor parameter, LSER parameter (dimensionless)
Γ surface concentration ($N\ L^{-2}$)

δ thickness of the boundary layer (L)

δ_I integration constant in breakthrough curve equations (dimensionless)
subscripts:
- F film diffusion
- S intraparticle (surface) diffusion

ε porosity (dimensionless)
subscripts:
- B bulk, bed
- P particle

ε adsorption potential (E N^{-1})

η parameter in the extended Redlich-Peterson isotherm (dimensionless)

η dynamic viscosity (M L^{-1} T^{-1})

η_A adsorber efficiency (dimensionless)

θ contact angle (degrees)

λ ratio of pore and surface diffusion, defined in eq. (5.108) (dimensionless)

λ first-order biodegradation rate constant (T^{-1})
subscripts:
- l liquid phase
- s solid (adsorbed) phase

superscript:
- * retarded (λ/R_d)

μ chemical potential (E N^{-1})

μ_{max} maximum biomass growth rate, Monod equation (T^{-1})

μ^*_{max} modified maximum biomass growth rate, Monod equation (T^{-1})

ν kinematic viscosity (L^2 T^{-1})

π spreading pressure (F L^{-1} or M T^{-2})

π^* polarity/polarizability parameter, LSER parameter (dimensionless)

ρ density (M L^{-3})
subscripts:
- B bulk, bed
- M material
- P particle
- W water

σ surface free energy (surface tension) (F L^{-1} or M T^{-2})
subscripts:
- as adsorbate solution-solid interface
- ws water-solid interface

σ_s surface charge density (I T L^{-2})

τ_P tortuosity (dimensionless)

Φ association factor of the Wilke-Chang correlation (dimensionless)

φ spreading pressure term, defined in eq. (4.26) (N M^{-1})

ψ electrical potential (U)
subscripts:
- s surface layer
- β beta layer

ω empirical parameter that describes the loading dependence of k^*_S and D_S (M M^{-1})

Abbreviations

ADE	advection-dispersion equation
BAC	biological activated carbon (process)
BET	Brunauer-Emmett-Teller (isotherm)
BTC	breakthrough curve
BV	bed volume
CD	constant diffusivity
CMBR	completely mixed batch reactor
CMFR	completely mixed flow-through reactor
CPHSDM	constant pattern approach to the homogeneous surface diffusion model
CUR	carbon usage rate
DA	Dubinin-Astakhov (isotherm)
DBP	disinfection by-product
DFPSDM	dispersed-flow, pore and surface diffusion model
DFT	density functional theory
DHS	dissolved humic substances
DOC	dissolved organic carbon
DOM	dissolved organic matter
DR	Dubinin-Radushkevich (isotherm)
EBC	equivalent background compound
EBCM	equivalent background compound model
EBCT	empty bed contact time
ECM	equilibrium column model
EfOM	effluent organic matter
GAC	granular activated carbon
GC/MS	gas chromatography/mass spectrometry
HSDM	homogeneous surface diffusion model
IAST	ideal adsorbed solution theory
IUPAC	International Union of Pure and Applied Chemistry
LCA	low-cost adsorbent
LDF	linear driving force (model)
LEM	local equilibrium model
LSER	linear solvation energy relationship
LUB	length of unused bed (model)
MC	Monte Carlo (method)
MP	micropollutant
MTZ	mass transfer zone
NF	nanofiltration
NOM	natural organic matter
PAC	powdered activated carbon
PBC	pore-blocking compound
PD	proportional diffusivity
RSSCT	rapid small-scale column test
SAT	soil-aquifer treatment
SBA	short bed adsorber
SEBCM	simplified equivalent background compound model
SOM	sorbent organic matter
SPDFR	surface-to-pore diffusion flux ratio

TRC	trace compound
TRM	tracer model
TVFM	theory of volume filling of micropores
UF	ultrafiltration
VST	vacancy solution theory

References

Altmann J, Zietzschmann F, Geiling E-L, Ruhl AS, Sperlich A, Jekel M. Impacts of coagulation on the adsorption of organic micropollutants onto powdered activated carbon in treated domestic wastewater. Chemosphere, 2015, 125, 198–204.

Amiri F, Börnick H, Worch E. Sorption of phenols onto sandy aquifer material: the effect of dissolved organic matter (DOM). Water Res, 2005, 39, 933–941.

Amiri F, Rahman MM, Börnick H, Worch E. Sorption behaviour of phenols on natural sandy aquifer material during flow-through column experiments: the effect of pH. Acta Hydrochim Hydrobiol, 2004, 32, 214–224.

Antelo J, Arce F, Fiol S. Arsenate and phosphate adsorption on ferrihydrite nanoparticles. Synergetic interaction with calcium ions. Chem Geol, 2015, 410, 53–62.

Anumol T, Sgroi M, Park M, Roccaro P, Snyder SA. Predicting trace organic compound breakthrough in granular activated carbon using fluorescence and UV absorbance as surrogates. Water Res, 2015, 76, 76–87.

Baker JR, Mihelcic JR, Luehrs DC, Hickey JP. Evaluation of estimation methods for organic carbon normalized sorption coefficients. Water Environ Res, 1997, 69, 136–145.

Bear J, Cheng AHD. Modeling Groundwater Flow and Contaminant Transport, Dordrecht: Springer, 2010.

Benjamin MM. Water Chemistry, New York: McGraw-Hill, 2002.

Boehm HP. Chemical identification of surface groups, Eley DD ed, Advances in Catalysis, Vol. 16, New York: Academic Press, 1966, 179–274.

Boehm HP. Funktionelle Gruppen an Festkörperoberflächen [Functional groups on solid surfaces]. Chem Ing Tech, 1974, 46, 716–719.

Boehm HP. Surface oxides on carbon and their analysis: a critical assessment. Carbon, 2002, 40, 145–149.

Boyd GE, Adamson AW, Meyers LS. The exchange adsorption of ions from aqueous solutions by organic zeolites. II. Kinetics. J Am Chem Soc, 1947, 69, 2836–2848.

Broughton DB. Adsorption isotherms for binary gas mixtures. Ind Eng Chem, 1948, 40, 1506–1508.

Brunauer S, Emmett PH, Teller E. Adsorption of gases in multimolecular layers. J Am Chem Soc, 1938, 60, 309–319.

Burwig G, Worch E, Sontheimer H. Eine neue Methode zur Berechnung des Adsorptionsverhaltens von organischen Spurenstoffen in Gemischen [A new method to describe the adsorption behaviour of organic trace compounds in mixtures]. Vom Wasser, 1995, 84, 237–249.

Butler JAV, Ockrent C. Studies in electrocapillarity. Part III. The surface tensions of solutions containing two surface-active solutes. J Phys Chem, 1930, 34, 2841–2859.

Cameron DR, Klute A. Convective-dispersive solute transport with a combined equilibrium and kinetic adsorption model. Water Resour Res, 1977, 13, 183–188.

Carter MC, Weber WJ Jr, Olmstead KP. Effects of background dissolved organic matter on TCE adsorption by GAC. J Am Water Works Assoc, 1992, 84, 81–89.

Chinn D, King CJ. Adsorption of glycols, sugars, and related multiple –OH compounds onto activated carbons. 2. Solvent regeneration. Ind Eng Chem Res, 1999, 38, 3746–3753.

Collins JJ. The LUB/equilibrium section concept for fixed-bed adsorption. Chem Eng Prog S Ser, 74, 1967, 63, 31–35.

Cornel P, Fettig J. Bestimmung des äußeren Stoffübergangskoeffizienten in durchströmten Sorbensschüttungen und Aussagen zum kinetischen Verhalten unbekannter Sorptivgemische [Determination of the external mass transfer coefficient in packed beds and conclusions on the kinetic behavior of unknown sorbate mixtures], Veröffentlichungen des Bereichs für Wasserchemie der Universität Karlsruhe [Publications of the Department of Water Chemistry of

https://doi.org/10.1515/9783110715507-012

the University Karlsruhe] Nr. Vol. 20, Karlsruhe: Bereich für Wasserchemie der Universität Karlsruhe, 1982.
Corwin CJ, Summers RS. Scaling trace organic contaminant adsorption capacity by granular activated carbon. Environ Sci Technol, 2010, 44, 5403–5408.
Crank J. The Mathematics of Diffusion, London: Clarendon Press, 1975.
Crini G, Lichtfouse E, Wilson LD, Morin-Crini N. Conventional and non-conventional adsorbents for wastewater treatment. Environ Chem Lett, 2019, 17, 195–213.
Crittenden JC, Berrigan JK, Hand DW. Design of rapid small-scale adsorption tests for a constant surface diffusivity. J Water Pollut Control Fed, 1986a, 58, 312–319.
Crittenden JC, Berrigan JK, Hand DW, Lykins BW Jr. Design of rapid fixed-bed adsorption tests for nonconstant diffusivities. J Environ Eng–ASCE, 1987a, 113, 243–259.
Crittenden JC, Hand DW, Arora H, Lykins BW Jr. Design considerations for GAC treatment of organic chemicals. J Am Water Works Assoc, 1987b, 79, 74–82.
Crittenden JC, Hutzler NJ, Geyer DG, Oravitz JL, Friedman G. Transport of organic compounds with saturated groundwater flow: model development and parameter sensitivity. Water Resour Res, 1986b, 22, 271–284.
Crittenden JC, Reddy PS, Arora H et al. , Predicting GAC performance with rapid small-scale column tests. J Am Water Works Assoc, 1991, 83, 77–87.
Crittenden JC, Sanongraj S, Bulloch JL et al. , Correlations of aqueous-phase adsorption isotherms. Environ Sci Technol, 1999, 33, 2926–2933.
De Wilde T, Mertens J, Šimunek J et al. , Characterizing pesticide sorption and degradation in microscale biopurification systems using column displacement experiments. Environ Pollut, 2009, 157, 463–473.
DiGiano FA, Baldauf G, Frick B, Sontheimer H. A simplified competitive equilibrium adsorption model. Chem Eng Sci, 1978, 33, 1667–1673.
Ding L, Mariñas, BJ, Schideman LC, Snoeyink VL. Competitive effects of natural organic matter: parameterization and verification of the three-component adsorption model COMPSORB. Environ Sci Technol, 2006, 40, 350–356.
Dittmar S, Zietzschmann F, Mai M, Worch E, Jekel M, Ruhl AS. Simulating effluent organic matter competition in micropollutant adsorption onto activated carbon using a surrogate competitor. Environ Sci Technol, 2018, 52, 7859–7866.
Driehaus W, Jekel M, Hildebrandt U. Granular ferric hydroxide – a new adsorbent for the removal of arsenic from natural water. J Water Supply Res T – AQUA, 1998, 47, 30–35.
Dubinin MM, Astakhov VA. Razvitie predstavlenii ob obemnom zapolnenii mikropor pri adsorbtsii gazov i parov mikroporistymi adsorbentam. Soobshchenie 1: Uglerodnie adsorbenti [Development of ideas of volume filling of micropores during adsorption of gases and vapors by microporous adsorbents. 1. Carbonaceous adsorbents]. Izv Akad Nauk SSSR Ser Khim 1971: 5–11.
Dubinin MM, Zaverina ED, Radushkevich LV. Sorbtsya i struktura aktivnykh uglei. 1. Issledovanie adsorbtsii organicheskykh parov [Sorption and structure of activated carbons. 1. Investigation of organic vapor adsorption]. Zh Fiz Khim, 1947, 21, 1351–1362.
Dwivedi PN, Upadhyay SN. Particle-fluid mass transfer in fixed and fluidized beds. Ind Eng Chem Proc DD, 1977, 16, 157–165.
Eppinger P. Aromatische Amine in der Elbe und ihr Verhalten bei der Trinkwasseraufbereitung [Aromatic amines in the Elbe River and their behavior during drinking water treatment]. PhD thesis, Dresden University of Technology; 2000.
Freundlich H. Über die Adsorption in Lösungen [On the adsorption in solutions]. Z Phys Chem, 1906, 57, 385–470.

Frick B. Adsorptionsgleichgewichte zwischen Aktivkohle und organischen Wasserinhaltsstoffen in Mehrstoffgemischen bekannter und unbekannter Zusammensetzung [Adsorption equilibria between activated carbon and organic water constituents in multisolute mixtures of known and unknown composition]. PhD thesis, University Karlsruhe; 1980.

Fritz W, Merk W, Schlünder EU. Competitive adsorption of two dissolved organics onto activated carbon. II. Adsorption kinetics in batch reactors. Chem Eng Sci, 1981, 36, 731–741.

Fritz W, Schlünder EU. Simultaneous adsorption equilibria of organic solutes in dilute aqueous solutions on activated carbon. Chem Eng Sci, 1974, 29, 1279–1282.

Fukuchi K, Kobuchi S, Arai Y. Application of vacancy solution theory to adsorption from dilute aqueous solutions. J Chem Eng Jpn, 1982, 15, 316–318.

Gawlik BM, Sotiriou N, Feicht EA, Schulte-Hostede S, Kettrup A. Alternatives for the determination of the soil adsorption coefficient, K_{oc}, of non-ionic organic compounds – a review. Chemosphere, 1997, 12, 2525–2551.

Gerstl Z. Estimation of organic chemical sorption by soils. J Contam Hydrol, 1990, 6, 357–375.

Girvin DC, Scott AJ. Polychlorinated biphenyl sorption by soils: measurement of soil-water partition coefficients at equilibrium. Chemosphere, 1997, 35, 2007–2025.

Glueckauf E. Theory of chromatography. Pt. 10: Formulae for diffusion into spheres and their application to chromatography. T Faraday Soc, 1955, 51, 1540–1551.

Glueckauf E, Coates JI. Theory of chromatography. Pt. 4. The influence of incomplete equilibrium on the front boundary of chromatograms and the effectiveness of separation. J Chem Soc, 1947, 149(II), 1315–1321.

Gnielinski V. Gleichungen zur Berechnung des Wärme- und Stoffaustausches in durchströmten Kugelschüttungen bei mittleren und großen Peclet-Zahlen [Equations for estimating the heat and mass transfer in packed beds of spherical particles at medium and high Peclet numbers]. Verfahrenstechnik, 1978, 12, 363–366.

Graham MR, Summers RS, Simpson MR, MacLeod BW. Modelling equilibrium adsorption of 2-methylisoborneol and geosmin in natural waters. Water Res, 2000, 34, 2291–2300.

Gupta VK, Carrott PJM, Ribeiro Carrott MML, Suhas. Low-cost adsorbents: growing approach to wastewater treatment – a review. Crit Rev Environ Sci Technol, 2009, 39, 783–842.

Halsey G. Physical adsorption of non-uniform surfaces. J Chem Phys, 1948, 16, 931–937.

Hand DW, Crittenden JC, Thacker WE. User-oriented batch reactor solutions to the homogeneous surface diffusion model. J Environ Eng – ASCE, 1983, 109, 82–101.

Hand DW, Crittenden JC, Thacker WE. Simplified models for design of fixed-bed adsorption systems. J Environ Eng – ASCE, 1984, 110, 440–456.

Harkins WD, Jura G. Surfaces on solids. XII. An absolute method for the determination of the area of a finely divided crystalline solid. J Am Chem Soc, 1944a, 66, 1362–1366.

Harkins WD, Jura G. Surfaces on solids. XIII. A vapor adsorption method for the determination of the area of a solid without the assumption of a molecular area, and the areas occupied by nitrogen and other molecules on the surface of a solid. J Am Chem Soc, 1944b, 66, 1366–1373.

Hassett JJ, Banwart WL, Griffen RA. Correlation of compound properties with sorption characteristics of nonpolar compounds by soil and sediments: concepts and limitations, Francis CW, Auerbach SI eds, Environmental and Solid Wastes: Characterization, Treatment and Disposal, Newton, MA: Butterworth Publishers, 1983, 161–178.

Haul R, Dümbgen G. Vereinfachte Methode zur Messung von Oberflächengrößen durch Gasadsorption [Simplified method for measuring the surface area by gas adsorption]. Chem Ing Tech, 1960, 32, 349–354.

Haul R, Dümbgen G. Vereinfachte Methode zur Messung von Oberflächengrößen durch Gasadsorption, 2. Mitteilung [Simplified method for measuring the surface area by gas adsorption. 2. Communication]. Chem Ing Tech, 1963, 35, 586–589.

Hayduk W, Laudie H. Prediction of diffusion coefficients for nonelectrolytes in dilute aqueous solutions. Am Inst Chem Eng J, 1974, 20, 611–615.

Heese C. Untersuchungen zur Kinetik und Dynamik der Adsorption organischer Wasserinhaltsstoffe [Investigations on the kinetics and dynamics of the adsorption of organic water constituents]. PhD thesis, Dresden University of Technology; 1996.

Hickey JP, Passino-Reader DR. Linear solvation energy relationships: "rule of thumb" for estimation of variable values. Environ Sci Technol, 1991, 25, 1753–1760.

Hoppe H, Worch E. Zur Anwendung des Adsorptionszonenmodells auf die Adsorption binärer Gemische. I. Berechnung von Durchbruchszeiten [On the application of the adsorption zone model to the adsorption of binary mixtures. I. Calculation of breakthrough times]. Chem Tech Leipzig, 1981a, 33, 572–575.

Hoppe H, Worch E. Zur Anwendung des Adsorptionszonenmodells auf die Adsorption binärer Gemische. II. Berechnung von Zonenwanderungsgeschwindigkeiten [On the application of the adsorption zone model to the adsorption of binary mixtures. II. Calculation of zone traveling velocities]. Chem Tech Leipzig, 1981b, 33, 644–646.

Jain JS, Snoeyink VL. Adsorption from bisolute systems on active carbon. J Water Pollut Control Fed, 1973, 45, 2463–2479.

Jaroniec M, Tóth J. Adsorption of gas mixtures on heterogeneous solid surfaces. 1. Extension of Tóth isotherm on adsorption from gas mixtures. Colloid Polym Sci, 1976, 254, 643–649.

Johannsen K, Worch E. Eine mathematische Methode zur Durchführung von Adsorptionsanalysen [A mathematical method for evaluation of adsorption analyses]. Acta Hydrochim Hydrobiol, 1994, 22, 225–230.

Jüntgen H, Seewald H. Charakterisierung der Porenstruktur mikroporöser Adsorbentien aus Kohlenstoff [Characterization of the pore structure of microporous carbonaceous adsorbents]. Ber Bunsenges Phys Chem, 1975, 79, 734–738.

Karickhoff SW, Brown DS, Scott TA. Sorption of hydrophobic pollutants on natural sediments. Water Res, 1979, 13, 241–248.

Kataoka T, Yoshida H, Ueyama K. Mass transfer in laminar region between liquid and packing material surface in the packed bed. J Chem Eng Jpn, 1972, 5, 132–136.

Kearns J, Dickensen E, Knappe D. Enabling organic micropollutant removal from water by full-scale biochar and activated carbon adsorbers using predictions from bench-scale column data. Environ Engin Sci, 2020, 37, 459–471.

Kenaga EE, Goring CAI. Relationship between water solubility, soil sorption, octanol–water partitioning, and concentration of chemicals in biota, Eaton JG, Parrish PR, Hendricks AC eds, Aquatic Toxicology ASTM STP 707, Philadelphia, PA: American Society for Testing and Materials, 1980, 79–115.

Kennedy AM, Reinert AM, Knappe DRU, Summers RS. Prediction of full-scale GAC adsorption of organic micropollutants. Environ Eng Sci, 2017, 34, 496–507.

Knappe DRU, Matsui Y, Snoeyink VL, Roche P, Prados MJ, Bourbigot MM. Predicting the capacity of powdered activated carbon for trace organic compounds in natural waters. Environ Sci Technol, 1998, 32, 1694–1698.

Kuennen RW, van Dyke K, Crittenden JC, Hand, DW. Predicting the multicomponent removal of surrogate compounds by a fixed-bed adsorber. J Am Water Works Assoc, 1989, 81, 46–58.

Kumar PS, Ejerssa WW, Wegener CC et al. , Understanding and improving the reusability of phosphate adsorbents for wastewater effluent polishing. Water Res, 2018, 145, 365–374.

Kunaschk M. Entwicklung eines Verfahrenskonzepts zur Entfernung von Phosphor in der dezentralen Abwasserbehandlung [Development of a process concept for the removal of phosphorus in decentralized wastewater treatment]. PhD thesis, Dresden University of Technology; 2020.

Kunaschk M, Schmalz V, Dietrich N, Dittmar T, Worch E. Novel regeneration method for phosphate loaded granular ferric (hydr)oxide – A contribution to phosphorus recycling. Water Res, 2015, 71, 219–226.

Kutzner S, Schaffer M, Börnick H, Licha T, Worch E. Sorption of the organic cation metoprolol on silica gel from its aqueous solution considering the competition of inorganic cations. Water Res, 2014, 54, 273–283.

Kutzner S, Schaffer M, Licha T, Worch E, Börnick H. Sorption of organic cations onto silica surfaces over a wide concentration range of competing electrolytes. J Colloid Interf Sci, 2016, 484, 229–236.

Langmuir I. The adsorption of gases on plane surfaces of glass, mica and platinum. J Am Chem Soc, 1918, 40, 1361–1403.

Li Q, Mariñas BJ, Snoeyink VL, Campos C. Three-component competitive adsorption model for flow-through PAC systems 1. Model development and verification with a PAC/membrane system. Environ Sci Technol, 2003, 37, 2997–3004.

Lowell S, Shields JE, Thomas MA, Thommes M. Characterization of Porous Solids and Powders: Surface Area, Pore Size and Density, Dordrecht: Kluwer Academic Publishers, 2010.

Lukchis GM. Adsorption systems. Pt.1: Design by mass transfer zone concept. Chem Eng New York, 1973, 80, 111–116.

Maraqa MA. Prediction of mass-transfer coefficient for solute transport in porous media. J Contam Hydrol, 2001, 53, 153–171.

Marczewski AW, Jaroniec M. A new isotherm equation for single-solute adsorption from dilute solutions on energetically heterogeneous solids. Monatsh Chem, 1983, 114, 711–715.

Mathews AP, Weber WJ Jr. Mathematical modeling of adsorption in multicomponent systems. ACS Symp Ser, 1980, 135, 27–53.

Michaels AL. Simplified method of interpreting kinetic data in fixed-bed ion exchange. Ind Eng Chem, 1952, 44, 1922–1930.

Myers AL, Prausnitz JM. Thermodynamics of mixed-gas adsorption. Am Inst Chem Eng J, 1965, 11, 121–127.

Najm IN, Snoeyink VL, Richard Y. Effect of initial concentration of a SOC in natural water on its adsorption by activated carbon. J Am Water Works Assoc, 1991, 83, 57–63.

Neretnieks I. Analysis of some adsorption experiments with activated carbon. Chem Eng Sci, 1976, 31, 1029–1035.

Newcombe G, Morrison J, Hepplewhite C, Knappe DRU. In the (adsorption) competition between NOM and MIB, who is the winner, and why?. Water Sci Technol Water Supply, 2002a, 2, 59–67.

Newcombe G, Morrison J, Hepplewhite C, Knappe DRU. Simultaneous adsorption of MIB and NOM onto activated carbon. II. Competitive effects. Carbon, 2002b, 40, 2147–2156.

Ogata A, Banks RB. A solution of the differential equation of longitudinal dispersion in porous media, U.S. Geological Survey Professional Paper 411-A. Washington, DC: U.S. Geological Survey, 1961.

Ohashi H, Sugawara T, Kikuchi KI, Konno H. Correlation of liquid-side mass transfer coefficient for single particles and fixed beds. J Chem Eng Jpn, 1981, 14, 433–438.

Pelekani C, Snoeyink VL. Competitive adsorption in natural water: role of activated carbon pore size. Water Res, 1999, 33, 1209–1219.

Pickens JF, Grisak GE. Scale-dependent dispersion in a stratified granular aquifer. Water Resour Res, 1981, 17, 1191–1211.

Polanyi M. Über die Adsorption vom Standpunkt des dritten Wärmesatzes [On adsorption from the viewpoint of the third law of thermodynamics]. Verh Dtsch Phys Ges, 1914, 16, 1012–1016.

Polson A. Some aspects of diffusion in solution and a definition of colloidal particle. J Phys Colloid Chem, 1950, 54, 649–652.

Qi S, Schideman L, Mariñas BJ, Snoeyink VL, Campos C. Simplification of the IAST for activated carbon adsorption of trace organic compounds from natural water. Water Res, 2007, 41, 440–448.

Rabolt B. Untersuchungen zur konkurrierenden Adsorption von Mikroverunreinigungen und natürlichen organischen Wasserinhaltsstoffen [Investigations on the competitive adsorption of micropollutants and natural organic water constituents]. PhD thesis, Dresden University of Technology; 1998.

Rabolt B, Heubach J, Schilling H, Worch E. Untersuchungen zur konkurrierenden Adsorption von chlorierten Kohlenwasserstoffen und organischen Hintergrundkomponenten der Elbe [Investigations into the competitive adsorption of chlorinated hydrocarbons and organic background compounds of the Elbe River]. Vom Wasser, 1998, 90, 135–151.

Radke CJ, Prausnitz JM. Thermodynamics of multi-solute adsorption from dilute liquid solutions. Am Inst Chem Eng J, 1972, 18, 761–768.

Raghavan NS, Ruthven DM. Numerical simulation of a fixed-bed adsorption column by the method of orthogonal collocation. Am Inst Chem Eng J, 1983, 29, 922–925.

Rebhun M, De Smedt F, Rwetabula J. Dissolved humic substances for remediation of sites contaminated by organic pollutants. Binding-desorption model predictions. Water Res, 1996, 30, 2027–2038.

Redlich O, Peterson DL. A useful adsorption isotherm. J Phys Chem, 1959, 63, 1024–1024.

Rosene MR, Manes M. Application of the Polanyi adsorption potential theory to adsorption from solution on activated carbon. VII. Competitive adsorption of solids from water solution. J Phys Chem, 1976, 80, 953–959.

Rosene MR, Manes M. Application of the Polanyi adsorption potential theory to adsorption from solution on activated carbon. IX. Competitive adsorption of ternary solid solutes from water solution. J Phys Chem, 1977, 81, 1646–1650.

Sabljiç A, Güsten H, Verhaar H, Hermans J. QSAR modelling of soil sorption, improvements and systematics of log K_{oc} vs. log K_{ow} correlations. Chemosphere, 1995, 31, 4489–4514.

Schay G, Fejes P, Szathmáry J. Studies on the adsorption of gas mixtures. I. Statistical theory of physical adsorption of the Langmuir-type in multicomponent systems. Acta Chim Acad Sci Hung, 1957, 12, 299–307.

Schideman LC, Mariñas BJ, Snoeyink VL, Campos C. Three-component competitive adsorption model for fixed-bed and moving-bed granular activated carbon adsorbers. Part I. Model development. Environ Sci Technol, 2006a, 40, 6805–6811.

Schideman LC, Mariñas BJ, Snoeyink VL, Campos C. Three-component competitive adsorption model for fixed-bed and moving-bed granular activated carbon adsorbers. Part II. Model parameterization and verification. Environ Sci Technol, 2006b, 40, 6812–6817.

Schoenheinz D. DOC als Leitparameter zur Bewertung und Bewirtschaftung von Grundwasserleitern mit anthropogen beeinflusster Infiltration [DOC as a guide parameter for assessment and management of aquifers with anthropogenically influenced infiltration]. PhD thesis, Dresden University of Technology; 2004.

Schreiber B. Untersuchungen zur Sorption aromatischer Kohlenwasserstoffe an mineralischen Bodenbestandteilen [Investigations on the sorption of aromatic hydrocarbons onto mineral soil components]. PhD thesis, Dresden University of Technology; 2002.

Schreiber B, Worch E. Untersuchungen zur konkurrierenden Sorption organischer Schadstoffe an Bodenmaterial [Investigations into competitive sorption of organic contaminants onto soil material]. Vom Wasser, 2000, 95, 279–292.

Schwarzenbach RP, Gschwend PM, Imboden DM. Environmental Organic Chemistry, New York: John Wiley & Sons, 1993.

Schwarzenbach RP, Westall J. Transport of nonpolar organic compounds from surface water to groundwater. Laboratory sorption studies. Environ Sci Technol, 1981, 15, 1360–1367.
Seewald H, Jüntgen H. Kinetik der nicht-isothermen Desorption organischer Stoffe von Aktivkohle [Kinetics of non-isothermal desorption of organic substances from activated carbon]. Ber Bunsenges Phys Chem, 1977, 81, 638–645.
Sheindorf C, Rebhun M, Sheintuch M. A Freundlich-type multicomponent isotherm. J Colloid Interf Sci, 1981, 79, 136–142.
Shimabuku KK, Paige J, Luna-Aguero M, Summers RS. Simplified modeling of organic contaminant adsorption by activated carbon and biochar in the presence of dissolved organic matter and other competing adsorbates. Environ Sci Technol, 2017, 51, 10031–10040.
Sigg L, Stumm W. Aquatische Chemie [Aquatic Chemistry], Zürich: vdf Hochschulverlag, 2011.
Sips R. On the structure of a catalyst surface. J Chem Phys, 1948, 16, 490–495.
Slavik I. Vorausberechnung des Adsorptionsverhaltens ausgewählter organischer Wasserinhaltsstoffe an Aktivkohle bei der Trinkwasseraufbereitung [Prediction of the adsorption behaviour of selected organic water constituents on activated carbon during drinking water treatment]. PhD thesis, Dresden University of Technology; 2006.
Slavik I, Uhl W, Börnick H, Worch E. Assessment of SOC adsorption prediction in activated carbon filtration based on Freundlich coefficients calculated from compound properties. RSC Adv, 2016, 6, 19587–19604.
Smith EH, Weber WJ Jr. Evaluation of mass transfer parameters for adsorption of organic compounds from complex organic matrices. Environ Sci Technol, 1989, 23, 713–722.
Sontheimer H, Crittenden JC, Summers RS. Activated Carbon for Water Treatment. Karlsruhe: DVGW-Forschungsstelle, Engler-Bunte-Institut, Universität Karlsruhe, 1988.
Sperlich A. Phosphate adsorption onto granular ferric hydroxide (GFH) for waste water reuse. PhD thesis, Technical University Berlin; 2010.
Sperlich A, Schimmelpfennig S, Baumgarten B et al. , Predicting anion breakthrough curves in granular ferric hydroxide (GFH) adsorption filters. Water Res, 2008, 42, 2073–2082.
Speth TF, Adams JQ. GAC and air stripping design support for the safe drinking water act, Clark RM, Summers RS eds, Strategies and Technologies for Meeting SDWA Requirements, Lancaster: Technomic Publishing Company, 1993, 47–89.
Stumm W. Chemistry of the Solid-Water Interface, New York: Wiley-Interscience, 1992.
Sudo Y, Misic DM, Suzuki M. Concentration dependence of effective surface diffusion coefficients in aqueous phase adsorption on activated carbon. Chem Eng Sci, 1978, 33, 1287–1290.
Sun N-Z. Mathematical Modeling of Groundwater Pollution, New York/Berlin/Heidelberg: Springer, 1996.
Sutikno T, Himmelstein KJ. Desorption of phenol from activated carbon by solvent regeneration. Ind Eng Chem Fundam, 1983, 22, 420–425.
Suwanayuen S, Danner RP. A gas adsorption isotherm equation based on vacancy solution theory. Am Inst Chem Eng J, 1980a, 26, 68–76.
Suwanayuen S, Danner RP. Vacancy solution theory of adsorption from gas mixtures. Am Inst Chem Eng J, 1980b, 26, 76–83.
Suzuki M, Kawazoe K. Concentration decay in a batch adsorption tank – Freundlich isotherm with surface diffusion kinetics. Seisan-Kenkyu [J Inst Ind Sci Univ Tokyo], 1974a, 26, 275–277.
Suzuki M, Kawazoe K. Concentration decay in a batch adsorption tank – Freundlich isotherm with pore diffusion kinetics. Seisan-Kenkyu [J Inst Ind Sci Univ Tokyo], 1974b, 26, 296–299.
Thompson KA, Shimabuku KK, Kearns JP, Knappe DRU, Summers RS, Cook SM. Environmental comparison of biochar and activated carbon for tertiary wastewater treatment. Environ Sci Technol, 2016, 50, 11253–11262.
Tóth J. State equations of the solid-gas interface layers. Acta Chim Acad Sci Hung, 1971, 69, 311–328.

Traegner UK, Suidan MT, Kim BR. Considering age and size distributions of activated-carbon particles in a completely-mixed adsorber at steady state. Water Res, 1996, 30, 1495–1501.
van Genuchten MT, Wierenga PJ. Mass transfer studies in sorbing porous media. I. Analytical solutions. Soil Sci Soc Am J, 1976, 40, 473–480.
van Gestel CAM, Ma W, Smit CE. Development of QSARs in terrestrial ecotoxicology: earth-worm toxicity and soil sorption of chlorophenols, chlorobenzenes and dichloroaniline. Sci Total Environ, 1991, 109/110, 589–604.
Vermeulen T, Klein G, Hiester NK. Adsorption and ion exchange, Perry RH, Chilton CH eds, Chemical Engineer's Handbook, 5th ed, New York: Mc Graw-Hill, 1973, 16/1–16/50.
Weber WJ Jr. Preloading of GAC by natural organic matter in potable water treatment systems: mechanisms, effects, and design considerations. J Water Supply Res T – AQUA, 2004, 53, 469–482.
Weber WJ Jr, DiGiano FA. Process Dynamics in Environmental Systems, New York: Wiley, 1996.
Weber WJ Jr, Liu KT. Determination of mass transport parameters for fixed-bed adsorbers. Chem Eng Commun, 1980, 6, 49–60.
Wilke CR, Chang P. Correlations of diffusion coefficients in dilute solutions. Am Inst Chem Eng J, 1955, 1, 264–270.
Williamson JE, Bazaire KE, Geankoplis CJ. Liquid-phase mass transfer at low Reynolds numbers. Ind Eng Chem Fundam, 1963, 2, 126–129.
Wilson EJ, Geankoplis CJ. Liquid mass transfer at very low Reynolds numbers in packed beds. Ind Eng Chem Fundam, 1966, 5, 9–14.
Worch E. Eine neue Gleichung zur Berechnung von Diffusionskoeffizienten gelöster Stoffe [A new equation for estimating diffusion coefficients of solutes]. Vom Wasser, 1993, 81, 289–297.
Worch E. Modelling the solute transport under non-equilibrium conditions on the basis of mass transfer equations. J Contam Hydrol, 2004, 68, 97–120.
Worch E. Fixed-bed adsorption in drinking water treatment: a critical review on models and parameter estimation. J Water Supply Res T – AQUA, 2008, 57, 171–183.
Worch E. Hydrochemistry, Basic Concepts and Exercises, Berlin/Boston: De Gruyter, 2015.
Worch E, Grischek T, Börnick H, Eppinger P. Laboratory tests for simulating attenuation processes of aromatic amines in riverbank filtration. J Hydrol, 2002, 266, 259–268.
Xu M, Eckstein Y. Use of weighted least-squares method in evaluation of the relationship between dispersivity and scale. Ground Water, 1995, 33, 905–908.
Yiacoumi S, Rao AV. Organic solute uptake from aqueous solutions by soil: a new diffusion model. Water Resour Res, 1996, 32, 431–440.
Yiacoumi S, Tien C. A model of organic solute uptake from aqueous solutions by soils. Water Resour Res, 1994, 30, 571–580.
Young DF, Ball WP. Effects of column conditions on the first-order rate modeling of nonequilibrium solute breakthrough. Water Resour Res, 1995, 31, 2181–2192.
Zhang Q, Crittenden J, Hristovski K, Hand D, Westerhoff P. User-oriented batch reactor solutions to the homogeneous surface diffusion model for different activated carbon dosages. Water Res, 2009, 43, 1859–1866.
Zietzschmann F, Altmann J, Ruhl AS et al. , Estimating organic micro-pollutant removal potential of activated carbons using UV absorption and carbon characteristics. Water Res, 2014, 56, 48–55.
Zietzschmann F, Mitchell R-L, Jekel M. Impacts of ozonation on the competition between organic micropollutants and effluent organic matter in powdered activated carbon adsorption. Water Res, 2015, 84, 153–160.
Zoschke K, Engel C, Börnick H, Worch E. Adsorption of geosmin and 2-methylisoborneol onto powdered activated carbon at non-equilibrium conditions: influence of NOM and process modelling. Water Res, 2011, 45, 4544–4550.

Index

https://doi.org/10.1515/9783110715507-013

Bei Fragen zur Produktsicherheit wenden Sie sich bitte an:
If you have any questions regarding product safety,
please contact:

Walter de Gruyter GmbH
Genthiner Straße 13
10785 Berlin
productsafety@degruyterbrill.com